FOAMGLAS® INDUSTRIAL INSULATION HANDBOOK

PC PITTSBURGH CORNING EUROPE N.V./S.A.

FOAMGLAS® INDUSTRIAL INSULATION HANDBOOK

1992

PITTSBURGH CORNING EUROPE NV/SA
WATERLOO OFFICE PARC
Drève RICHELLE 161 c
B-1410 WATERLOO
BELGIUM

© Pittsburgh Corning Europe NV/SA, Waterloo, 1992

D/1992/6331/1
ISBN 2-930049-00-6

Printed in France
by Imprimerie Louis Jean, 05000 Gap.

PREFACE

Pittsburgh Corning Corporation was founded in 1937, started production of glass blocks in 1938 and of FOAMGLAS® cellular glass in 1942. *Pittsburgh Corning Europe* was founded in 1963 and launched production of FOAMGLAS® insulation in Belgium in 1965.

Since that time, a considerable experience of FOAMGLAS® insulation production and industrial applications has been developed, particularly with the insulation system approach which was used, right from the beginning. More generally speaking, data have been gathered on thermal insulation.

A few years ago, a decision was made to re-assemble and re-evaluate all this information, and a team with members from both sides of the Atlantic ocean was constituted. Only factual, serious and well established information was considered. It implies substantial work including consultation of more than 700 reference papers, written by outstanding authorities in the insulation field and a very large number of internal reports. This information was summarised in a document aimed exclusively at *Pittsburgh Corning*'s staff and personnel.

Considering the uniqueness of the previous document, it was decided to prepare a new version to share this information with professionals interested in industrial insulation. The *FOAMGLAS® INDUSTRIAL INSULATION HANDBOOK* is the result of this vast, team work.

It is aimed at owners, designers and contractors who are in charge of industrial insulation.

In this book, the term industrial insulation incorporates the insulation of refineries, petrochemical plants, gas processing plants, chemical and offshore installations.

It also includes applications in food industry, such as breweries, dairies, meat producing plants and chilled water lines, above ground and underground pipe-

PREFACE

lines, air ducts, refrigeration plants, ice rinks, paper mills, digesters, water treatment plants, power and nuclear plants, and industrial chimneys.

The following subjects are covered:

   I. Where to use cellular glass
  II. Properties of FOAMGLAS® cellular glass and its accessories
 III. How to design and install FOAMGLAS® insulation for industrial applications
 IV. Case studies
  V. Why to choose FOAMGLAS® insulation.

Although fairly complete, Part I is not necessarily exhaustive.

The properties quoted in Part II mainly concern FOAMGLAS® cellular glass.

Since the application of cellular glass is rather specific and implies well-chosen adhesives, sealers, coatings, etc. the main accessory products marketed by *Pittsburgh Corning* are shortly described and data on the FOAMGLAS® compatibility with most usual accessory products coming from other sources are presented.

Part III proposes the answers to the question "How to design and install FOAMGLAS®".

In the first chapter a critical review of the general requirements that should be met by an insulation system is presented. Then a chapter is devoted to the determination of the thickness and another one to the available fabricated elements needed to fit the equipment shape. In the fourth chapter, many applications are described, indicating in each particular case the requirements, the recommended insulation system with an evaluation of the need for each layer, the thickness, the number of layers, the need for sealers and the suggested type of coating and jacket. It should be understood as specification concepts with some explanations, but not as detailed specifications for which the reader should consult more specific documents issued by *Pittsburgh Corning*.

General design recommendations are then presented for particular points such as expansion/contraction joints, insulation of pipelines flanges, supports, sphere legs, heat traced pipelines, underground lines.

Finally, guidance is provided for the application on the jobsite and for the maintenance. Although the first goal is to present the FOAMGLAS® case, many of the recommendations could be useful for other insulation systems.

In Part IV, Case studies, short descriptions of a few typical existing references are given, including an "old" one for which an evaluation has been made of the current situation compared to the initial one, many years ago.

In Part V, the question "Why" is handled, based on comparison between the possible performance of various insulation systems. This is carried out from several significant standpoints which are important for the owners, designers and contractors.

They are focused on the main problems an insulation system can be exposed to: fire and absorption of combustible liquids, water in liquid and vapour form, thermal efficiency in relation to ageing, corrosion under insulation, chemical durability, dimensional stability, mechanical strength and resistance to vermin. The goal is to base the evaluations and comparisons between the various insulation systems on factual and documented data coming from independent authors or competitive literature, thus remaining as objective as possible. In a way, this part presents the exercise that an independent decision maker should carry out at the design stage.

Moreover, the book is organised to make each part more or less independent from the other ones to facilitate reading on a specific subject. This implies some repetition of ideas and quoted references.

CONTENTS

Preface V

Part I
WHERE TO USE CELLULAR GLASS

I.1 CRITERIA FOR USING CELLULAR GLASS 3

    I.1.1 Cold 3
    I.1.2 Cryogenic 4
    I.1.3 High Load-bearing Insulation Material 4
    I.1.4 Protection against Corrosion through Water Ingress 5
    I.1.5 Fire Risk 5
    I.1.6 Absorption of Combustible Liquids or Gases 6
    I.1.7 Bacterial Risks 7

I.2 HYDROCARBON CHEMISTRY 9

    I.2.1 Introduction to Hydrocarbon Chemistry 9
    I.2.2 Types of Hydrocarbons 10
        I.2.2.1 Hydrocarbons with Open Carbon Chains 10
        I.2.2.2 Hydrocarbons with Closed Carbon Chains 10
    I.2.3 Classification 12
        I.2.3.1 Numbers of Carbons 12
        I.2.3.2 Boiling Temperatures 12

I.3 REFINERY 15

I.3.1 The Destructive Transformation Processes 15
I.3.1.1 The Catalytic Reforming Process.................................. 16
I.3.1.2 The Isomerisation Process.. 16

I.3.2 Processes which Convert Excess Products into Lighter Products.. 16
I.3.2.1 The Catalytic Cracking Process 16
I.3.2.2 The Thermal Cracking Process..................................... 16

I.3.3 The Hydrocracking... 17
I.3.4 The Synthesis Processes 17
I.3.4.1 The Alkylation Process 17
I.3.4.2 The Dimersol Process... 17

I.3.5 The Purification Processes... 17
I.3.6 Separation Processes 18
I.3.6.1 The First Distillation Cuts Fractionation Processes...... 18
I.3.6.2 The Fractionation Processes 18

I.3.7 Miscellaneous Processes.. 18
I.3.8 General Refinery Scheme 18
I.3.9 Recommendations for the Use of Cellular Glass in Refining Processes... 18

I.4 NATURAL GAS PROCESSING.. 21

I.4.1 Introduction.. 21
I.4.1.1 Field Processing.. 21
I.4.1.2 Plant Processing.. 21

I.4.2 General Natural Gas Processing Scheme....................................... 21
I.4.2.1 H_2S, CO_2 Removal Processes 21
I.4.2.2 Dehydration... 23
I.4.2.3 LPG-NGL Recovery... 23
I.4.2.4 LNG Liquefaction Processes.. 23
I.4.2.5 LNG Receiving Terminals 27

I.5 GAS PROCESSING 29

I.5.1 Synthesis Gases 29
I.5.2 Industrial Gases 29

I.6 PETROCHEMISTRY ... 31

I.6.1 Introduction ... 31

I.6.2 General Petrochemistry Scheme ... 32

- I.6.2.1 Ammonia ... 32
- I.6.2.2 Carbon Monoxide ... 35
- I.6.2.3 Ethylene and Propylene by Steam Cracking ... 35
- I.6.2.4 P-Xylene. Arco Two-step Process ... 44
- I.6.2.5 Ethylene Oxide ... 44
- I.6.2.6 Vinyl Chloride (From Pyrolysis of Ethylene Dichloride) ... 46
- I.6.2.7 MDI (Diphenyl Methane 4,4-Diisocyanate) and TDI (Tolylene Diisocyanate) ... 46
- I.6.2.8 Polypropylene. Mitsui Montedison Technology ... 46
- I.6.2.9 PVC ... 46
- I.6.2.10 Polybutadiene ... 46
- I.6.2.11 Methanol ... 46
- I.6.2.12 MTBE ... 52

I.6.3 Miscellaneous ... 52

- I.6.3.1 Vertical Storage Tanks ... 52
  - A. Base Insulation ... 52
  - B. Walls and Roofs ... 53
- I.6.3.2 Spherical Storage Tanks ... 53
- I.6.3.3 Offshore Installations ... 54
- I.6.3.4 Bitumen Tanks and Barges/Ships ... 54
- I.6.3.5 Heat Transfer Lines ... 54
- I.6.3.6 Sulphur Pits ... 54
- I.6.3.7 Pool Fire Control of Liquefied Gas ... 55

I.7 CHEMICAL – PHARMACEUTICAL – COSMETIC ... 57

I.8 OTHER INDUSTRIAL APPLICATIONS ... 59

I.8.1 Food ... 59

- I.8.1.1 Breweries ... 59
- I.8.1.2 Malt Houses ... 59
- I.8.1.3 Dairies. Cheese Dairies ... 60
- I.8.1.4 Meat. Poultry-Farms. Farms ... 60

I.8.2 Cold Stores.. 60
    I.8.2.1 Wall 60
    I.8.2.2 Floor 60
    I.8.2.3 Roof 61
    I.8.2.4 Freezing Lines 61

I.8.3 Heating. Ventilation. Air Conditioning 62
    I.8.3.1 Chilled Water 62
    I.8.3.2 Mining Air Conditioning 62
    I.8.3.3 Air Ducts 63
    I.8.3.4 Heating 63
        A. Central Heating 63
        B. Underground District Heating 63

I.8.4 Others 63
    I.8.4.1 Ice Rinks 63
    I.8.4.2 Paper Mills 64
    I.8.4.3 Digesters. Water Treatment 64
    I.8.4.4 Power Plants. Nuclear Plants 65
    I.8.4.5 Chimneys 65
    I.8.4.6 Road Tunnels 66
    I.8.4.7 Architecture. Building 66

I.9 SUMMARY 67

Part II
PROPERTIES OF FOAMGLAS® CELLULAR GLASS AND ITS ACCESSORIES

II.1 FOAMGLAS® CELLULAR GLASS 73
    II.1.1 Production Process 73
    II.1.2 General Properties 73
    II.1.3 Dimensions of the Slabs 75
    II.1.4 Fire Behaviour 76

II.1.5	Water and Water Vapour	77
II.1.6	Constant Thermal Efficiency	78
II.1.7	Corrosion	79
II.1.8	Chemical Durability	79
II.1.9	Dimensional Stability	81
II.1.10	Coefficient of Thermal Expansion	81
II.1.11	Ecological Standpoint: Pay Back in Energy	82
II.1.12	Ecological Standpoint: CO_2 in the Cells	83
II.1.13	Different Types of FOAMGLAS® Cellular Glass	84
II.1.14	Thermal Conductivity	86
II.1.15	Compressive Strength	86
II.1.16	Flexural Strength	89
II.1.17	Shear Strength	90
II.1.18	Tensile Strength	91
II.1.19	Temperature Influence on the Properties of Cellular Glass	91
II.1.20	Limitations to the Use of Cellular Glass	92

II.2	**ACCESSORY PRODUCTS**	**95**
II.2.1	Adhesives	95
	II.2.1.1 Bitumen	96
	II.2.1.2 Two Component Adhesives	97
	II.2.1.3 Contact Adhesives	97
	II.2.1.4 Hydraulic Setting Cement Adhesives	97
II.2.2	Sealers	98
II.2.3	Coatings and Jackets	99
II.2.4	Mastics	102
II.2.5	Reinforcement Fabrics	102
II.2.6	Metal Jackets	103
II.2.7	Anti-abrasive Coatings	107
II.2.8	Choice of Accessory Products	107
	II.2.8.1 Points to Be Considered when Selecting Accessory Products	108
	II.2.8.2 Properties for Particular Applications	109

XIII

II.2.9 Accessory Products Marketed by *Pittsburgh Corning* for Industrial Applications 109

    II.2.9.1   Adhesives 110
                  A. PC® 56 ADHESIVE 110
                  B. PC® 88 ADHESIVE 110
                  C. PC® 86T 111
                  D. PC® 80 M MORTAR 111

    II.2.9.2   Sealer 112
                  A. PITTSEAL® 444 112

    II.2.9.3   Coatings and Jackets 113
                  A. PITTCOTE® 300 113
                  B. PITTCOTE® 404 113
                  C. PC® 74A 114
                  D. PC® 85 POWDER 114
                  E. PC® 18 114
                  F. PITTWRAP® 115

    II.2.9.4   Reinforcement Fabrics 115

    II.2.9.5   Anti-abrasives 116
                  A. PC® ANTI-ABRASIVE COMPOUND 2A 116
                  B. PC® HIGH TEMPERATURE ANTI-ABRASIVE 116

    II.2.9.6   Chemical Resistance Table for Coatings 117

II.2.10 Some other Accessory Products Compatible with FOAMGLAS® Insulation 117

    II.2.10.1   Adhesives 117
                  A. Foster® 81-80 and 81-82 117
                  B. Foster® 82-10 and 81-33 120

    II.2.10.2   Sealers 120
                  A. Foster® Foamseal® 30-45 120
                  B. Foster® 95-50 120

    II.2.10.3   Coatings 120
                  A. Foster® 60-25 120
                  B. Foster® 60-75 120
                  C. Foster® Lagtone® 30-70 121
                  D. Foster® Monolar® 121
                  E. Encacel T 121

    II.2.10.4   Anti-abrasive 121
                  A. Foster® Anti-abrasive 30-16 121

Part III
HOW TO DESIGN AND INSTALL FOAMGLAS® INSULATION FOR INDUSTRIAL APPLICATIONS

Introduction 125

III.1 APPLICATION REQUIREMENTS 127

- III.1.1 Thermal Requirements 127
  - III.1.1.1 Heat Transfer Limitation 127
  - III.1.1.2 Surface Condensation Limitation 128
  - III.1.1.3 Protection from Burns 129
  - III.1.1.4 Surface Temperature Limitation 129
  - III.1.1.5 Temperature Process Control 130
  - III.1.1.6 Rapid Temperature Variations 130
  - III.1.1.7 Protection of Structural Material against Excessive Temperature 130
  - III.1.1.8 Fire Protection Requirements 131
  - III.1.1.9 Non Absorption of Flammable Liquids 133
  - III.1.1.10 Compatibility with Liquid Oxygen 133
  - III.1.1.11 Nuclear Applications 133
- III.1.2 **Moisture Resistance Requirements** 134
- III.1.3 **Corrosion Resistance Requirements** 135
- III.1.4 **Chemical Resistance Requirements** 136
- III.1.5 **Mechanical Requirements** 138
- III.1.6 **Important Notes** 138

III.2 THICKNESS DETERMINATION 141

- III.2.1 **Cold Applications** 141
- III.2.2 **Hot Applications** 154
- III.2.3 **Dual Temperature Applications** 161
- III.2.4 **Overfit/Retrofit on High Temperature Applications** 162
- III.2.5 **Insulated Underground Pipings** 163
- III.2.6 **Tank Insulation** 166

XV

III.3 FOAMGLAS® CELLULAR GLASS FABRICATED ELEMENTS 167

III.3.1 Number of Layers 167
III.3.2 Geometry of the FOAMGLAS® Cellular Glass Elements to Be Selected as a Function of the Surface to Be Insulated 168
   III.3.2.1 General Principles 168
   III.3.2.2 Fabrication Method 170
   III.3.2.3 Definition of some Available FOAMGLAS® Cellular Glass Elements 171

III.4 REVIEW OF THE MAIN FOAMGLAS® INSULATION CONCEPTS 183

III.4.1 Preliminary Conditions before Insulation Application 183
III.4.2 Cold Applications: Piping and Equipment 185
III.4.3 High Temperature Applications: Piping and Equipment 189
III.4.4 Dual Temperature Applications: Piping and Equipment 192
III.4.5 Chilled Water Pipelines 194
III.4.6 Overfit/Retrofit 197
III.4.7 Underground Lines 199
III.4.8 Low Temperature Tank Bottom 202
III.4.9 Hot Tank Base 210
III.4.10 Low Temperature and Cryogenic Tank Wall and Roof 216
III.4.11 Low Temperature Tank Wall and Roof 221
III.4.12 Cryogenic Tank Wall 225
III.4.13 Low Temperature Sphere 229
III.4.14 Digester 234
III.4.15 Industrial Chimney 238
III.4.16 Offshore 242
III.4.17 Stainless Steel Support 244
III.4.18 Nuclear 247
III.4.19 Particular Points 249
   III.4.19.1 Expansion/Contraction Joints on Insulated Pipelines 249
   III.4.19.2 Expansion/Contraction Joints of Vessels and Tank Walls 256
   III.4.19.3 Insulation of Pipeline Flanges 257

III.4.19.4	Insulation of Pipe Hangers and Support Assembly	259
III.4.19.5	Vessel Supports	262
III.4.19.6	Sphere Legs	264
III.4.19.7	Heat Traced Pipelines	265
III.4.19.8	Particular Points of Underground Insulated Pipelines	269
III.4.20	**Table of Standardisation of Fabricated Products, Inside, Joints and Outside Area**	274
III.4.21	**Application Guide**	278
III.4.22	**Maintenance**	287

Annex
CALCULATION METHOD FOR ABOVE GROUND INSULATED PIPES 292

Part IV
CASE STUDIES

IV.1 COOL IT! 297

IV.2 THERMAL INSULATION ON CHILLED WATER PIPES AT THE LURGI GmbH COOLING SYSTEM IN FRANKFURT 301

IV.3 WRAPPING UP AT MOSSMORRAN 303

IV.4 DURABLE INSULATION FOR LIQUID GAS STORAGE TANKS 307

IV.5 LAGGING THE DRAGON 311

IV.6 FOAMGLAS® INSULATION 32-YEAR PERFORMANCE 315

IV.7 THE NETHERLANDS: INSULATING THE BASE OF A TANK (110°C) AT THE SUIKERUNIE, DINTELOORD 317

IV.8 SCHWEDENECK-SEE OFFSHORE PROJECT 319

Part V
WHY TO CHOOSE
FOAMGLAS® INSULATION

V.1 FIRE AND SMOKE 323

    V.1.1 The Problem: Importance of Fires 323

        V.1.1.1 Insulation Flammability and Wicking 323
        V.1.1.2 Consequences of Fire 324

    V.1.2 Misleading Terminology 325

        V.1.2.1 Historical Terminology 325
        V.1.2.2 The Development of Current Terminology 326

    V.1.3 Test Methods 327

    V.1.4 Influence of Insulating Materials 329

        V.1.4.1 Generalities 329
        V.1.4.2 Examples of Large Fires 331

    V.1.5 Review of Insulating Material Manufacturers' Literature 333

    V.1.6 The FOAMGLAS® Cellular Glass Solution 335

        V.1.6.1 Combustibility 335
        V.1.6.2 Heat Evolution 336
        V.1.6.3 Smoke and Toxic Gases 336
        V.1.6.4 Fire Resistance Tests 338
            A. Systems Involving Adhesives 338
            B. Pipe Tests 339
            C. Petrochemical Fire Tests 340
            D. Penetrations 341
            E. Panels 341
        V.1.6.5 FOAMGLAS® Cellular Glass Fire Experience 341
            A. Ethylene Plant Fire 342

B. LNG System	342
C. Metal Decks	343
D. Cold Stores	343
E. Insurances	343
F. LNG Pool Fires	343

V.2 COMBUSTIBLE LIQUIDS IN INSULATION ... 345

V.2.1 The Problem ... 345
    V.2.1.1 Explanation of the Problem ... 346
    V.2.1.2 Sources of Leaks ... 348

V.2.2 The Solution ... 348

V.2.3 Installations Using FOAMGLAS® Insulation ... 349

V.2.4 High Temperature Systems ... 350

V.2.5 Low Temperature Considerations ... 350
    V.2.5.1 Hydrocarbons ... 350
    V.2.5.2 Oxygen ... 351

V.3 LIQUID WATER ABSORPTION AND RETENTION IN INSULATION ... 353

V.3.1 The Problem ... 353

V.3.2 Water in Low Temperature Insulation ... 354

V.3.3 Water in High Temperature Insulation ... 356
    V.3.3.1 High Temperature Insulations ... 356
        A. Calcium Silicate ... 357
        B. Mineral Wool ... 357
        C. Water Repellent Insulations ... 358
    V.3.3.2 Water Entry ... 358
    V.3.3.3 Can Insulation Dry out? ... 358
        A. Low Temperature ... 358
        B. High Temperature ... 358
        C. Mechanism of Water Accumulation in Permeable High Temperature Insulation ... 359

V.3.4 FOAMGLAS® Insulation ... 360
    V.3.4.1 Water Resistance ... 360
    V.3.4.2 Examples of Water Resistance ... 360
    V.3.4.3 Cautions ... 361
        A. Low Temperatures ... 361
        B. High Temperatures ... 362

XIX

V.3.5 FOAMGLAS® Insulation Systems 362
    V.3.5.1 Total FOAMGLAS® Insulation Systems 362
    V.3.5.2 FOAMGLAS® Cellular Glass/Mineral Fibre Composite
           Insulation System 363
    V.3.5.3 Overfit/Retrofit 364

Annex
PATHS FOR WATER ENTRY 365

V.4 WATER VAPOUR TRANSPORT AND CONDENSATION 369

V.4.1 The Problem 369

V.4.2 Explanation of the Problem 370
    V.4.2.1 Absolute Humidity 370
    V.4.2.2 Relative Humidity 370
    V.4.2.3 Partial Pressure of Water Vapour in the Air 371
    V.4.2.4 Condensation and Dew Point Temperature 371
    V.4.2.5 Partial Pressure Difference 371
    V.4.2.6 Permeance and Permeability. Water Vapour Diffusion Resistance
           Number 372
    V.4.2.7 Specific Resistance against Water Vapour Diffusion 374
    V.4.2.8 Water Vapour Flow 375

**V.4.3 Examples of Water Vapour Transmission through Low
Temperature Insulation Systems** 375
    V.4.3.1 Water Vapour Transmission through an Insulating Material 375
    V.4.3.2 Water Vapour Transmission through an Insulating System of a
           Low Temperature Application 377
    V.4.3.3 The Water Vapour Barrier or Water Vapour Retarder 379

**V.4.4 Water Absorption in Medium and High Temperature
Insulation** 381

V.4.5 The Overfit/Retrofit System 384

V.4.6 Insulating Materials and Water Vapour Transmission 386

V.4.7 Conclusion 388

Annex
TABLES 389

V.5 THERMAL EFFICIENCY IN REAL LIFE 395

V.5.1 The Problem 395
    V.5.1.1 Operating Costs 395
    V.5.1.2 Process Control 396
    V.5.1.3 Personnel Protection 396

  V.5.1.4 Damage 397
  V.5.1.5 Corrosion 397
V.5.2 Mechanisms for Loss of Efficiency 397
  V.5.2.1 Loss of Efficiency in the Insulation Material 397
    A. The Effect of Water and Water Vapour 397
    B. Ageing of Materials 400
    C. Absorption of Liquid Chemicals 401
    D. Compression and/or Vibrations 401
  V.5.2.2 Loss of Efficiency in the Insulating System 401
    A. Convection 401
    B. Thermal Bridges 403
  V.5.2.3 Misleading Concepts 404
V.5.3 Conclusion 404

V.6 CORROSION UNDER INSULATION 405

V.6.1 The Problem 405
  V.6.1.1 Economics 406
  V.6.1.2 Safety 406
V.6.2 Conditions 406
  V.6.2.1 Water 407
  V.6.2.2 Chemistry 407
  V.6.2.3 Temperature 408
V.6.3 Solutions 409
V.6.4 Insulations 410
  V.6.4.1 Cellular Plastics 410
    A. Polyurethane 410
    B. Phenolics 412
  V.6.4.2 High Temperature Insulations 412
    A. Calcium Silicate 413
    B. Fibrous Insulation 413
    C. Stress Corrosion Cracking under High Temperature Insulation . 414
  V.6.4.3 FOAMGLAS® Insulation 414
    A. Chemical Properties 415
    B. Carbon Steel Compatibility 416
    C. Stainless Steel Compatibility 417

V.7 CHEMICAL DURABILITY 419

V.7.1 The Problem 419

V.7.2	**The Solution**	420
	V.7.2.1 Glass Durability	421
	V.7.2.2 FOAMGLAS® Insulation is 100% Glass	422
	V.7.2.3 FOAMGLAS® Insulation is Impermeable	423
V.7.3	**Measures of Chemical Durability**	423
	V.7.3.1 FOAMGLAS® Cellular Glass Insulation	423
	A. Water Durability	424
	B. Durability in Other Reagents	424
	C. The Role of Accessory Materials	424
	D. Insulation Specialist's Recommendations	425
	V.7.3.2 Competitive Insulation Materials	425
	A. Polyurethane, Polyisocyanurate, Polystyrene	425
	B. Phenolics	426
	C. Mineral Fibre	426
	D. Perlite	427
V.7.4	**Conclusions**	428

V.8 DIMENSIONAL STABILITY . . . 429

V.8.1	**The Problem**	429
	V.8.1.1 Reversible Changes in the Dimensions of Insulations	429
	V.8.1.2 Non Reversible Changes of Dimensions	430
	V.8.1.3 The Combined Effect of Temperature and Humidity	431
	V.8.1.4 The Combined Effect of Temperature and Loading	432
V.8.2	**Consequences of a Lack of Dimensional Stability: Different Types of Problems**	433
	V.8.2.1 Warping and other Types of Distortion	433
	V.8.2.2 Open Joints or Cracks	434
	V.8.2.3 Stresses between Insulation and Coatings or Waterproofing Materials	434
V.8.3	**FOAMGLAS® Cellular Glass Has an Exceptional Dimensional Stability**	435
	V.8.3.1 Reversible Dimensional Changes Due to Temperature Variations Are Minimal	435
	V.8.3.2 Non Reversible Changes Due to the Influence of Temperatures in the Range of +430°C to –200°C Are not Known	436
	V.8.3.3 Humidity Does not Cause Dimensional Changes to FOAMGLAS® Cellular Glass	436
V.8.4	**Conclusion**	436

V.9 COMPRESSIVE STRENGTH . . . 437

V.9.1	**The Problem**	437
	V.9.1.1 Low Temperature Tank Base Insulation	438
	V.9.1.2 Semi-underground Conical Digesters. Base Insulation	439
	V.9.1.3 High Temperature Tank Base Insulation	439
	V.9.1.4 Industrial Floors	440
	V.9.1.5 Underground Pipe and Vessel Insulation	440
	V.9.1.6 Pipe Supports and Hangers	441
	V.9.1.7 Living Loads	441
	V.9.1.8 Industrial Metal Decks	443
	V.9.1.9 Self Supporting FOAMGLAS® Insulation Walls	443
V.9.2	**Definition of Compressive Strength Characteristics**	443
	V.9.2.1 Definition of Compressive Strength Characteristics	444
	V.9.2.2 The Deformation Issue	444
	V.9.2.3 Testing Procedure for FOAMGLAS® Cellular Glass	447
	V.9.2.4 The Capping Influence	447
	V.9.2.5 Consequences of Important Deformation	447
	V.9.2.6 Influence of Temperature and Time	448
	V.9.2.7 Influence of the Load Direction	450
V.9.3	**Engineering Design Information**	450
	V.9.3.1 Types of FOAMGLAS® Insulation	450
	V.9.3.2 The Safety Factor	451
	V.9.3.3 Quality Control Procedure: Internal and by Third Party	451
V.9.4	**Performance of FOAMGLAS® Insulation Compared to Alternative Insulation**	452
V.10	**VERMIN RESISTANCE OF INSULATION MATERIALS**	455
V.10.1	**The Problem**	455
V.10.2	**What Types of Creatures are Dangerous to Insulating Materials?**	455
	V.10.2.1 Rodents	455
	V.10.2.2 Insects	456
	V.10.2.3 Microorganisms	456
V.10.3	**The Different Types of Damage Caused by Rodents and Vermin**	457
	V.10.3.1 Insulating Efficiency	457
	V.10.3.2 Bad Hygienic Conditions	457
	V.10.3.3 Economic Losses	457
V.10.4	**Different Insulation Materials Show Different Behaviour against Rodents and Vermin**	458
	V.10.4.1 FOAMGLAS® Insulation	458
	V.10.4.2 Other Insulation Materials	459

XXIII

V.10.5 Installations Using FOAMGLAS® Cellular Glass as the Insulation Material 459

CONCLUSION 461

SUBJECT INDEX 463

I
WHERE TO USE CELLULAR GLASS

I.1
CRITERIA FOR USING CELLULAR GLASS

Before going into detail regarding the question "Where to use cellular glass?", it is useful to remember the specific and sometimes unique characteristics of cellular glass.

Each of these characteristics will be thoroughly examined in the following chapters but their mere statement clarifies, from the very start, the reasons for using cellular glass in various applications:

- Totally incombustible
- Does not absorb inflammable or dangerous liquids or gases
- Waterproof
- Impervious to water vapour
- No deterioration of thermal conductivity
- Dimensional stability
- Resistant to most acids, gases and solvents
- Chemically neutral
- High compressive strength, without crushing (hard material)
- Resistant to insects, rodents.

Of the characteristics mentioned above, one or any combination of them, makes cellular glass an ideal choice in many different situations such as:

I.1.1 COLD

Equipment operating at temperatures lower than the ambient temperature is subject to condensation. Traditional insulation materials are more or less pervious to gases and are thus sensitive to humidity in the air that may condense within the insulation system.

Only cellular glass has fully tight cells, which perfectly protect it against moisture absorption due to condensation.

This characteristic is a vital requirement in cold environments.

I.1.2 CRYOGENIC

Besides the condensation, another phenomenon can be observed: the vacuum created beneath the vapour barrier of a cryogenic insulation system. Its importance increases with the decrease in temperature according to the law of gases:

$$PV = nRT$$

where pressure considerably reduces with the decrease in absolute temperature.

The following approximate underpressures can be found in the insulation beneath the vapour barriers:

Product	Temperature	Under pressure
Ammonia	to $-33°C$	$+/-0.07 \times 10^{-5}$ Pa
Propane	to $-45°C$	$+/-0.10 \times 10^{-5}$ Pa
Ethylene	to $-104°C$	$+/-0.20 \times 10^{-5}$ Pa
Methane	to $-161°C$	$+/-0.30 \times 10^{-5}$ Pa
Oxygen	to $-183°C$	$+/-0.34 \times 10^{-5}$ Pa

As a consequence, the air and its humidity are literally sucked into the insulant through all the imperfections of the vapour barrier, which unfortunately is rarely perfect (Fig. I.1).

The phenomenon is even more disastrous, when piping and equipment in intermittent service such as loading lines or tanks whose loading levels vary, operate at variable temperatures. In this case, a real air pump, loaded with humidity that will condense during the next cooling cycle, will be created.

I.1.3 HIGH LOAD-BEARING INSULATION MATERIAL

This characteristic is particularly interesting for:

Fig. I.1. Underpressure created under the vapour barrier at –23°C.

- Tank bottoms, hot and cold
- Equipment exposed to traffic, even occasional
- Pipe supports.

I.1.4 PROTECTION AGAINST CORROSION THROUGH WATER INGRESS

Certain types of equipment, particularly in humid, polluted or coastal regions, will be rapidly attacked by water, especially in case of salty sea water or acid rain. Deluge tests have disastrous consequences, particularly when carried out with gutter water. Using cellular glass at certain well-chosen spots will considerably decrease the corrosion risk that can have very serious consequences.

I.1.5 FIRE RISK

Cellular glass contributes very positively to the fire protection of the equipment (Fig. I.2). It does not spread the fire nor does it generate smoke or transmit

Fig. I.2. Fire risk with inflammable insulations.

heat quickly. Hydrocarbon fire tests were run with 2 or 3 layers and resulted in up to 2 hours' resistance against fire.

I.1.6 ABSORPTION OF COMBUSTIBLE LIQUIDS OR GASES

Penetrations of combustible liquids or gases in insulants dramatically increase the risk of fire or explosion, because of the larger surface of contact with oxygen. As a result, the flash point or spontaneous combustion temperature decreases considerably. This explains many oil or ethylene oxide fires or explosions.

For instance, different non combustible insulation materials were placed in 250 ml graduated cylinders that were filled with kerosene to the same level. Only FOAMGLAS® cellular glass insulation did not wick the kerosene and did not burn (Fig. I.3).

Fig. I.3. Possible wicking effect with non combustible products.

I.1.7 BACTERIAL RISKS

The food industry is eager to avoid any bacterial proliferation. Many insulation materials offer an excellent environment for the development of bacteria, parasites and rodents.

Cellular glass, due to its mineral composition and waterproof structure, offers the best guarantees against this danger.

I.2

HYDROCARBON CHEMISTRY

I.2.1 INTRODUCTION TO HYDROCARBON CHEMISTRY

Hydrocarbons are products whose molecules chiefly comprise carbon combined in open or closed chains, and hydrogen atoms.

They are found in nature in two forms:

- **In liquid phase**:
  they constitute the crude oils, which are refined. The processes consist in separating the multiple molecules and in modifying some of them.
- **In vapour phase**:
  they constitute natural gas that is treated in processes consisting mainly of separating the molecules.

Refining and Natural Gas processes yield raw materials, which may be used just as they are or modified in Petrochemical processes to become more elaborate products, for everyday use.

These conversion processes (Refining, Natural Gas Processing and Petrochemistry) often use very sophisticated techniques and expensive equipment operating at varying temperatures and frequently require thermal insulation materials that are efficient, reliable and safe.

The latter part of this chapter, after a simple recall of basic data on hydrocarbons, will examine where FOAMGLAS® cellular glass can be of use in bringing a solution to these problems.

I.2.2 TYPES OF HYDROCARBONS

The family of hydrocarbons includes molecules consisting of carbon (C), hydrogen (H) and possibly other atoms like oxygen (O), chlorine (Cl), fluorine (F), etc.

I.2.2.1 Hydrocarbons with Open Carbon Chains

$$—C—C—C…$$

or

$$—C=C—…$$

which consists of:

- **Saturated** hydrocarbons (alkanes or paraffins) with low reactivity
- **Unsaturated** hydrocarbons, for instance a carbon compound – carbon with a double bond C=C (alkenes or olefins) or triple C≡C (alkynes or acetylenic hydrocarbons) which are more or less reactive due to the possibility of addition of other atoms or molecules to the double or triple bond.

I.2.2.2 Hydrocarbons with Closed Carbon Chains

which consists of:

- **Saturated** hydrocarbons (cyclanes or naphtenes) with low reactivity

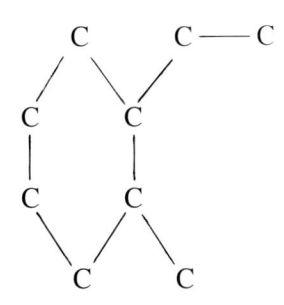

- **Unsaturated** hydrocarbons, often including a benzene nucleus

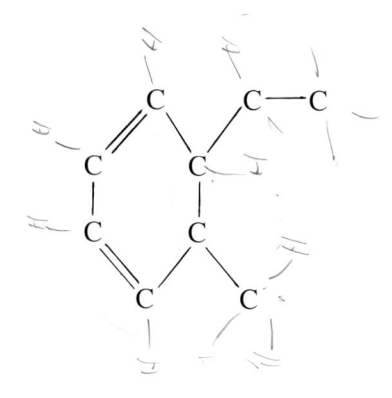

which are more or fairly reactive (Fig. I.4).

Name	Synonyms	Family of words	Molecular formula	Structure	Reactivity
Alkanes (saturated)	paraffined hydrocarbons **paraffins**	...ANE	C_nH_{2n+2}	$-C-C-$	– very low – possible substitution of atoms or molecules
Cyclanes (saturated)	cyclanic hydrocarbons **naphthene** naphthenic hydrocarbons	Cyclo...ANE	C_nH_{2n}	$(CH_2)_n$	– very low. – possible substitution of atoms or molecules
Alkenes (unsaturated)	olefinic hydrocarbons **olefins**	...ENE	C_nH_{2n}	$-C=C-$	– strong – easy addition of atoms and molecules
Alkynes (unsaturated)	**acetylenic hydrocarbons**	...YNE	C_nH_{2n-2}	$-C\equiv C-$	– strong – easy addition
Aromatics	**benzenoid hydrocarbons**				– possible substitution and addition

Fig. I.4. Recapitulatory table.

I.2.3 CLASSIFICATION

I.2.3.1 Numbers of Carbons

The names of the prefixes characterising the most common hydrocarbon chains according to the number of carbon atoms are the following:

- C   Meth
- C—C   Eth...
- C—C—C   Prop...
- 4 x C   But...
- 5 x C   Pent...
- 6 x C   Hex...
- 7 x C   Hept...
- 8 x C   Oct...

I.2.3.2 Boiling Temperatures

The boiling temperatures of hydrocarbons mainly depend on the number of carbon atoms in its molecule, but also to a lesser extent on the structure of the chains (Fig. I.5).

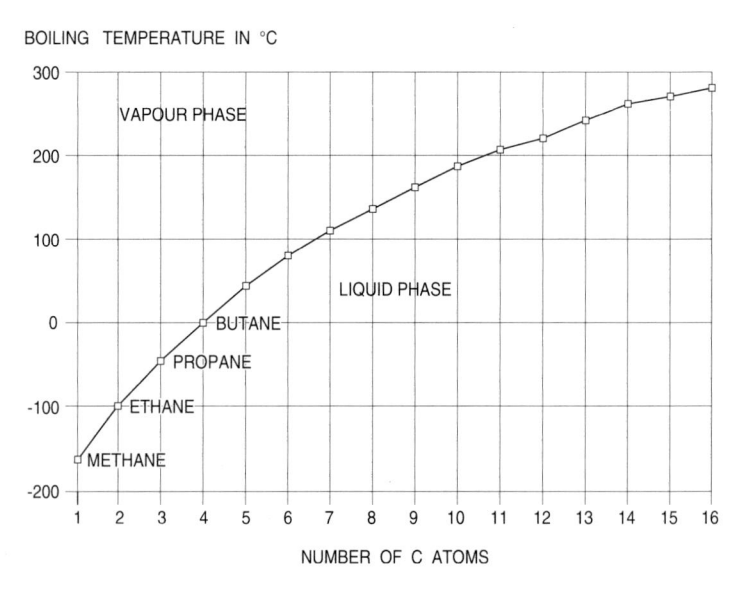

Fig. I.5. Boiling temperature curve of alkanes.

Example: Isomerism of n-Hexane

- Hexane: C—C—C—C—C—C
  Boiling temperature = 68.7°C

- 2-methylpentane:
  $$\begin{array}{c} \text{C} \\ | \\ \text{C—C—C—C—C} \end{array}$$
  Boiling temperature = 60.3°C

- 3-methylpentane:
  $$\begin{array}{c} \text{C} \\ | \\ \text{C—C—C—C—C} \end{array}$$
  Boiling temperature = 63.3°C

- 2.2. dimethylbutane:
  $$\begin{array}{c} \text{C} \\ | \\ \text{C—C—C—C} \\ | \\ \text{C} \end{array}$$
  Boiling temperature = 49.7°C

- 2.3. dimethylbutane:
  $$\begin{array}{c} \text{C—C—C—C} \\ | \quad | \\ \text{C} \quad \text{C} \end{array}$$
  Boiling temperature = 58°C

One understands the importance of the latter one for the fineness of product obtained in the refinery distillation columns.

In practice, refineries classify products from 1 to 5 carbon atoms.

For products exceeding this classification, the operation becomes complex and onerous and the separation is carried out from complex mixtures or "hydrocarbon fractions".

The table I.1 below indicates the boiling temperature at atmospheric pressure of the different chemical products, encountered.

TABLE I.1

Products	°C
Helium	−269
Hydrogen	−253
Nitrogen	−196
Oxygen	−183
Methane	−161
Ethylene	−104
Ethane	−88
CO_2	−78
Propylene	−47
Propane	−44
Ammonia	−33
Iso-butane	−10
n-Butane	−1
Gas-oil	54-77
Residual # 6 fuel	77-88
Asphalt	93-149

I.3
REFINERY

The first stage of a modern refinery is the distillation plant which in general includes:

- An atmospheric distillation unit that fractionates the crude oil into different cuts: LPG gasoline, kerosene, gas oil and atmospheric residue
- A LPG gasoline fractionation unit that achieves the separation of this cut into fuel gas, LPG (Liquefied Petroleum Gases: a mixture of propane and butanes), light and heavy gasoline
- A vacuum distillation unit that achieves the separation of the atmospheric residue into vacuum gas oil and vacuum residue.

All these products are the bases of the refinery finished products. They will undergo either molecular transformations or a further physical separation.

In fact, the situation at the outlet of this first stage, in terms of qualities and quantities of products, shows that:

- No product, except the atmospheric residue, which can be sold as heavy fuel oil, can be considered as finished product
- The quantities of the different cuts do not correspond to the quantities required by the market. There is an excess of heavy products and a shortage of light ones

It is then necessary to use transformation and separation processes to improve the quality of the products and decrease the excess of heavy products while producing more gasoline and gas oil.

I.3.1 THE DESTRUCTIVE TRANSFORMATION PROCESSES

Two processes are modifying the molecular structure and improving the quality. They are:

I.3.1.1 The Catalytic Reforming Process

The catalytic reforming process increases the octane number of heavy gasoline. Hydrogen, fuel gas and LPG are also produced.

I.3.1.2 The Isomerisation Process

The isomerisation process consists in the transformation of normal paraffin into isoparaffins that have a high octane number.

I.3.2 PROCESSES WHICH CONVERT EXCESS PRODUCTS INTO LIGHTER PRODUCTS

In order to obtain a larger amount of light products, heavy molecules can be modified by the following processes:

I.3.2.1 The Catalytic Cracking Process

The catalytic cracking of vacuum gas oil or atmospheric residue aims at producing gasoline with high octane number. This process produces also fuel gas, LPG, gas oil and fuel oil.

I.3.2.2 The Thermal Cracking Process

This process consists in the thermal cracking of atmospheric or vacuum residues which according to the severity of the cracking can be:
- The visbreaking process that is a mild cracking. It produces gas oil and a residue of lower viscosity than the feed. This process produces also fuel gas, LPG and gasoline
- The coking process that is a deep cracking. It produces gas oil, but also gas, LPG, gasoline and coke.

I.3.3 THE HYDROCRACKING

The hydrocracking process of vacuum gas oils consists in producing kerosene and gas oil but also fuel gas, LPG and gasoline. The hydrogen presence allows to get products of good quality with low sulphur content and good stability. Kerosene and gas oil obtained through these units are in general of very good quality.

I.3.4 THE SYNTHESIS PROCESSES

These processes use LPG and especially unsaturated LPG to produce gasoline. The main processes of this type are:

I.3.4.1 The Alkylation Process

The alkylation process produces gasoline by recombination of an olefin and a paraffin.

I.3.4.2 The Dimersol Process

The Dimersol process produces gasoline by combination of two olefins.

I.3.5 THE PURIFICATION PROCESSES

Although classified in the molecular transformation process group, these processes do not transform, or only slightly, the molecular structure of the product. They allow either to destroy the impurities such as sulphur, nitrogen and organo-metallic compounds in the presence of hydrogen – in this case these processes are called hydrotreatment processes – or to chemically transform the sulphur compounds especially mercaptans into less undesirable products such as disulphides – in this case these processes are called chemical treatment processes.

I.3.6 SEPARATION PROCESSES

This family of processes includes:

I.3.6.1 The First Distillation Cuts Fractionation Processes

The solvent deasphalting process is included in this category. It allows to separate, using a solvent like propane or butane for instance, the asphalt from a vacuum residue. The deasphalted oil can then be sent to a conversion unit.

I.3.6.2 The Fractionation Processes

This category comprises the distillation of all the effluents of the destructive and synthesis processes.

The amine treating process used to remove H_2S from the refinery fuel gas or unit gas hydrotreatment can also be included in this family.

I.3.7 MISCELLANEOUS PROCESSES

The Claus process used to transform into sulphur the hydrogen sulphide recovered from refinery gases in the amine treating unit belongs to this group of processes.

I.3.8 GENERAL REFINERY SCHEME

Figure I.6 shows how in a refinery, the above described processes are combined.

I.3.9 RECOMMENDATIONS FOR THE USE OF CELLULAR GLASS IN REFINING PROCESSES

In refining processes the temperatures are often high and a thermal insulation is required. FOAMGLAS® cellular glass can be recommended in case of:

- Risk of fire due to the absorption of combustible liquids caused by spillage or leakage from valves, flanges, outlets or overflows

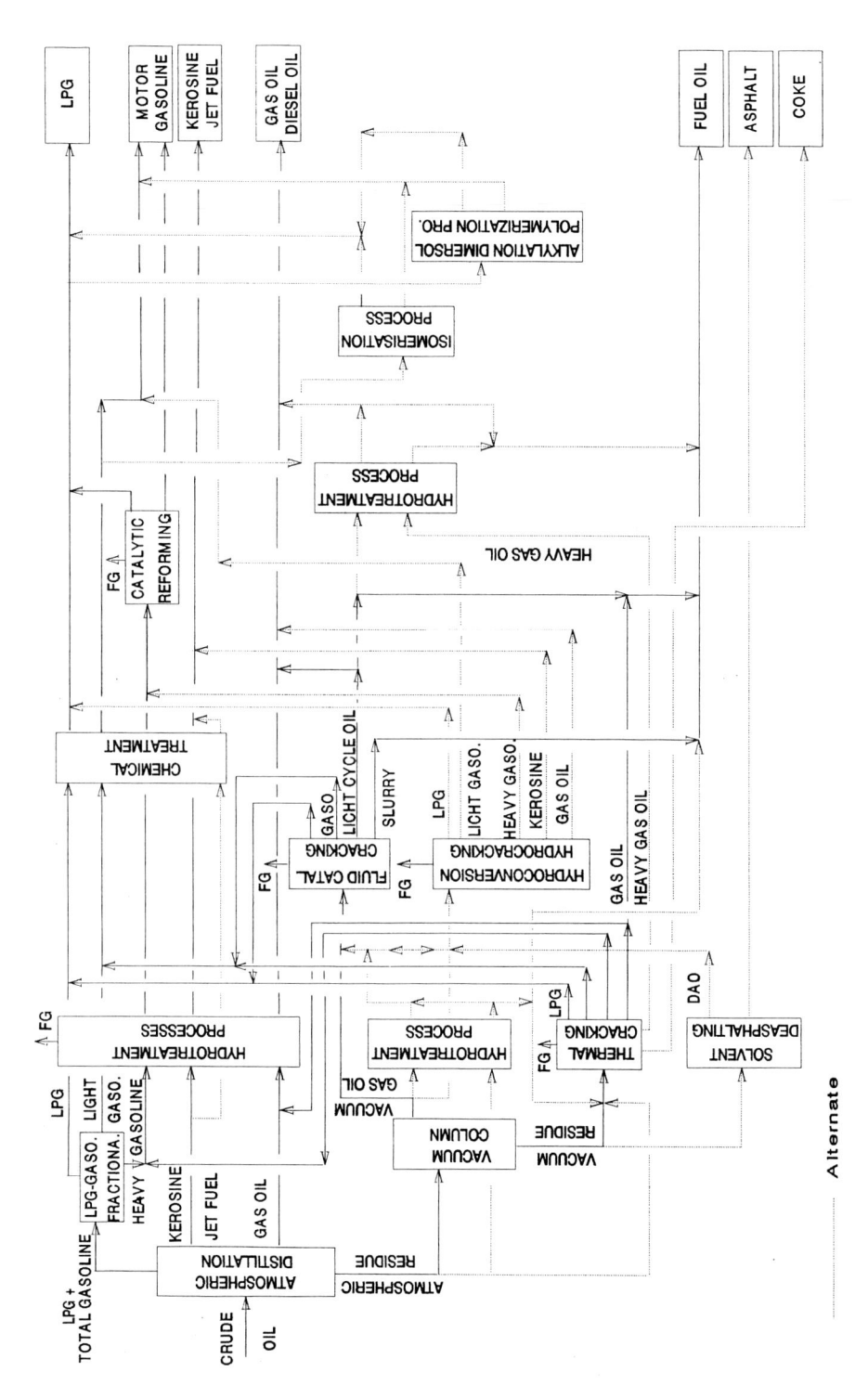

Fig. I.6. General refinery scheme *(BEICIP Document)*.

PART I. WHERE TO USE CELLULAR GLASS

- Risk of corrosion, especially around rings, supports, equipment and pipe connections
- Traffic for maintenance or control particularly at the upper part of some pipes, exposed vessels or tank roofs where the deterioration of the weather protection will lead to water ingress
- Risk of moisture ingress and retention in the external part of the insulation for equipment working above 100°C. There FOAMGLAS® cellular glass can be used as external layer where it will act as a self drying agent with the overfit system. This will be explained in V.4.5
- Pipes and vessels where temperature control is essential for the process or to avoid crystallisation, high viscosity, chemical reaction, icing etc. Often these lines and equipment are heat traced as in the following processes:
  - Atmospheric distillation: on the steam traced lines for the atmospheric residue
  - Vacuum distillation: for the same steam traced atmospheric residue lines and for the heavy vacuum gas oil and vacuum residue
  - Sulphur unit: where the temperature control of the sulphur is critical. If too high it increases the viscosity and if too low it crystallises the sulphur
- High temperature tank bases such as heavy fuel, bitumen, sulphur
- Inner insulation of bitumen tank
- Exposed steam lines
- Heating oil lines around flanges, valves, tees.

For sections where the temperatures are cold, which is seldom the case in refineries, cellular glass can be strongly recommended for its resistance to moisture ingress.

The cold areas where FOAMGLAS® cellular glass is recommended in refineries are mainly the LPG storage and the sulphuric acid alkylation process.

In the MTBE production unit, which will be described in I.6.2.12, cellular glass is particularly recommended around flanges or valves for safety reasons, due to the danger of leaks in this process.

I.4

NATURAL GAS PROCESSING

I.4.1 INTRODUCTION

Natural gas processing is a general term implying that natural gas is being manipulated.

I.4.1.1 Field Processing

Field processing is a process necessary to make the gas marketable. It consists in removing the water excess of hydrocarbon liquids, and impurities.

I.4.1.2 Plant Processing

Plant processing can be similar to field processing but also includes the separation of products contained in the gas stream, such as methane, ethane, propane, butane or natural gasoline, to increase the value of well production, and the removal of hydrogen sulphide and carbon dioxide.

I.4.2 GENERAL NATURAL GAS PROCESSING SCHEME (Fig. I.7)

I.4.2.1 H_2S, CO_2 Removal Processes

There are various licences of H_2S, CO_2 removal processes.

The Rectisol process that uses methanol and the Fluor Solvent process that uses propylene carbonate operate at negative temperatures, where cellular glass insulation can be recommended.

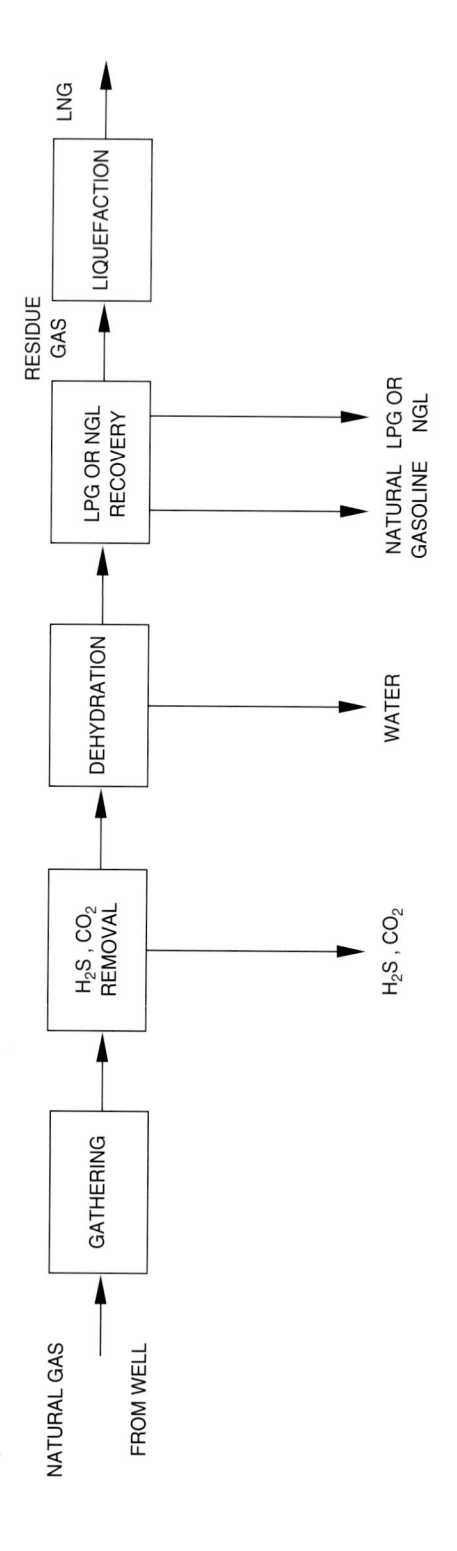

Fig. I.7. General natural gas processing scheme (*BEICIP Document*).

I.4.2.2 Dehydration

There are two main processes:

- The absorption that uses a liquid: triethylene glycol
- The adsorption that uses a solid: activated alumina or silica gel.

I.4.2.3 LPG–NGL Recovery

There are two types of processes used to recover the LPG or LNG liquids from natural gases:

- The cold oil absorption process from –20°C to –30°C (Fig. I.8)
- The cold process using either:
  - The Joule-Thompson effect by pressure drop through a valve from –10°C to –30°C
  - The external refrigeration at –35°C
  - The turbo expander from –100°C to –130°C (Fig. I.9).

Cellular glass insulation is recommended for all cold pipes, vessels and equipment.

I.4.2.4 LNG Liquefaction Processes

These are various commercial processes available to liquefy natural gas but all of them are based on the refrigeration of the gas down to –161°C to liquefy it under atmospheric pressure.

Whatever process is used, the operating temperatures range from ambient to –161°C.

Produced LNG is stored in double-wall storage tanks.

Boil-off gas from the storage tanks and tanker loading is compressed and sent to the fuel gas system or to the flare.

The process flow scheme of the storage tanks and loading facilities of a liquefaction plant is very similar to the receiving terminal hereafter depicted (Fig. I.10).

PART I. WHERE TO USE CELLULAR GLASS

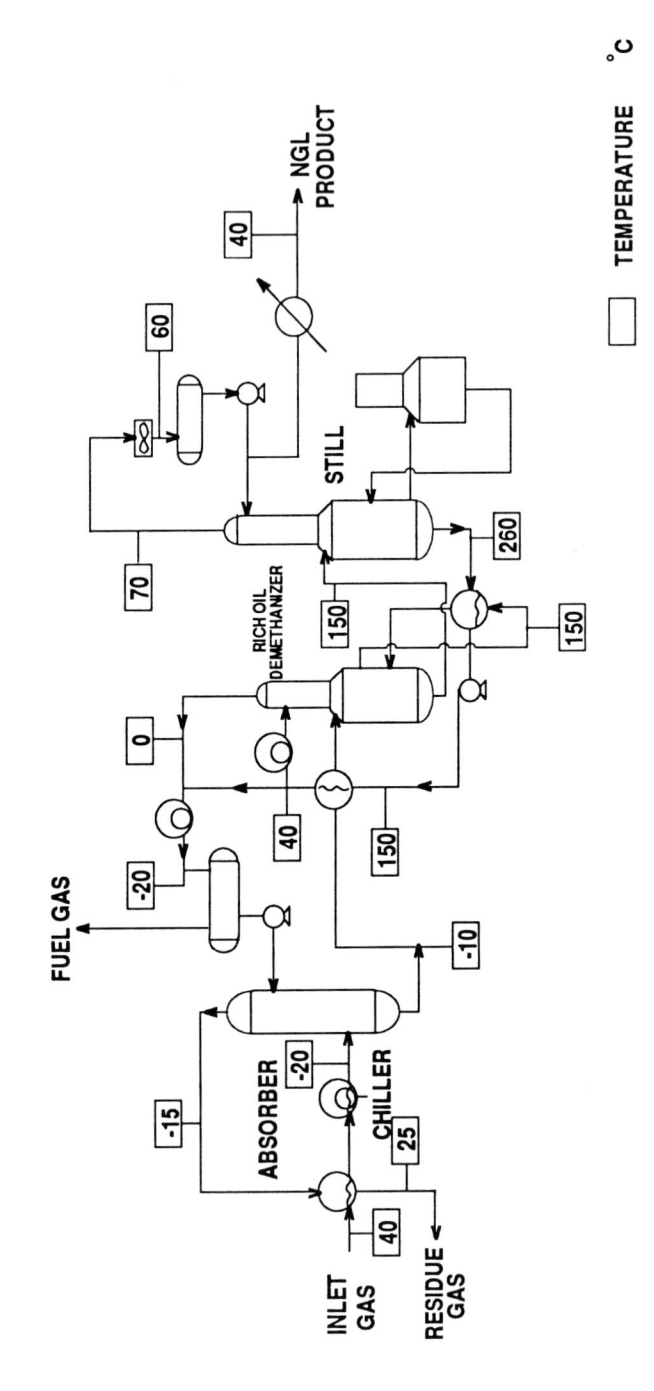

Fig. I.8. NGL recovery. Cold oil absorption process (*BEICIP Document*).

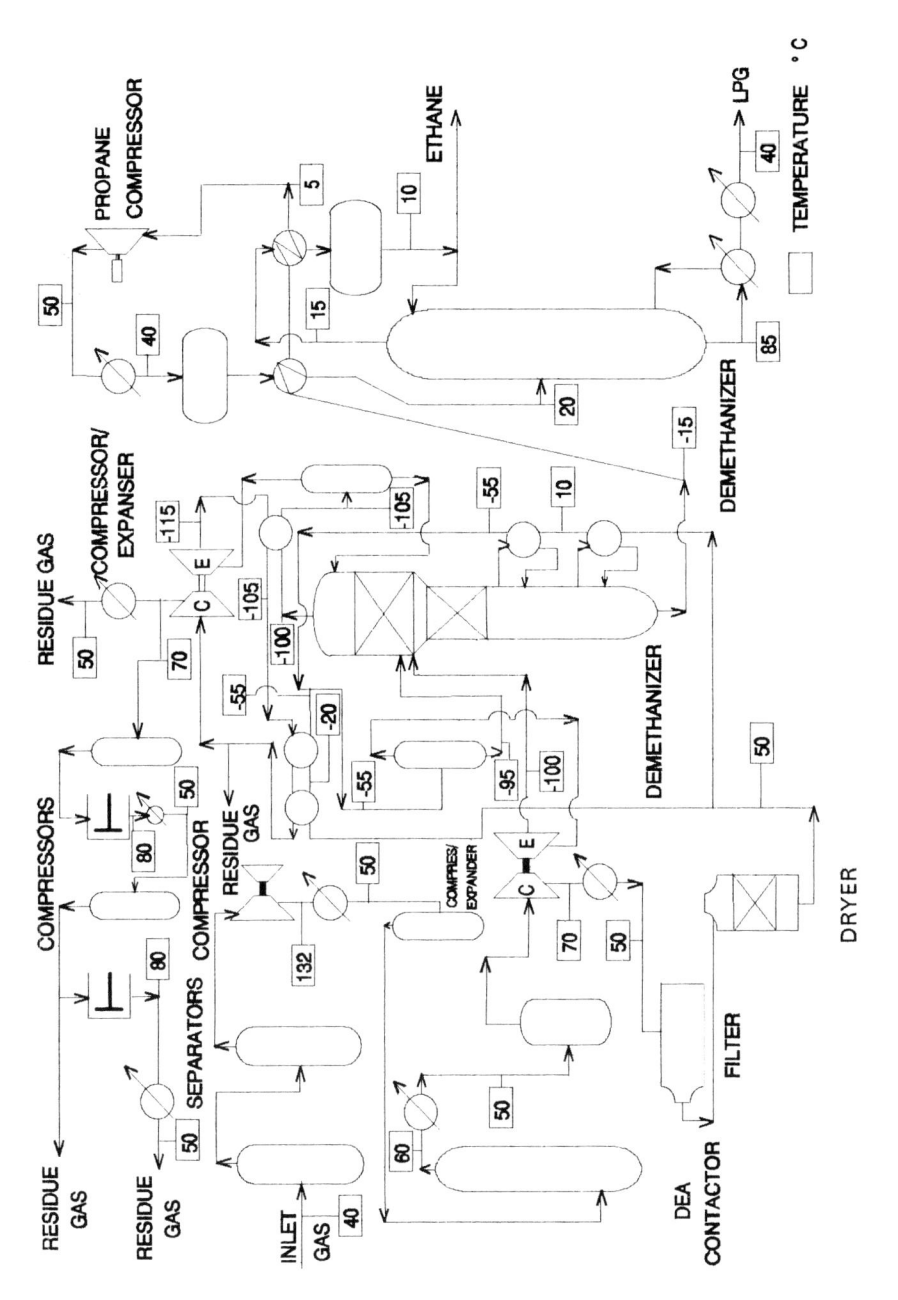

Fig. I.9. NGL recovery. Turbo expander process (*BEICIP Document*).

PART I. WHERE TO USE CELLULAR GLASS

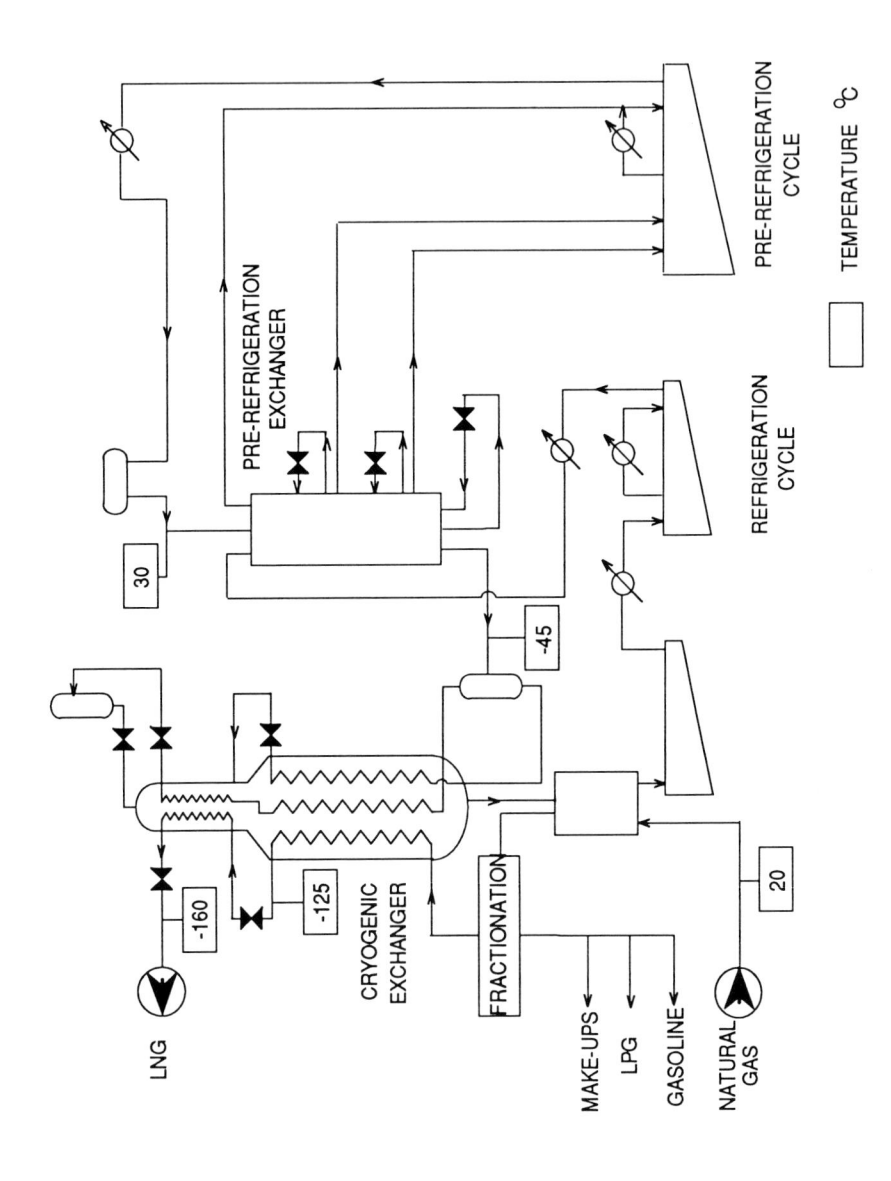

Fig. I.10. Liquefaction of natural gas. Tealarc process (*BEICIP Document*).

I.4.2.5 LNG Receiving Terminals

The receiving terminal includes pumps and vaporisers required to vaporise LNG before sending it to the transportation and distribution pipeline (Fig. I.11).

All equipment of the LNG storage, loading facilities and receiving terminal, except vaporisers and downstream equipment, operate at a temperature of about −161°C and require appropriate insulation.

Loading, vapour return and unloading lines are kept at low temperature, even when not in use, by circulating LNG or N_2.

The quality, the longevity and the productivity of LNG plants are very much relying on the qualities of the thermal insulation.

Many LNG plants in the world have been insulated with FOAMGLAS® cellular glass, combined with other insulation materials.

The development of new types of cellular glass, with a lower thermal conductivity at low or cryogenic temperature, has allowed new insulation techniques, using 100% cellular glass.

Advantages of these new techniques are numerous:

- Reduced number of layers
- Improved fire resistance
- Lower cost
- Reduced maintenance
- Improved longevity etc.

When revamping existing plants and insulating new plants these new techniques should be considered.

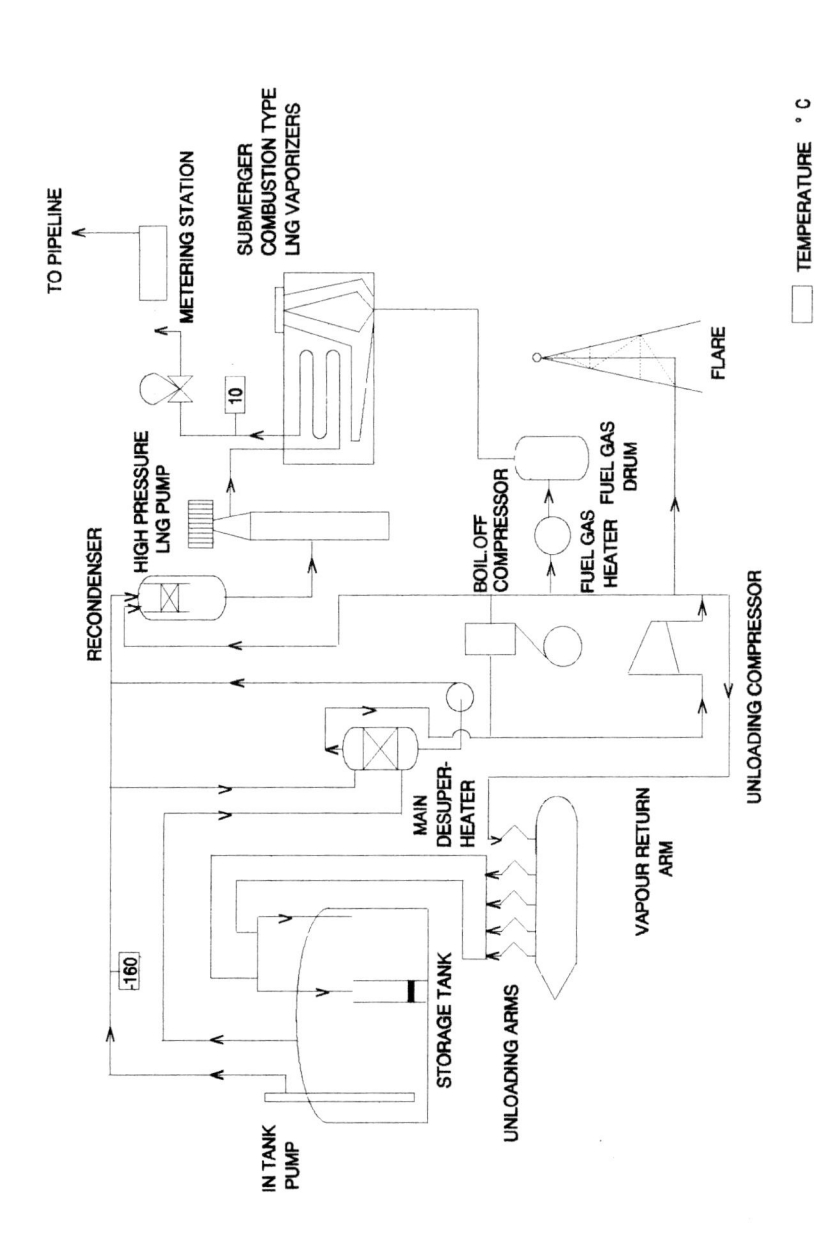

PART I. WHERE TO USE CELLULAR GLASS

Fig. I.11. LNG receiving and vaporising terminal (*BEICIP Document*).

I.5
GAS PROCESSING

I.5.1 SYNTHESIS GASES

Many gases are produced from coal, kerosene, naphtha, fuel, solids.

Some processes are using fractionation at low or cryogenic temperature.

I.5.2 INDUSTRIAL GASES

Gases, such as oxygen, nitrogen, argon, rare gases, hydrogen and helium, are often produced by air distillation at cryogenic temperature.

Organic materials are reacting dangerously with liquid oxygen and the only safe way to insulate is to use either a gas sealed (and expensive) system or cellular glass, totally inorganic.

I.6
PETROCHEMISTRY

I.6.1 INTRODUCTION

Petrochemistry consists in modifying and transforming the products obtained in refining and in gas treatment, sometimes in several operations, to obtain final materials such as:

- Urea, fertilizers
- Plastics like PVC or polyethylene
- Textile fibres
- Neoprenes
- Polyurethanes
- Polystyrenes
- Resins
- Raw materials for the chemical industry.

The initial step is to manufacture various types of highly reactive chemical products from petroleum and gas hydrocarbons.

These products are the first generation intermediates:

- Hydrogen, ammonia, methanol
- Olefinic and dienic hydrocarbons: ethylene, propylene, butadiene, isoprene, etc.
- Aromatic hydrocarbons: benzene, toluene, styrene, xylene, etc.
- Acetylene.

At a second stage, a new series of chemical operations is conducted to introduce various hetero-atoms into the final molecule, including oxygen, nitrogen, chlorine and sulphur. This leads to the formation of the so-called second generation intermediates.

A final operation is needed to obtain the target product by determining its formulation, so that its properties correspond to the intended use. These products include plastics, synthetic fibres, fertilizers, solvents, elastomers, insecticides and detergents, etc.

The three-step classification shows however many exceptions:

- A large number of operations are sometimes necessary to complete one step or another
- The basic reactants must also be used in suitable form: this may require additional operations
- The manufacture of some products doesn't need so many steps.

Despite these exceptions, the foregoing classification allows to identify the three major types of petrochemical complexes:

- Complexes based on synthesis gas chemistry and culminating in the production of fertilizers and resins. These processes are centred on steam reforming
- Olefinic and diolefinic complexes, of which the basic reaction consists in steam cracking
- Aromatic complexes, more specific to the refining industry for the production of high octane motor fuels, centred around catalytic reforming. With respect to petrochemicals these are mainly intended to obtain benzene, ethylbenzene, ortho and paraxylene.

I.6.2 GENERAL PETROCHEMISTRY SCHEME (Fig. I.12)

I.6.2.1 Ammonia

Ammonia production processes are mainly hot, but the storage of ammonia is effected cold: under pressure in spherical tanks at temperature below ambient, or at atmospheric pressure in cylindrical tanks at $-33°C$ (Fig. I.13).

Cellular glass insulation is recommended in the cooling process and the thermal insulation of:

- The spherical tanks according to the so-called "fully adhered compact system"

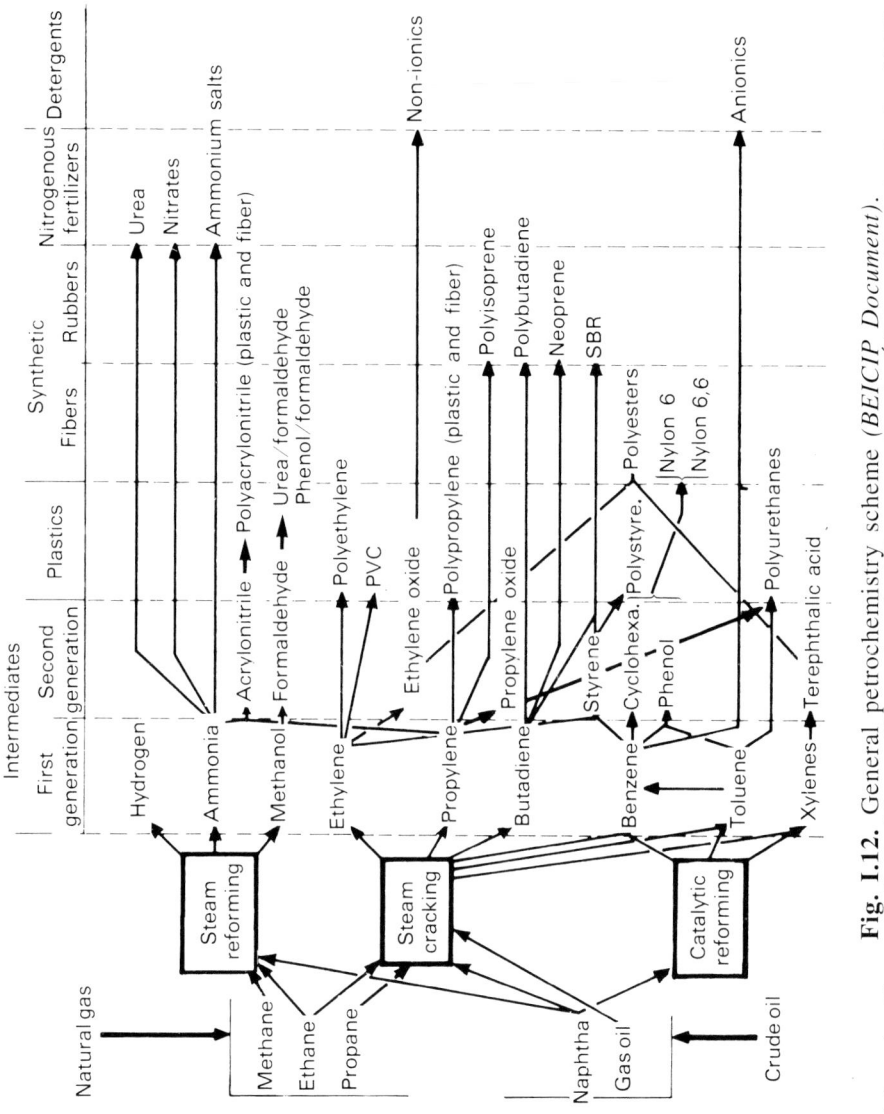

Fig. I.12. General petrochemistry scheme (*BEICIP Document*).
(After A. Chauvel, G. Lefebvre, "Petrochemical Processes", t.1, *Éditions Technip*, Paris, 1989)

PART I. WHERE TO USE CELLULAR GLASS

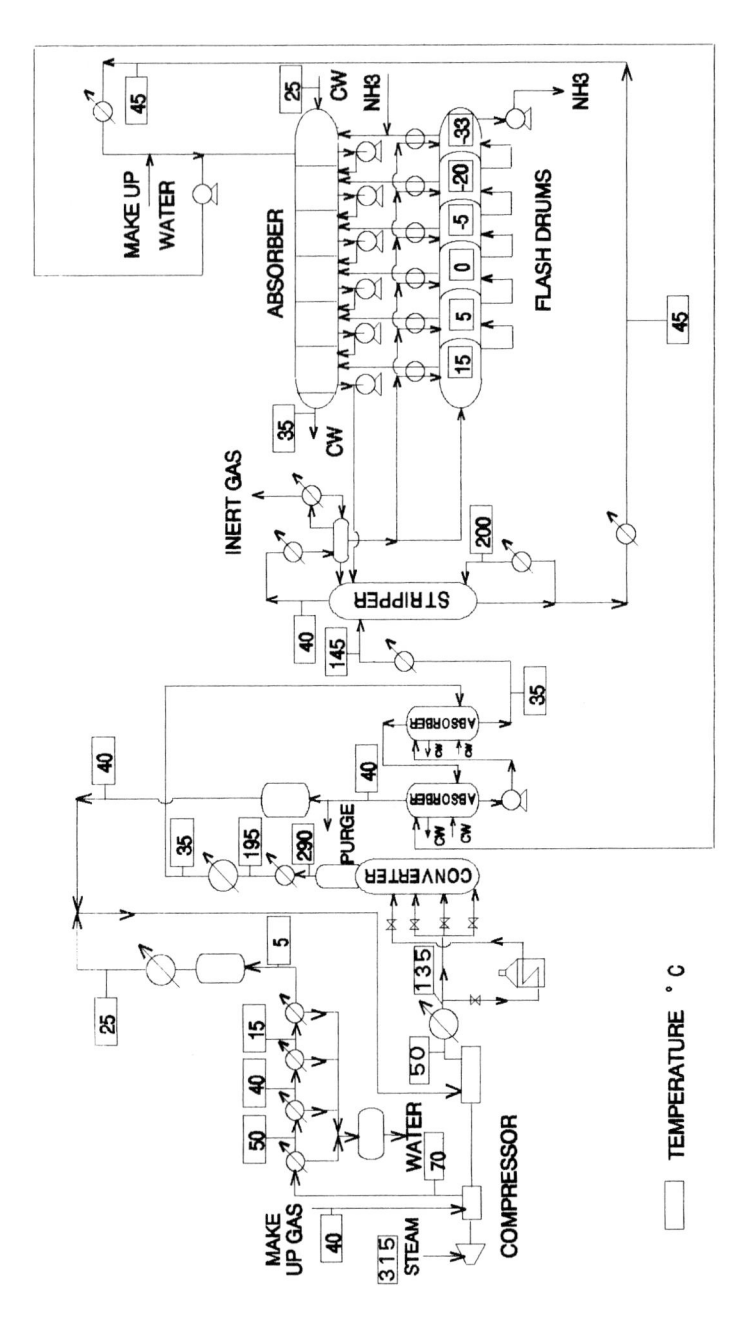

Fig. I.13. Ammonia synthesis. Low pressure. M.W. Kellogg. Type process *(BEICIP Document)*.

- The cylindrical tanks where FOAMGLAS® cellular glass is used for the insulation of tank bottom as well as of the walls and the roof with the fully adhered compact system.

I.6.2.2 Carbon Monoxide (Fig. I.14)

The Carbon Monoxide produced by partial condensation process is obtained at low or very low temperature (–120°C to –185°C) and cellular glass is particularly recommended as thermal insulation for the:

- Pipes
- Vessels
- Separators
- Heat exchangers.

The main licensers using low temperature or cryogenic separation are *Air Products*, *Air Liquide*, *Petrocarbon Development*, *Uhde* and *Union Carbide*.

I.6.2.3 Ethylene and Propylene by Steam Cracking

Steam cracking is a process that produces primarily ethylene but also propylene and as secondary products, depending on the feedstock employed, a C_4 cut, rich in butadiene and a C_5^+ cut with a high content of aromatics, particularly benzene.

Ethylene, C_2H_4,

$$\begin{array}{c} H \quad H \\ | \quad | \\ C = C \\ | \quad | \\ H \quad H \end{array}$$

highly reactive because of the double carbon link, is the basic molecule used for the production of most of the polymers (PVC, polyesters, polystyrenes, etc.).

The feedstocks can be ethane, LPG, naphtha or gas oils, but the most commonly used in the world is naphtha as shown in the following table (in % weight):

PART I. WHERE TO USE CELLULAR GLASS

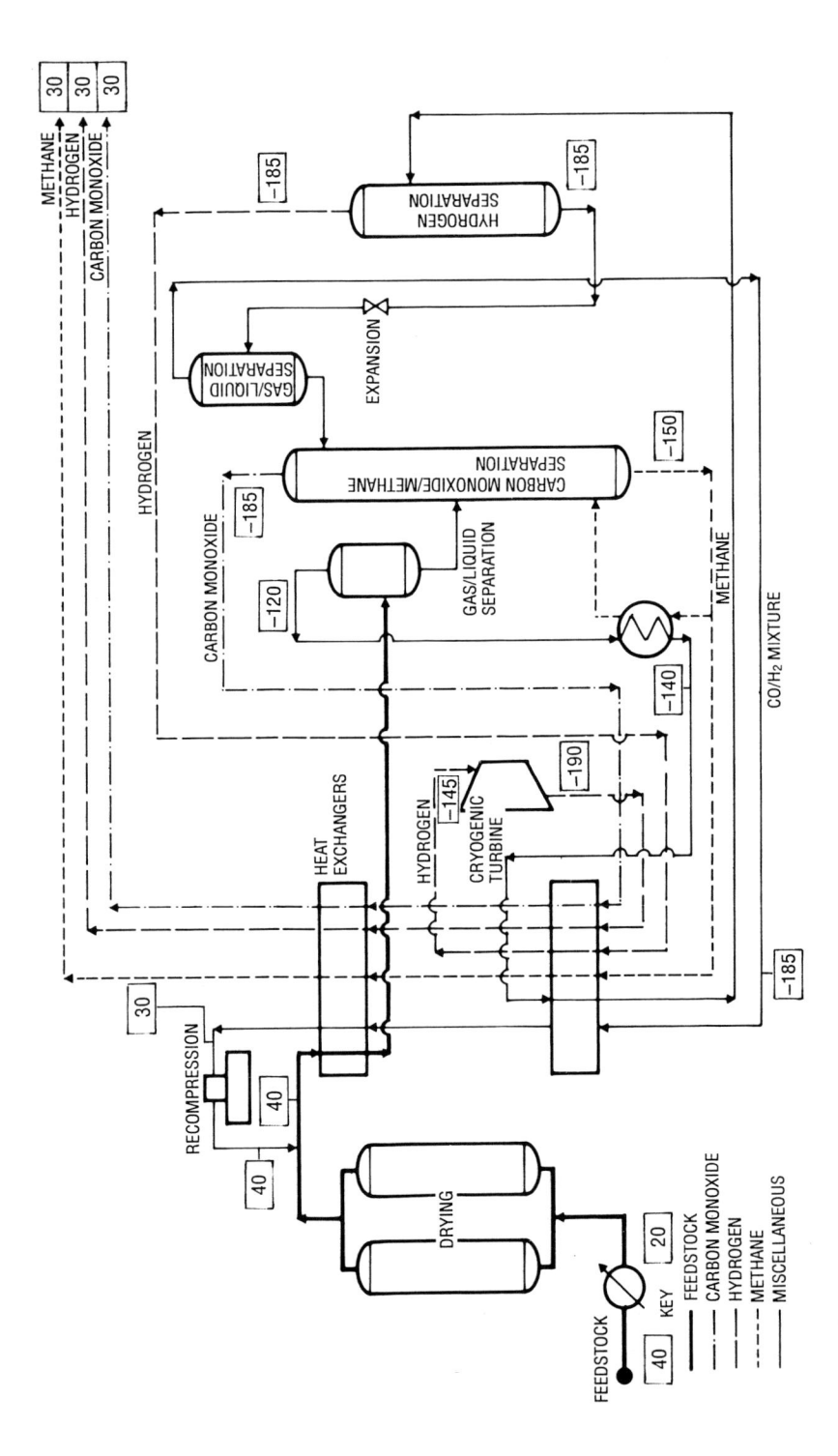

Fig. I.14. Cryogenic manufacture of carbon monoxide. Partial condensation (*BEICIP Document*). (After A. Chauvel, G. Lefebvre, "Petrochemical Processes", t.1, *Éditions Technip*, Paris, 1989)

Feedstock	Western Europe	United States	Japan	World
Ethane	8.0	57.5	–	30.5
LPG	11.0	19.0	7.5	11.0
Naphtha	69.0	9.5	92.5	49.0
Gas oils	12.0	14.0	–	8.5
Miscellaneous	–	–	–	1.0
Total	100.0	100.0	100.0	100.0

The ethylene is produced, together with propylene, in the vapour cracker furnace at +/–850°C in a less than one second reaction time and then cooled very quickly to +/–400°C to avoid polymerisation of the products.

A simple field diagram of a naphtha cracker shows the different steps of a complete unit (Fig. I.15 and Fig. I.16).

Separation of the different components is carried out first at high pressure. The purified products are finally recovered at low temperature (Fig. I.17).

Many low temperature steps are required to recover C_4-C_5 cuts, propane, propylene, ethane, ethylene, residual methane and hydrogen (Fig. I.18 and Fig. I.19).

The refrigeration circuits are using propylene and ethylene (Fig. I.20).

For the good working and the high efficiency of such a plant, it is very important to have excellent thermal insulation of the cold equipment, pipes, reactors or separators.

This thermal insulation may be submitted to fluctuating temperatures dependant upon the production, the liquid level, cleaning and shut downs, etc. and is therefore subject to a hostile environment.

Non-vapour proof insulation materials are fully dependant on the quality and the good performance during a number of years of the vapour barrier applied over the insulation. It is almost impossible to obtain this quality at places such as supports, elbows, flanges and tees. Moreover, moisture may saturate the insulation.

The demethanisation, the fractionation and the separation of ethylene and propylene are processes at low and very low temperature (down to –170°C) where non-absorbent and moisture resistant insulation is required.

PART I. WHERE TO USE CELLULAR GLASS

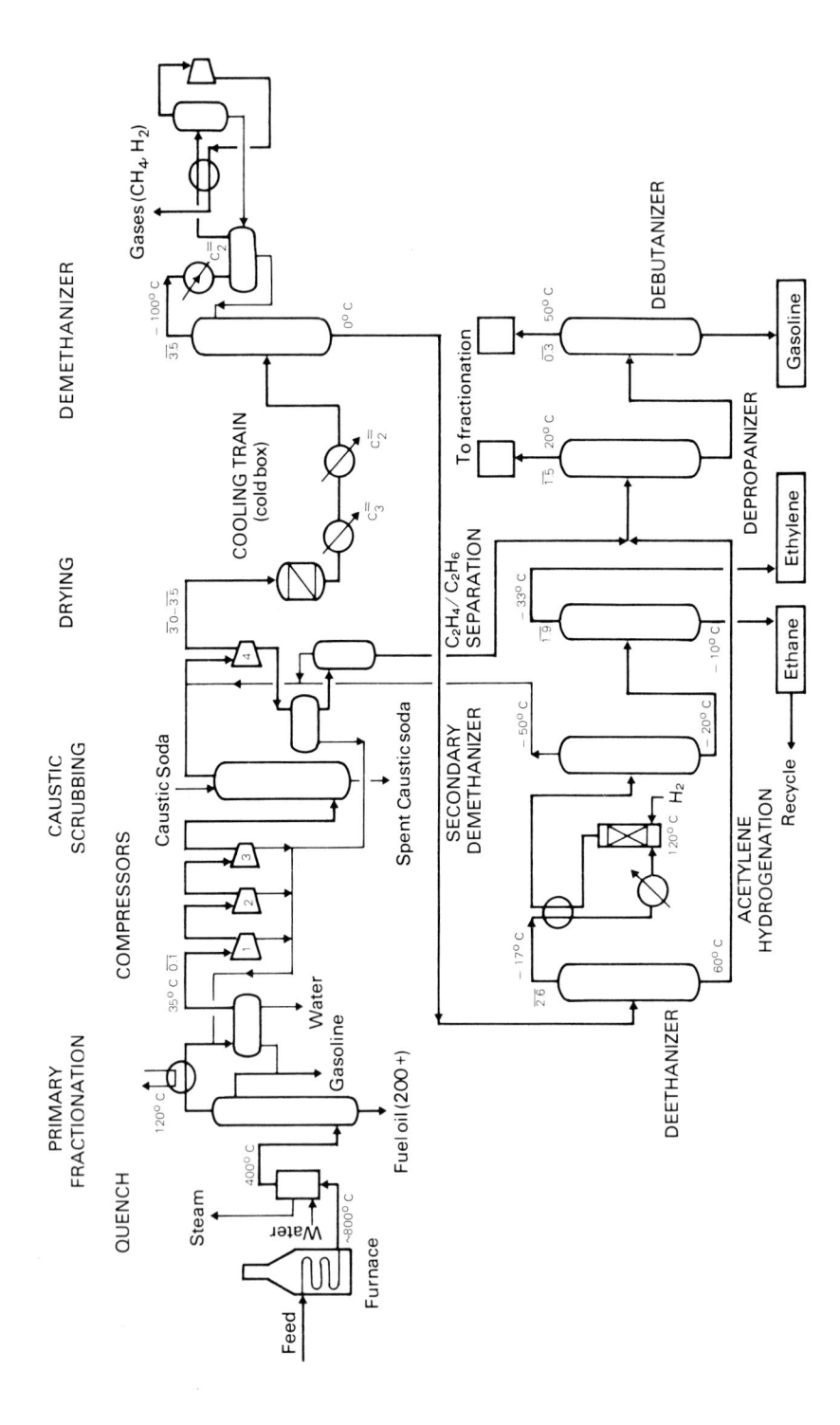

Fig. I.15. Flow diagram: naphtha cracker (*BEICIP Document*).

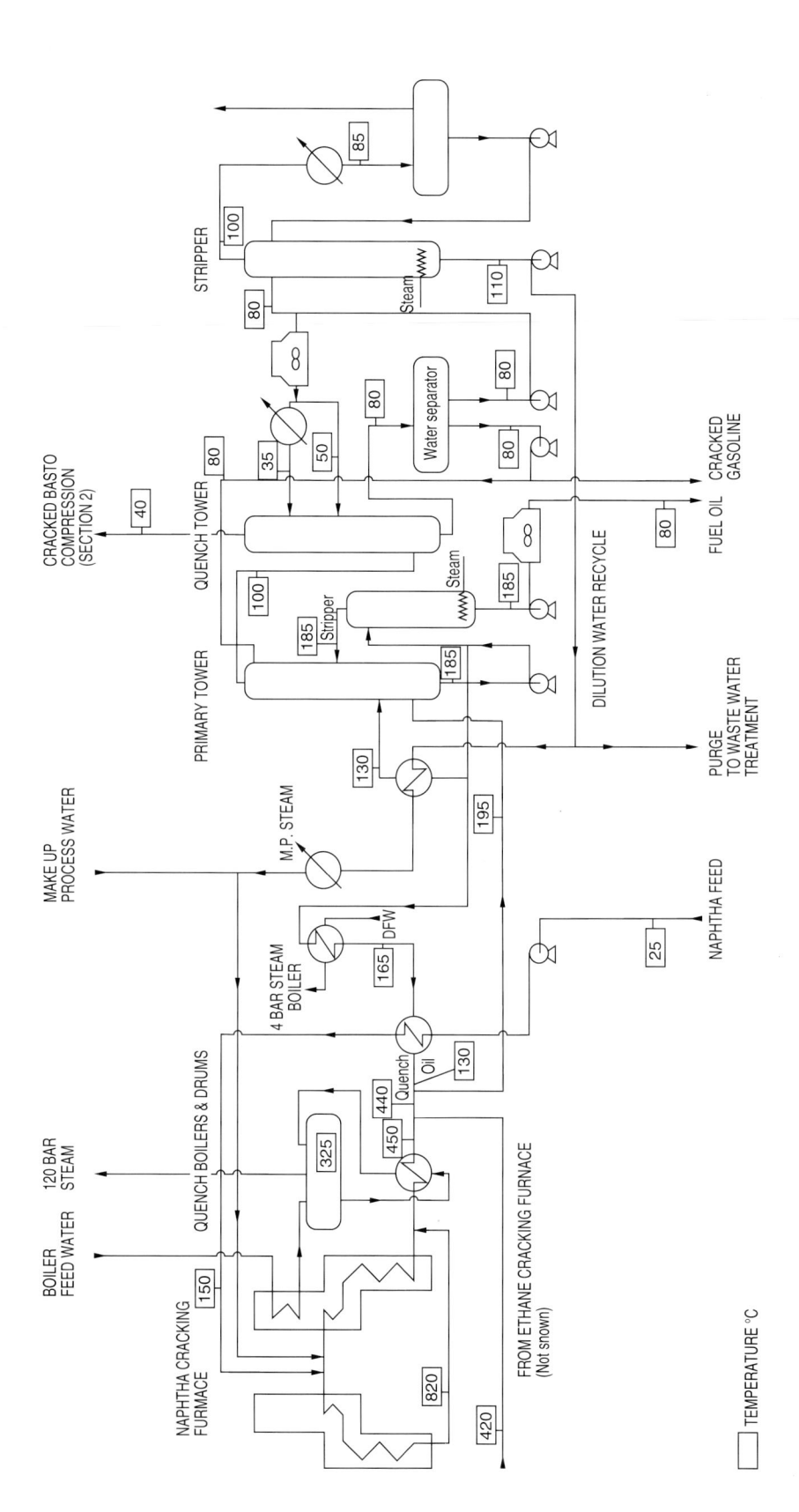

Fig. I.16. Flow diagram: cracking and primary fractionation naphtha cracking, section 1 (*BEICIP Document*).

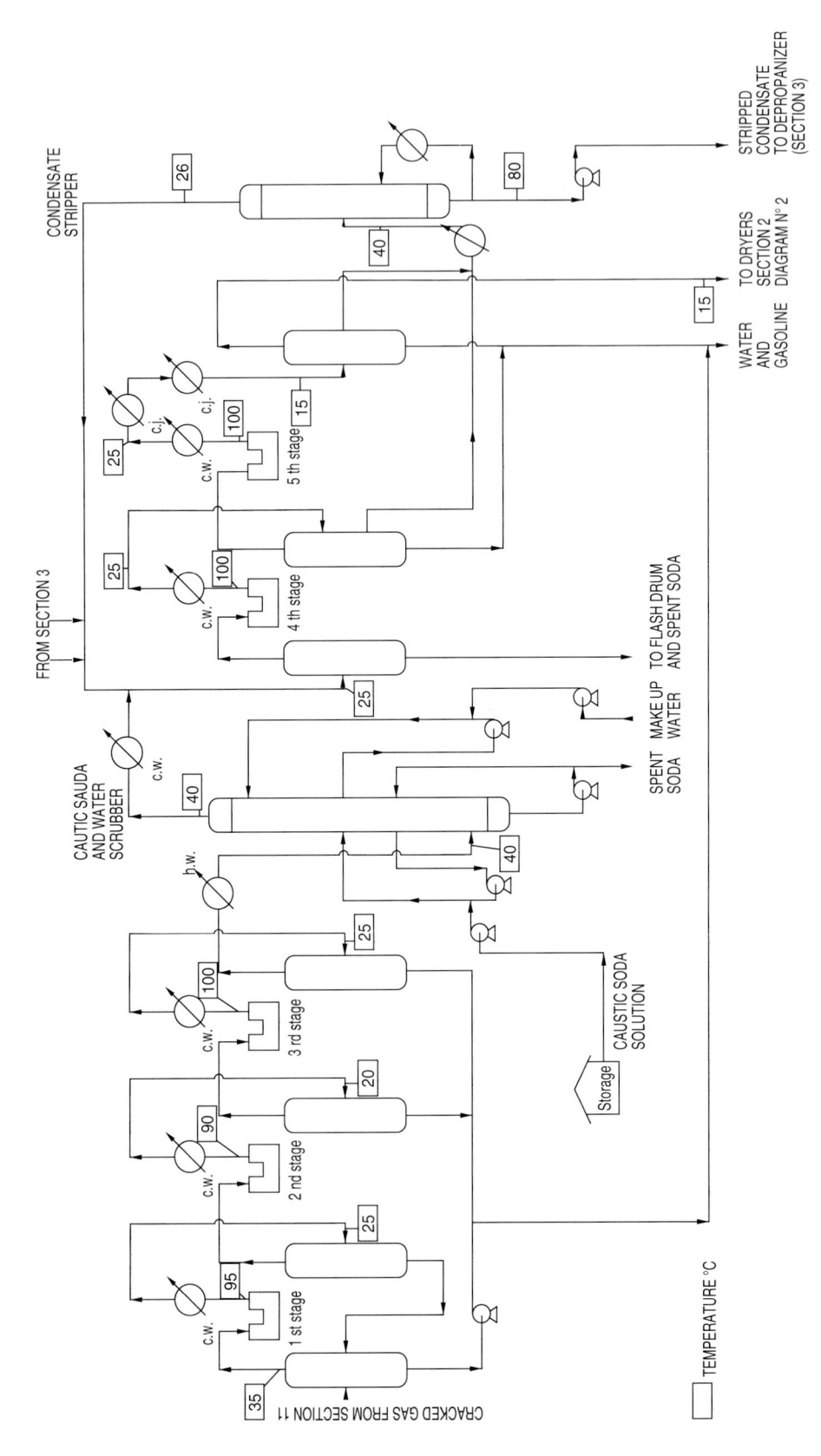

Fig. I.17. Flow Diagram: compression naphtha cracking, section 2.1 (*BEICIP Document*).

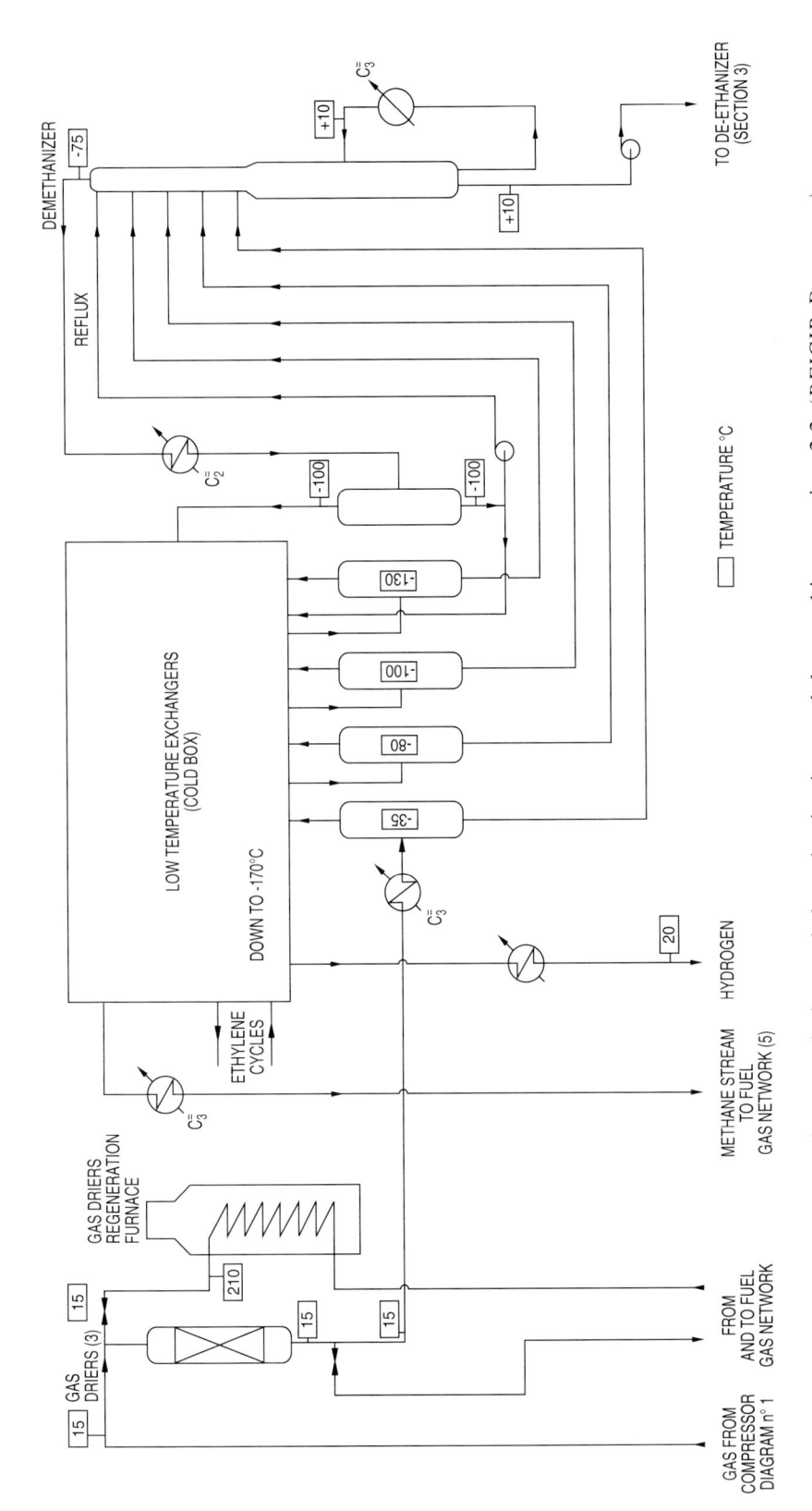

Fig. 1.18. Flow diagram: drying and demethanization naphtha cracking, section 2.2 (*BEICIP Document*).

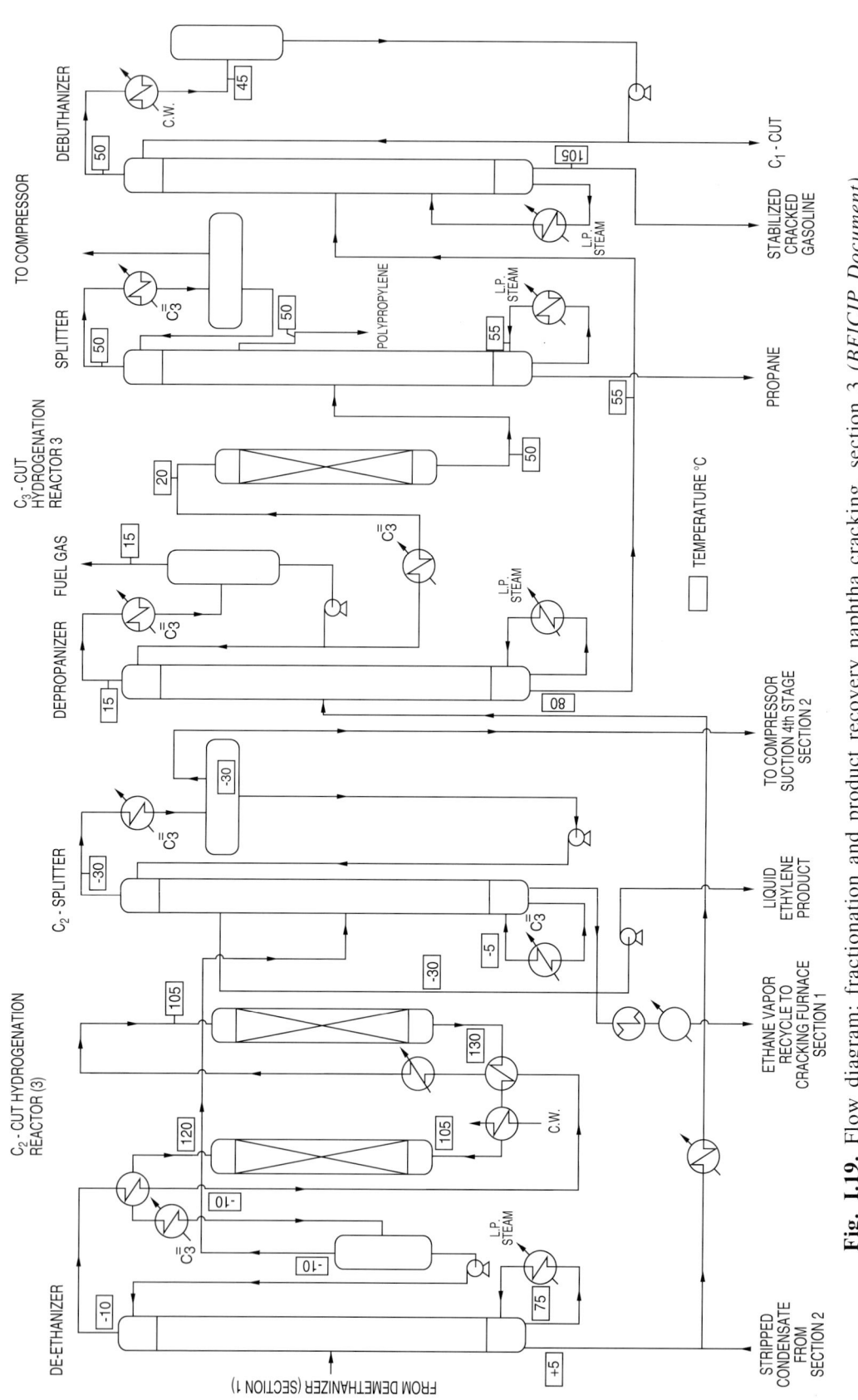

Fig. 1.19. Flow diagram: fractionation and product recovery naphtha cracking, section 3 (*BEICIP Document*).

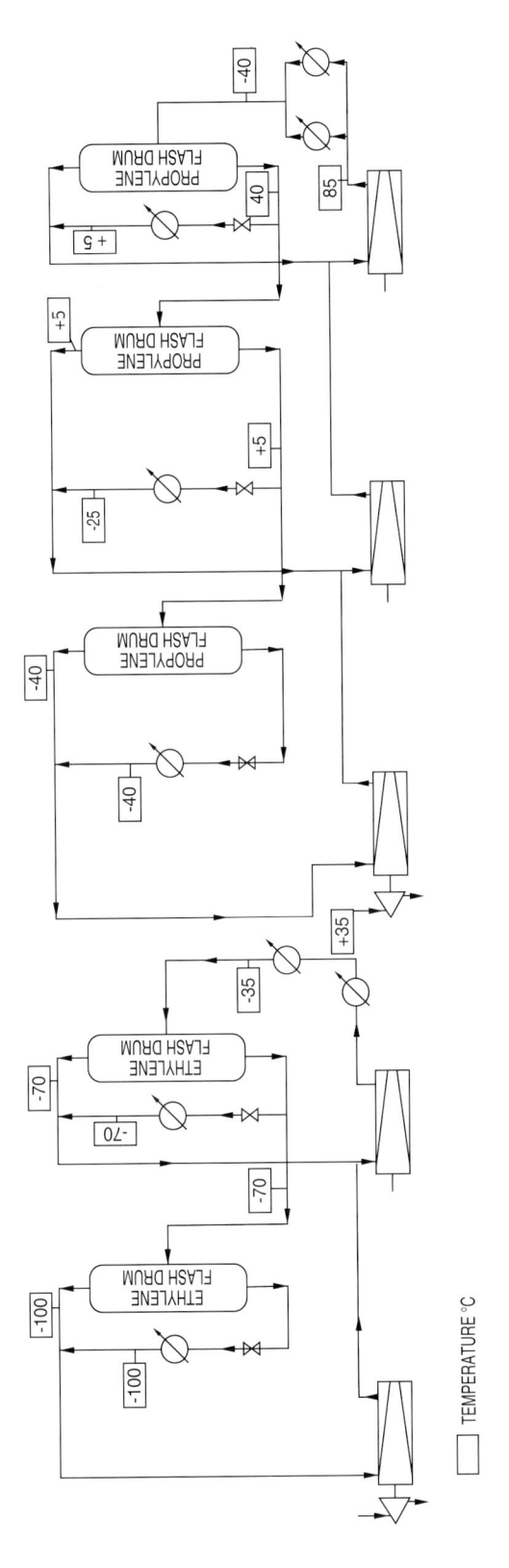

Fig. 1.20. Flow diagram: refrigeration naphtha cracking (*BEICIP Document*).

The majority of the ethylene plants in the world have been insulated, at least partially, with cellular glass.

The main licensors are *ABB Lummus Crest, M.W. Kellogg, Linde, Lurgi*.

I.6.2.4 P-Xylene. Arco Two-Step Process

P-Xylene can be separated from its isomers by crystallisation (Fig. I.21).

The temperatures in the Arco two-step process are low or very low at the first stage crystallisers (–45°C to –75°C) and cold at the second stage (–5°C).

I.6.2.5 Ethylene Oxide

Ethylene oxide leaks and the absorption of glycols in the insulation have caused serious explosions during the last years.

The reports on these accidents by experts have resulted in the examination of the risks related to the absorption of combustible liquids in porous insulation materials.

The case of EO is particularly critical because there is a SIT risk (Self Ignition Temperature) at a temperature as low as 60°C in some insulation materials.

The choice of cellular glass, which is fully impervious, is widely recommended for this application nowadays.

Though, to improve the time of resistance to hydrocarbon fire, some experts have recommended insulation of the pipes or the equipment with cellular glass and to cover it with an external layer of mineral wool.

We should be aware that this solution is less appropriate regarding the risk of water vapour absorption.

It will fall to the engineers to determine which pipes and installations require greater fire protection.

An other solution consists of applying an all cellular glass insulation, the system being reinforced and protected by a stainless steel jacket. Hydrocarbon fire resistance tests have been run with these systems and have given very good results (see V.1.6.4.C).

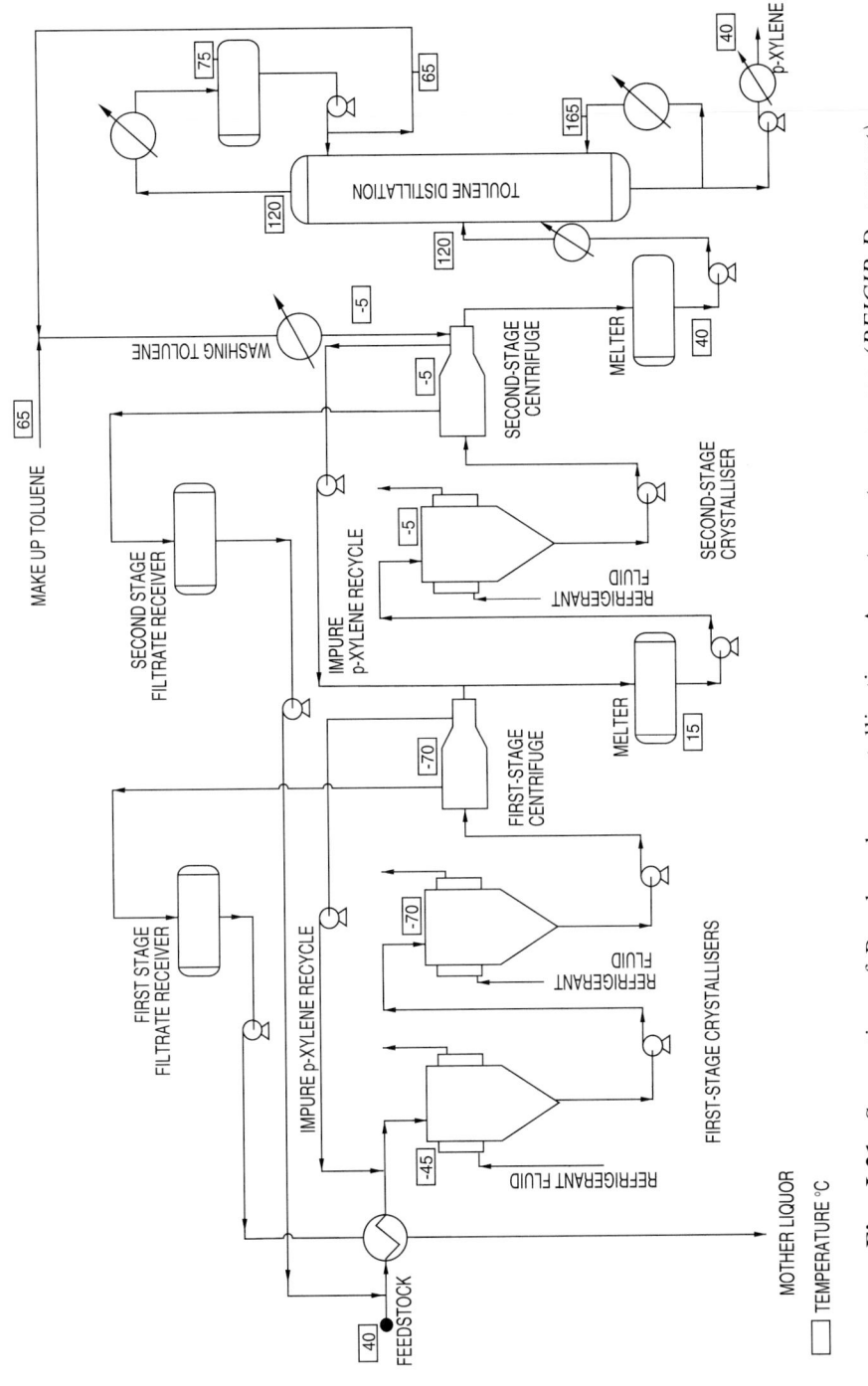

Fig. I.21. Separation of P-xylene by crystallisation. Arco two-step process (*BEICIP Document*). (After A. Chauvel, G. Lefebvre, "Petrochemical Processes", t.1, *Editions Technip*, Paris, 1989)

I.6.2.6 Vinyl Chloride (From Pyrolysis Of Ethylene Dichloride)

This process has a section working at temperatures down to –30°C at the head of the HCl column (Fig. I.22).

I.6.2.7 MDI (Diphenyl Methane 4,4 – Diisocyanate) And TDI (Tolylene Diisocyanate)

Some sections of the process, after the Phosgene dissolver (–18°C) and the Phosgene absorber (–20°C) and where the HCl is recovered, have to be insulated due to their negative operating temperatures (Fig. I.23 and Fig. I.24).

I.6.2.8 Polypropylene. Mitsui Montedison Technology

Equipment and pipes in the drying section of the polypropylene are operating at negative temperatures down to –15°C and must be insulated with a suitable product such as cellular glass (Fig. I.25).

I.6.2.9 PVC

The reactor in the polymerisation in suspension process operates at temperatures ranging from 5 to 80°C.

The insulation and the equipment are exposed to condensation and corrosion.

Cellular glass thermal insulation should be considered.

I.6.2.10 Polybutadiene

The raw material (Butadiene, Toluene) preparation section and the reactor are at low temperatures (–15°C, –5°C, 0°C) (Fig. I.26).

A non-absorbent insulation such as cellular glass should be considered.

I.6.2.11 Methanol

Depending on the catalyst used for the reaction, different technologies have been developed by the licensors.

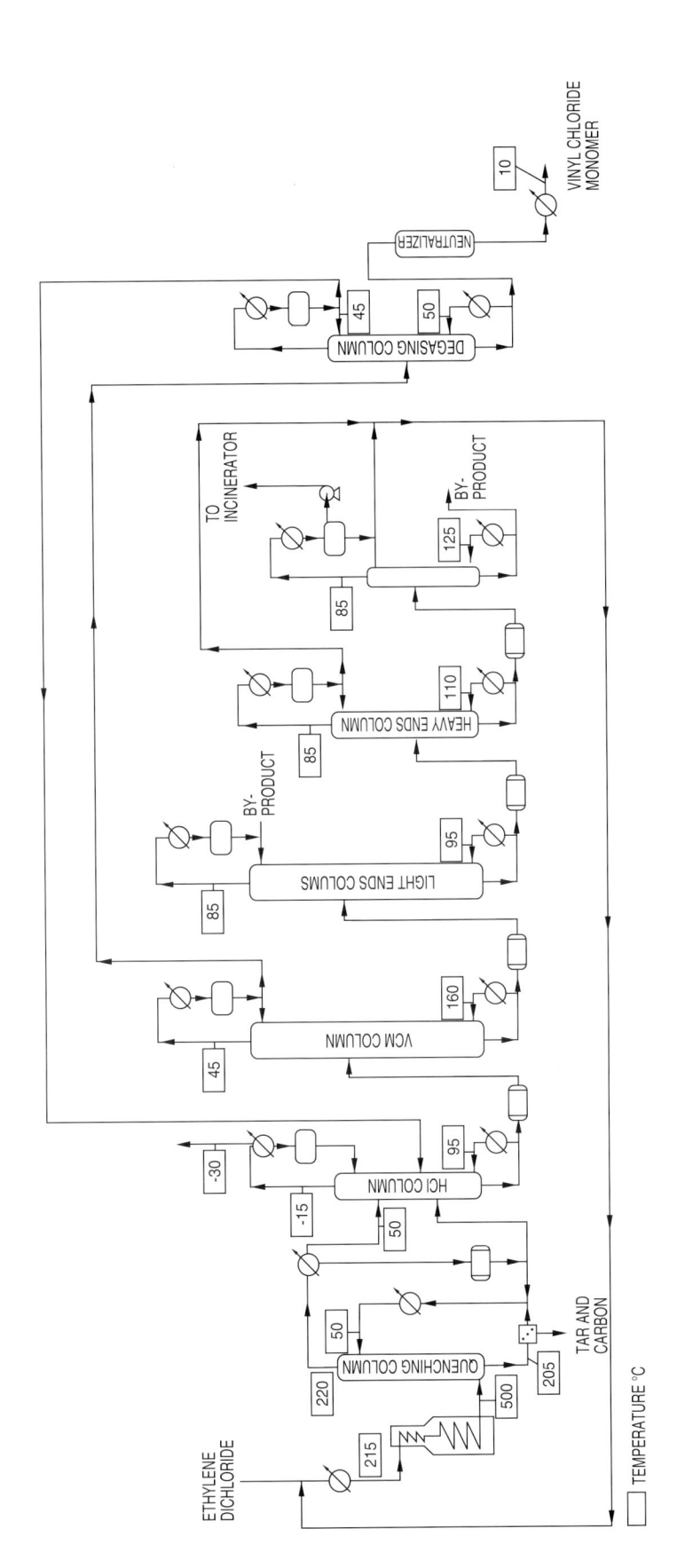

Fig. I.22. Vinyl chloride from ethylene dichloride by pyrolysis (*BEICIP Document*). (After A. Chauvel, G. Lefebvre, "Petrochemical Processes", t.2, *Editions Technip*, Paris, 1989)

PART I. WHERE TO USE CELLULAR GLASS

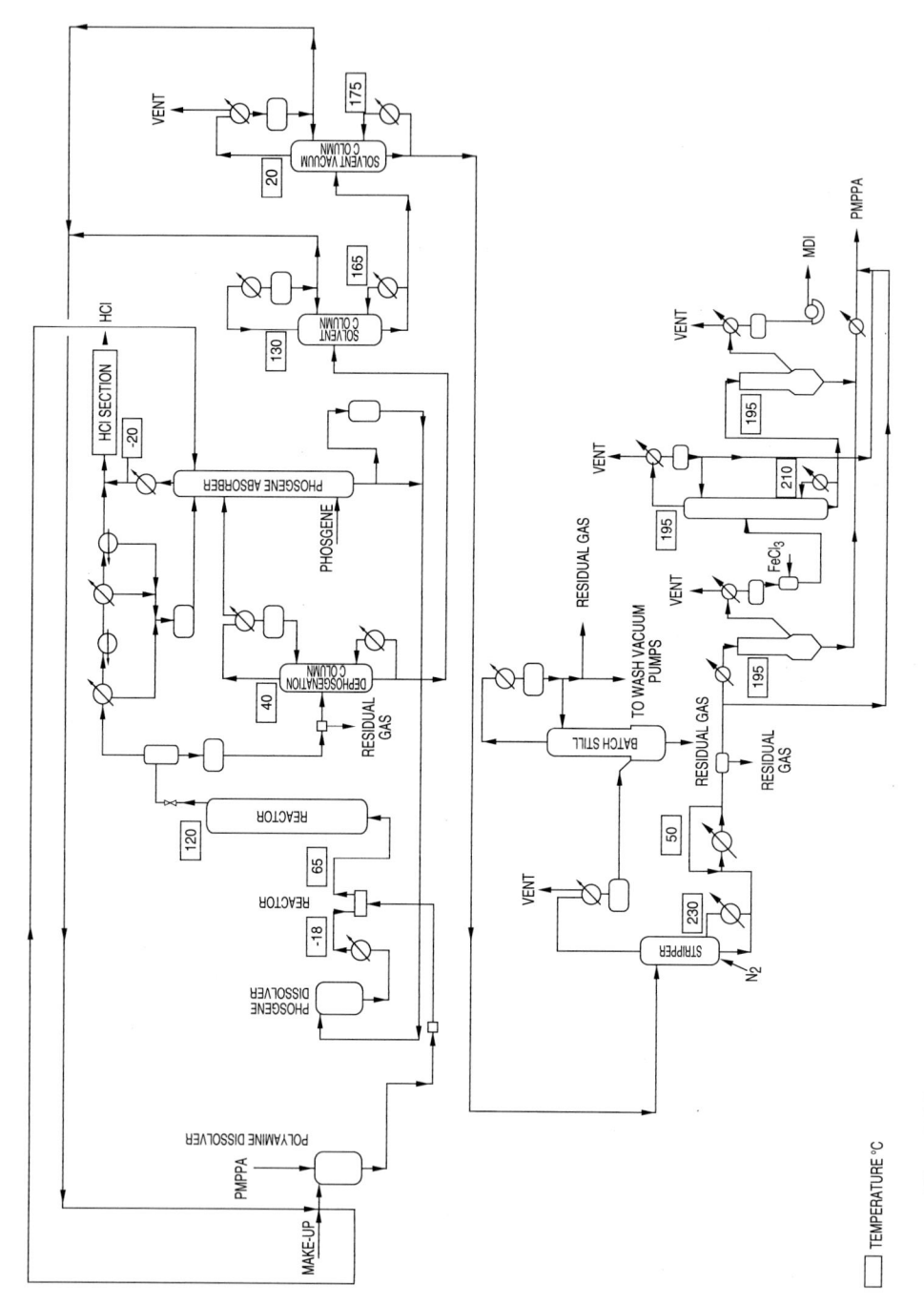

Fig. I.23. MDI by phosgenation of polymethylene polyphemylene amine (*BEICIP Document*).

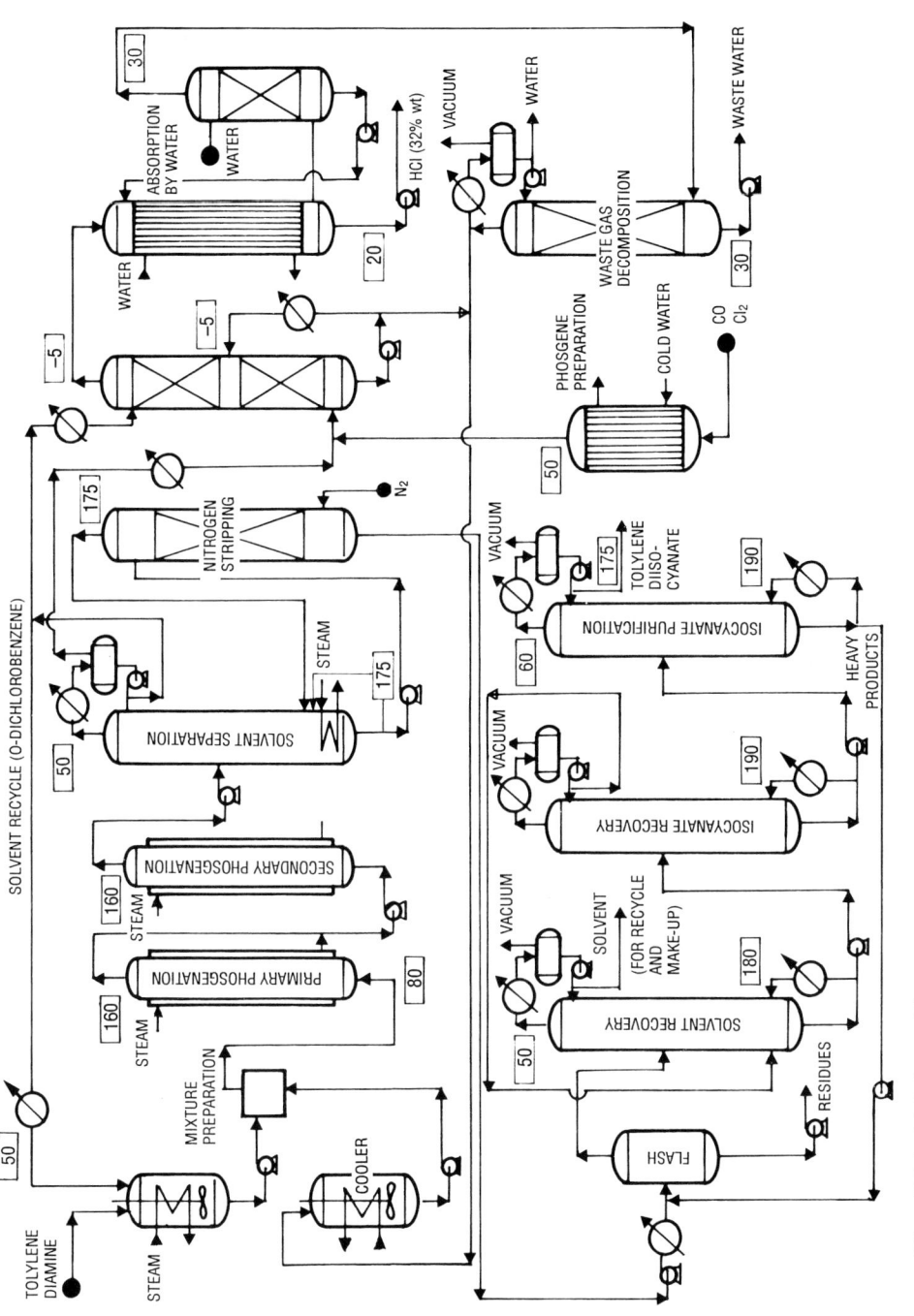

Fig. I.24. Tolylene diisocyanate production by phosgenation of tolylene diamine (*BEICIP Document*). (After A. Chauvel, G. Lefebvre, "Petrochemical Processes", t.2, *Editions Technip*, Paris, 1989)

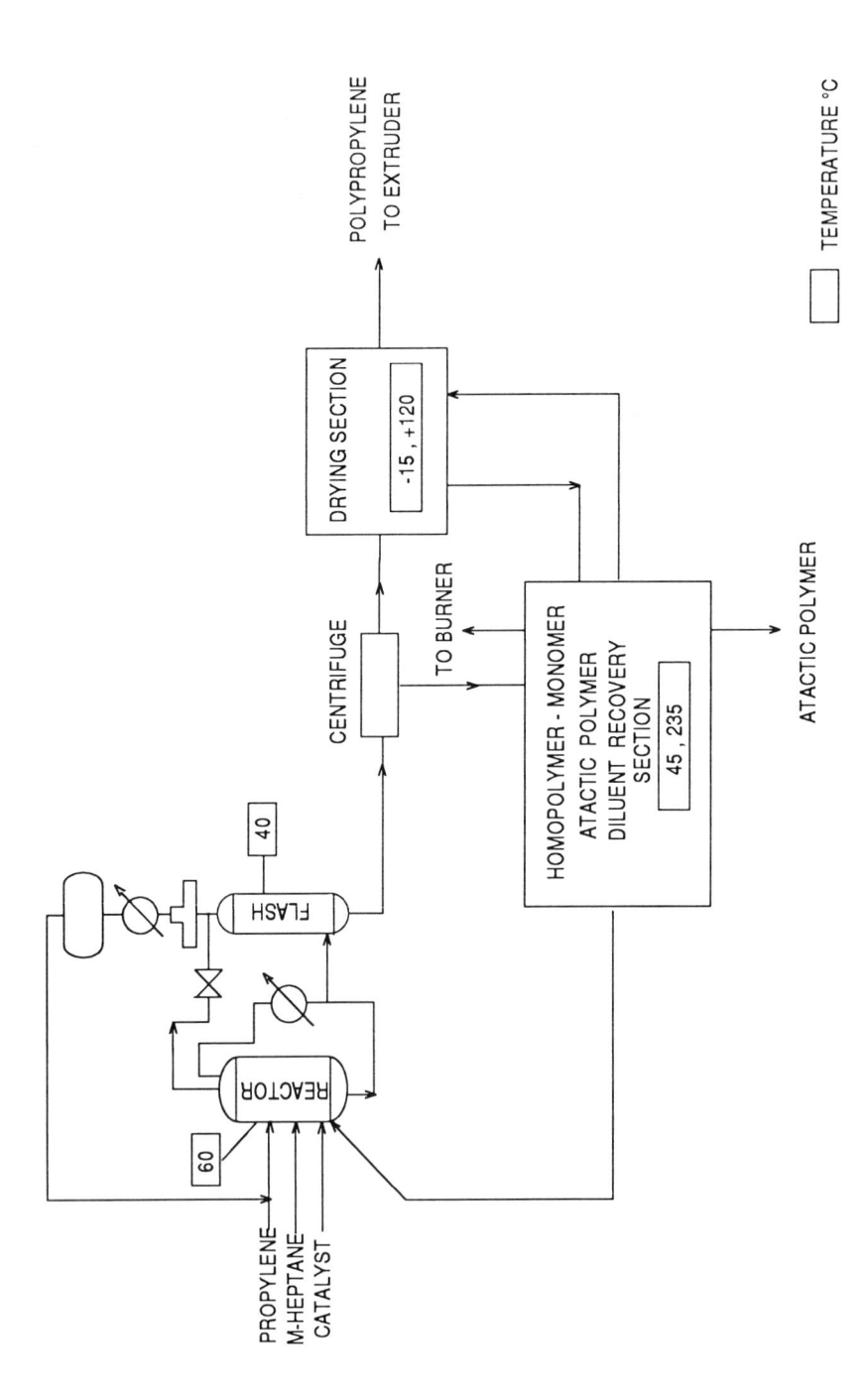

Fig. I.25. Polypropylene by the Mitsui petrochemical. Montedison technology (*BEICIP Document*).

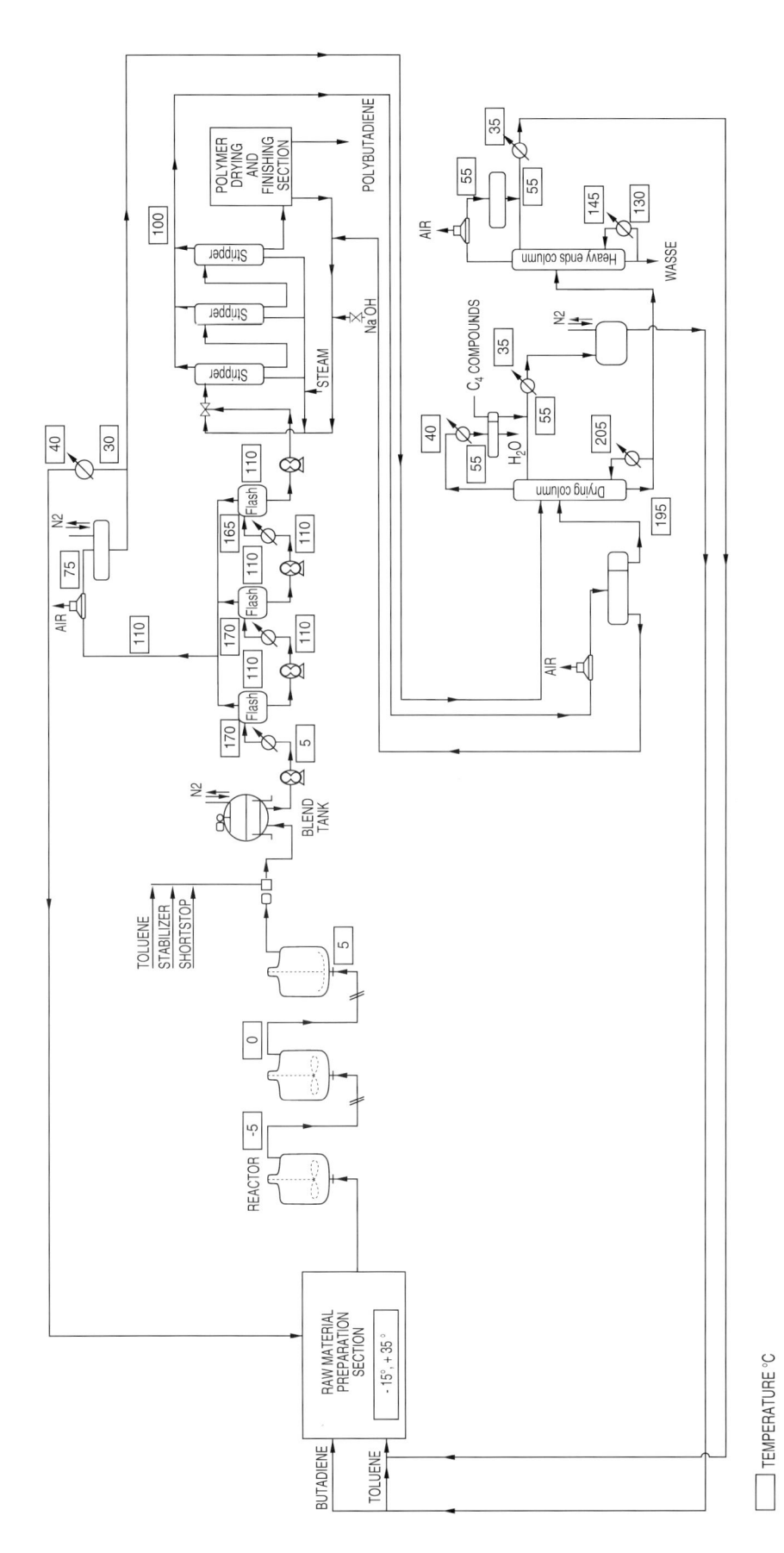

Fig. I.26. Polybutadiene by an iodine. Ziegler catalyst process (*BEICIP Document*).

The latest processes developed operate at low pressure and the major parts of the plant are at temperature below +150°C.

Some specific problems with methanol are leakages through flanges or valves and fire risks when methanol saturates the thermal insulation.

FOAMGLAS® cellular glass is therefore very widely used for safety reasons because of its non-absorbent properties.

I.6.2.12 MTBE

MTBE (Methyl Ter Butyl Ether), thanks to its high octane value, has become the refiners preferred gasoline additive, to replace the lead that is not considered environmentally safe.

The total world capacity is expected to double between 1990 and 1995.

MTBE is obtained by the reaction of methanol on isobutene in a C_4 cut.

As with methanol, MTBE attacks joint sealers. Leakage of the product is a big concern in the plants.

The risk is increased when the thermal insulation absorbs this highly combustible liquid.

Closed cell insulation such as FOAMGLAS® cellular glass is widely used in these plants.

I.6.3 MISCELLANEOUS

I.6.3.1 Vertical Storage Tanks

A. Base Insulation

Thanks to its high compressive strength, FOAMGLAS® cellular glass – specially HLB (High Load Bearing) types – has been used for over 35 years for the insulation of the base of low temperature and cryogenic storage tanks. In particular, it is used for the following liquefied gas storage tanks:

- Butane – butadiene
- Ammonia
- Propane – propylene – styrene
- CO_2
- Ethane – ethylene
- LNG
- Oxygen
- Nitrogen.

FOAMGLAS® cellular glass is also used in hot tanks:

- Bitumen
- Waxes
- Sulphur.

B. Walls and Roofs

For low temperature tanks, from –50°C and upper, cellular glass can be applied on the vertical walls and on the roof with the "PC® 88 ADHESIVE compact system". Many tanks around the world have an experience of over 20 years with this technique.

For hot applications, FOAMGLAS® cellular glass can be applied dry and kept in place with bands.

I.6.3.2 Spherical Storage Tanks

As for vertical storage tanks the fully adhered cellular glass compact system with PC® 88 ADHESIVE, presents many advantages, in particular no moisture penetration, for the insulation of spherical tanks.

The following products can be stored:

- Ammonia
- CO_2
- Propane – butane
- Butadiene
- Propylene
- Nitro-paraffins
- Ethylene oxide
- Waxes

- Pharmaceutical products
- Beer.

I.6.3.3 Offshore Installations

The particular problems with offshore installations are as follows:

- Fire risk
- Absorption of combustible liquids in the insulation
- High corrosion risk
- Freezing
- High ambient air humidity
- Consequences of regular test of the fire extinguishing equipment using sea water (deluge test)
- Vibrations
- Wind and weather.

I.6.3.4 Bitumen Tanks and Barges/Ships

Cellular glass is highly recommended for the internal insulation of bitumen tanks as it does not absorb bitumen. It is applied dry and kept in place during the installation. When first loaded, the liquid bitumen will migrate around cellular glass elements to the cold walls where it will adhere the cellular glass for the future life of the tank. Solvent, that could dissolve the cold bitumen, should not be stored in the same tank (Fig. I.27).

I.6.3.5 Heat Transfer Lines

For efficiency and safety reasons, due to its imperviousness to combustible liquids, cellular glass is recommended as insulation by heating oil manufacturers (*Dow*: Dowtherm®, *Monsanto*: Therminol®, Santotherm® or *Mobil*: Mobiltherm®), at least around valves, flanges and everywhere a risk of leakage may occur.

I.6.3.6 Sulphur Pits

Sulphur pits must be insulated to protect the structure with a mechanically strong and chemically resistant insulation material such as FOAMGLAS® cellular glass.

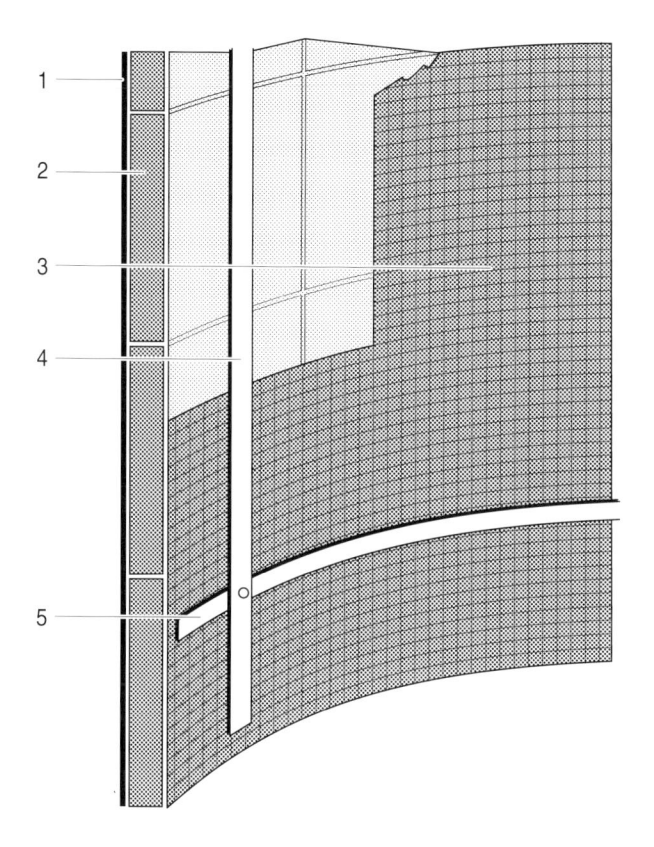

Fig. I.27. Inside wall insulation of vertical hot bitumen storage tanks.

1. Tank wall (inside).
2. FOAMGLAS® slabs dry applied.
3. Wire netting.
4. Flat steel bands 40 × 5 mm.
5. Flat steel bands 40 × 5 mm.

I.6.3.7 Pool Fire Control of Liquefied Gas

In case of spillage of liquid gas (LNG, ethylene, propane, etc.) foaming equipment is usually installed to keep the fire under control.

This equipment relies on its good functioning and is effective as long as the wind does not blow the foam away from the surface of the liquid gas.

Several companies including *Shell Research*, *Snam* and *British Gas* have contributed to the development of a purely static system by putting small blocks of cellular glass, packed in plastic bags, on the ground of the gas retention area in a +/− 20 cm thick layer. In case of fire, these rigid blocks will float on the surface and smother the fire.

This system, purely static, does not rely on mechanic, and is effective, even with high wind (which can be initiated by the fire itself).

I.7
CHEMICAL – PHARMACEUTICAL – COSMETIC

The use of cellular glass is advised for the insulation of chemical and pharmaceutical processes where the following characteristics may be met:

- Negative temperatures
- Thermal oils
- Steam
- High corrosion risks
- Tanks
- Dual temperatures
- Heat traced lines.

In the case of pharmaceutical industry the chemical neutrality and the particular resistance of cellular glass to bacteria, micro-organisms and fungi must be added and emphasised.

I.8

OTHER INDUSTRIAL APPLICATIONS

Other industrial applications where cellular glass can be used are the following:

I.8.1 FOOD

I.8.1.1 Breweries

Because of its chemical and bacteriological neutrality and its constant thermal conductivity, cellular glass is suited for the insulation of:

- Fermentation tanks
- Storage tanks
- Cold pipes
- Storage rooms
- Bottling rooms.

Due to its chemical neutrality it is particularly recommended for the insulation of stainless steel equipment.

I.8.1.2 Malt Houses

The germination of barley for malt production takes place in a high temperature and high relative humidity environment.

Therefore it is necessary to insulate the walls of the germination rooms with an insulation resistant to humidity, bacteria, insects, rodents, etc. also presenting a high degree of dimensional stability.

I.8.1.3 Dairies. Cheese Dairies

Because of its chemical and bacteriological neutrality, its resistance to lactic acid, to rodents and insects, cellular glass is an ideal choice for the insulation of tanks, pipes, fermentation and cold rooms of dairies and cheese production.

I.8.1.4 Meat. Poultry-Farms. Farms

Resistance to bacteria, micro-organisms, insects are characteristics that should be taken into account when choosing insulation materials for the meat and poultry industry.

I.8.2 COLD STORES

Numerous cold rooms have been constructed with cellular glass using the "black box" technique. This technique consists in applying the insulation around the bearing structure and the metallic framework to avoid thermal bridges that can cause many problems.

Due to the production of self-supporting panels which are more economical and easier to apply, the "black box" technique is less used.

These new techniques, incorporating FOAMGLAS® cellular glass, are as follows:

I.8.2.1 Wall

Sandwich boards can be manufactured with cellular glass as a core material. This ensures their dimensional stability, their resistance to fire and imperviousness to humidity.

I.8.2.2 Floor

The high compressive strength of FOAMGLAS® cellular glass allows it to be used for the insulation of floors especially those where goods are stacked or where the traffic of handling trolleys is high.

The high compressive modulus of elasticity and the strong compressive strength make it possible to realise lighter and thinner reinforced floor slabs.

I.8.2.3 Roof

The technique of adhering cellular glass to the metal deck by dipping the slabs in hot bitumen, has numerous advantages:

- Stiffening of the roof
- Fire resistance
- No thermal bridges
- No mechanical fixings
- High wind lift-up resistance
- Resistance to moisture
- Use of the compact insulation technique (all the elements are fully adhered together):
  - Cellular glass to the steel plate
  - The joints between slabs
  - The roofing membrane on the insulation.

This provides that an accidental leak would be confined to a small area.

I.8.2.4 Freezing Lines

Piping in refrigeration systems is sensitive to humidity. The insulation must therefore be kept dry to remain efficient.

Cellular glass made of closed cells and fully impervious to water vapour is the required insulation.

The application according to the semi-compact technique avoids the spreading of condensation along the pipes, should an accident occur.

The equipment is protected against moisture, fire (the insulation being incombustible) and resists corrosion better.

I.8.3 HEATING. VENTILATION. AIR CONDITIONING

I.8.3.1 Chilled Water

Like freezing lines, chilled water pipes are subject to moisture. Moreover, being often installed inside important and public buildings, the insulation should be incombustible and should not give off toxic gases or fumes in case of fire.

More and more, many authorities, often after having experienced a dramatic fire, impose the use of insulation materials, totally incombustible and not giving off toxic fumes or gases in case of fire.

Cellular glass perfectly answers these requirements while totally ensuring the insulation of the pipes.

Its constant insulation efficiency will prevent dripping due to surface condensation after a certain period of time.

I.8.3.2 Mining Air Conditioning

Working conditions in mines are already very hard at normal temperatures.
Mines and particularly gold mines are going deeper and deeper. A depth of over 3 000 m has been reached in South-African gold mines.
The temperature of the rock rises to 60°C. The work performance of the miners decreases very fast at temperature over 25°C.

Cooling the air at a distance of several kilometres, both vertically and horizontally, from the cooling plant located at the surface, needs sophisticated techniques and is very energy consuming.

Safety is, of course, of prime importance. The materials used should be incombustible and should not release dense or toxic smoke in case of fire.

Cellular glass presents all the required characteristics:

- Does not absorb moisture
- Is incombustible
- Does not release any smoke or gas during a fire
- Does not induce corrosion.

I.8.3.3 Air Ducts

Air ducts have to be insulated and the insulation material has to be resistant to:

- Fire
- Moisture
- Rodents, insects, bacteria.

Cellular glass perfectly meets these requirements.

I.8.3.4 Heating

A. Central Heating

Cellular glass is rarely used for central heating pipes except in the following cases:

- High risk of corrosion or water absorption
- Underground pipes.

B. Underground District Heating

Cellular glass has frequently been used for the insulation of district heating networks with underground pipes or gutter-pipes with the following advantages:

- Resistance to water
- Can be used in direct burial within banked/backfilled trenches
- If used in gutters, resistance to accidental flood
- Resistance to rodents, vermin.

I.8.4 OTHERS

I.8.4.1 Ice Rinks

The material chosen for the insulation of ice rinks must be resistant to mechanical actions, present an excellent dimensional stability and a high compression strength. It must not be attacked by rodents or insects.

For several years now, numerous ice rinks have been insulated with cellular glass.

I.8.4.2 Paper Mills

The atmosphere in paper mill is hot, humid, caustic and/or acid.

The building, the machines and the equipment suffer considerably from these conditions. Energy consumption is high.
Therefore, the choice of the materials and the method of installing them is very important.

Cellular glass will show excellent results in the following cases:

ROOFS: roofs are subject to heavy condensation. Insulating according to the compact roof technique avoids condensation in the insulation and under the surface and prevents the formation of drips, often acidic, that can fall on the equipment.

PIPES AND EQUIPMENT: because of its excellent resistance to water, it is also used for the insulation of pipes and tanks containing wood pulp, warm water and steam, in particular in the following processes:

- The Kraft pulping process for the insulation of:
  - Digesters (175°C)
  - Blow tanks (105°C)
  - Dryers (150°C)
  - Evaporators (115°C)
  - Recovery boilers (120-400°C)
  - Smelt dissolving tanks (105°C)
  - Green liquor clarifiers (87°C)
  - Slakers/causticisers (98°C)
  - Boilers (120-480°C)
- The Sulphite process for the insulation of:
  - Digesters (175°C)
  - Blend tanks (105°C)
  - Evaporators (105°C)
  - Recovery boilers (120-400°C)
  - Cleaners (95°C).

I.8.4.3 Digesters. Water Treatment

Fermentation of biodegradable residues in digesters has become more and more frequent in:

Photo 1. Overfit system.

Photo 2. Natural gas liquids separation plant at Mossmorran, Scotland.

Photo 3. Distrigas N.V. LNG terminal, Zeebrugge, Belgium.

Photo 4. Offshore platform.

Photo 5. Pool fire control. Small elements of FOAMGLAS® cellular glass packed in plastic bags.

Photo 6. Refrigerated storage vessels at Heineken Brewery.

Photo 7. Cargill Malt-House.

Photo 8. Chilled water pipes.

Photo 9. Air ducts.

Photo 10. Direct buried pipe.

Photot 11. Ice rink at Paris-Bercy Sports Centre.

Photo 12. 27 000 m^2 FOAMGLAS® insulation on the metal deck roof of a paper mill at Lanaken, Belgium.

Photo 13. Water treatment plant at the Gileppe Dam, Belgium.

Photo 14. FOAMGLAS® insulation for industrial chimneys.

Photo 15. Bottom of a digester.

Photo 16. Cone insulation.

- Water treatment stations
- Bio-mass installations to recover methane gas.

The atmosphere is particularly aggressive (moisture, acidity, rodents, vermin, bacteria, etc.).

Cellular glass is well suited to these conditions and has a high compressive strength that makes it ideal for the insulation of the bottom of digester tanks.

I.8.4.4 Power Plants. Nuclear Plants

In these plants, insulation is mainly hot and as corrosion risk is highly considered, cellular glass is strongly recommended.

It can be of use for particular problems such as:

- Cold pipes, chilled water
- Protection against frost
- Insulation inside the containement
- Underground insulation
- Building insulation
- Chimneys in power plants (see Chimneys)
- Exposed steam lines with the overfit system.

I.8.4.5 Chimneys

Cellular glass makes an excellent complementary insulation to the refractory bricks of industrial chimneys.

Indeed it resists the temperatures met in chimneys as well as the acids, and offers high dimensional stability.

Numerous chimneys have been insulated with cellular glass, in particular:

- Power plants
- Industrial chimneys
- Cement-works
- Glass-works
- Waste incineration plants.

I.8.4.6 Road Tunnels

To avoid slippery roads due to frost in tunnels, the spreading on the road of percolating water from the rock must be prevented. At the same time, the drain system for this water around the tunnel must not freeze.

Moreover, for security reasons, the insulation must be incombustible and must not give off toxic fumes or smoke when involved in a fire within the tunnel.

Cellular glass has all of these requirements and offers security and resistance.

I.8.4.7 Architecture. Building

The purpose of this book is to study the different fields of application for cellular glass in industry.

The different applications of cellular glass in architecture are numerous and are further developed in other books.

But the main applications should be mentioned here:

- **Flat roofs:**
    - On concrete bearing structures: according to the 'compact roof' technique, where cellular glass is bonded with hot bitumen directly to the concrete and is thus fully adhered to the roof
    - On metal deck: where cellular glass is applied, after having been dipped in a hot bitumen bath, or with cold adhesive
    - According to the 'tapered FOAMGLAS® roof system', where cellular glass is cut with an integrated fall that ensures water drainage even on a roof without slope or with counter-slopes
- **Floors:**
    - Under the screed
    - Under the foundation slab
- **Vertical walls:**
    - In cavity wall
    - Under light or heavy claddings
    - Internal insulation
    - Exterior insulation
- **Pitched roofs**
- **Insulation of Foundations.**

I.9
SUMMARY

Process and application temperatures have been classified in three ranges where the main reasons to use FOAMGLAS® cellular glass are:

- Below ambient:
  - Perfect moisture resistance
  - Protection against the corrosion risk
  - Protection against freezing
  - Dimensional stability
- Ambient to +120°C:
  - Protection against corrosion risk
  - Fire hazard in case of leakage of combustible product
  - No water absorption
  - Dimensional stability
  - Compressive strength
- 120 to 300°C:
  - Fire hazard with combustible liquids
  - Compressive strength
  - Protection against corrosion
  - Dimensional stability.

In the next pages, many FOAMGLAS® cellular glass applications are mentioned with an appreciation of its use:

- (A) = very limited
- (B) = limited
- (C) = moderate
- (D) = moderate to high
- (E) = high
- (F) = very high.

PART I. WHERE TO USE CELLULAR GLASS

Process	Reference	Cold <20°C	Warm 20 to 120°C	Hot >120°C	Comments
\multicolumn{6}{c}{Refinery}
Atmospheric distillation	I.3			A	
Gas, LPG, Gasoline Fractionation	I.3		E	B	Fire
Vacuum distillation				C	Heat tracing lines
Catalytic reforming	I.3.1.1		A		
Isomerisation	I.3.1.2		A		
Catalytic cracking	I.3.2.1		A		
Visbreaking	I.3.2.2				
Coking processes	I.3.2.2			A	
Hydroconversion	I.3.3		C	A	
Alkylation (HF)	I.3.4.1		A	A	
Alkylation (H_2SO_4)	I.3.4.1	E	A		
Dimersol	I.3.4.2		A		
Polymerisation			A		
Hydrotreatment processes	I.3.5		A		
Chemical treatment processes	I.3.5				Very limited insulation required
Solvent desasphalting	I.3.6.1		C	B	
Amine unit	I.3.6.2		D		
Sulphur unit	I.3.7		D	E	On traced lines, crystallisation risk
\multicolumn{6}{c}{Gas Processing}
H_2S, CO_2 removal processes	I.4.2.1	F	C		Depending on process used
LPG, LNG recovery processes	I.4.2.3	F	C		
LNG liquefaction processes	I.4.2.4	F			Very low temperature processes
LNG receiving terminal	I.4.2.5	F			All FOAMGLAS® cellular glass to be considered
Synthesis gases	I.5.1	F			
Industrial gases	I.5.2	F			LOX/LIN - no organic products

A = limited application...... F = FOAMGLAS® highly recommended.

68

CHAPTER I.9 SUMMARY

Petrochemistry

Ammonia	I.6.2.1	F	C	A	Storage (F)
Carbon monoxide	I.6.2.2	F			Cryogenic process: (F)
Methanol	I.6.2.11		C		Leakage
MTBE	I.6.2.12		C		Leakage
Urea			C	A	
Butadiene		F	B	A	
Steam cracking	I.6.2.3	F	E	B	
Isobutene		F	E		
Aromatics			D	D	
Cumene			B	B	Traced
Phenol/Acetone			B	B	
Ethylbenzene			B	A	
Styrene			E		Fractionation section
Nitrobenzene			B		
Aniline			B	A	
Paraxylene: (Crystallisation)	I.6.2.4	E	B		Crystallisation process
Paraxylene: (Parex)			B	A	
Ethylene oxide	I.6.2.5		F	F	Explosion risk
Ethylene glycol			F	F	
Propylene oxide			F	F	
Propylene glycol			F	F	
N. Butanol			C		
Isopropanol			D		
Higher alcohols			D	D	
2-Ethyl hexanol			D	D	
Acetaldehyde			B		
Acetic acid			B	B	
Acrylates			B		
Acrylic acid			B	B	
Acrylonitrile		F	C	B	
Vinyl acetate			B		
Vinyl chloride		F	D	B	
Adipic acid			B	A	
Maleic anhydride			B	A	
DMT			C	B	
TPA			B	B	
MDI	I.6.2.7	E	C	C	
TDI	I.6.2.7	F	C	C	
Polyether-polyols			B		
Polyethylene (LDPE)			B	A	
Polyethylene (HDPE)		F	E		
Polyethylene (LINEAR LDPE)			E		
Polypropylene	I.6.2.8	F	D	D	
Polystyrene			C	B	
PVC	I.6.2.9		C		
LAB			B	B	
Polybutadiene	I.6.2.10	E	D	D	
SBR			D		

A = limited application...... F = FOAMGLAS® highly recommended.

69

Application	Reference	Cold <20°C	Warm 20 to 120°C	Hot >120°C	Comments
Storage tanks					
Vertical tank: bottom	I.6.3.1	F	F	F	Compressive strength
Vertical tank: walls and roofs	I.6.3.1	F	D	A	
Spherical tank	I.6.3.2	F	D		
Miscellaneous					
Offshore	I.6.3.3	F	F	F	Fire-corrosion
Bitumen tanks	I.6.3.4		F	F	Fire
Heat transfer	I.6.3.5			F	Fire-explosion
Sulphur pits	I.6.3.6			F	
Pharmaceutical	I.7	F	F	F	Corrosion - bacteriological cycling temperature
Cosmetics	I.7	F	E		
Food					
Breweries	I.8.1.1	F	E		Stress corrosion on stainless steel bacteriological resistance temperature control
Malt house	I.8.1.2		F		High humidity
Dairies-cheese	I.8.1.3	E	D		Refrigeration lines
Meat-poultry	I.8.1.4	E			Refrigeration lines
Cold stores					
Walls	I.8.2.1	D			
Floor	I.8.2.2	F			High load
Roof	I.8.2.3	D			Steel deck: F
Freezing lines	I.8.2.4	F			Freon (-40°C) NH_3 (-30°C)
Heating- Ventilation - Air conditioning					
Chilled Water	I.8.3.1	E			Moisture-fire
Mining	I.8.3.2	E			Fire
Air ducts	I.8.3.3	E	D		
Heating	I.8.3.4		D		
Underground	I.8.3.4	D	B	E	
Others					
Ice rinks	I.8.4.1	F			
Paper mills: Roofs	I.8.4.2		F		
Paper mills: Kraft process	I.8.4.2		D		
Paper mills: Sulphite process	I.8.4.2		D		
Digester	I.8.4.3		F		Bacteriological
Water treatment	I.8.4.3		D		
Power plants	I.8.4.4		D		Corrosion
Nuclear plants	I.8.4.4		E	F	Corrosion
Chimneys	I.8.4.5			F	H_2SO_4
Road tunnels	I.8.4.6	F			Fire risk
Building: Flat roofs	I.8.4.7		F		Compact roof technique
Building: Pitched roofs	I.8.4.7		C		
Building: Walls	I.8.4.7		D		
Building: Floors	I.8.4.7		E		

A = limited application...... F = FOAMGLAS highly recommended.

II

PROPERTIES OF FOAMGLAS® CELLULAR GLASS AND ITS ACCESSORIES

II.1

FOAMGLAS® CELLULAR GLASS

II.1.1 PRODUCTION PROCESS

Before reviewing the main properties of FOAMGLAS® cellular glass, it is worth summarising the production process.

Sand and various additives are automatically weighted, mechanically mixed and heated to produce a special glass that is then extruded and crushed to form a glass powder. A mixture of the powder and carbon is placed into moulds and put into an oven heated to approximately 1000°C. In this oven, the carbon oxidises and forms gas bubbles that begin the expansion.

The small amount of carbon gives FOAMGLAS® cellular glass its characteristic black colour. When the expansion process is complete, the product is taken out of the mould and passed through a long annealing oven. The slabs are then cut on all faces to the required dimensions, tested and packed (Fig. II.1).

II.1.2 GENERAL PROPERTIES

From this method of production, we can deduce a number of properties that have often been verified by tests. First of all, FOAMGLAS® cellular glass is made exclusively of glass. There is no binder or other material in its composition (Fig. II.2). Consequently, the product is strictly non combustible and presents an excellent resistance towards water and water vapour, just as plain glass. The only difference is that the cell glass walls have a thickness of about 10 microns. Also like glass, FOAMGLAS® cellular glass is dimensionally stable and has a low coefficient of thermal expansion: $\alpha = 9 \times 10^{-6} K^{-1}$ at ambient temperature.

It also has another of the desirable properties of glass: the constancy of the properties with time. Finally it resists vermin, rodents and many chemicals.

PART II. PROPERTIES OF FOAMGLAS ® CELLULAR GLASS AND ITS ACCESSORIES

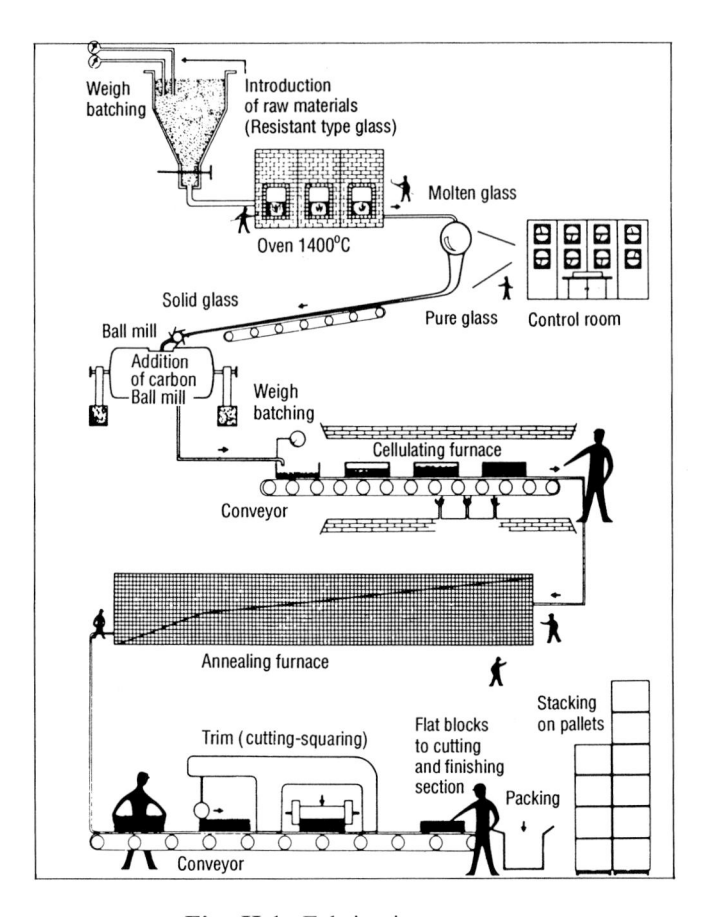

Fig. II.1. Fabrication process.

These points will be detailed in this chapter and in the corresponding chapters of part V.

Some properties are unaffected or very moderately affected by density, for instance:

Temperature limits [1] .... from –260°C to +430°C
Glass softening point .... about 730°C

---

[1] Applications have been successfully done in this temperature field. The design has to accomodate for possible hairline cracks resulting from thermal shocks. When they occurred, they are generally due to rapid temperature changes of restrained elements.

74

CHAPTER II.1 FOAMGLAS® CELLULAR GLASS

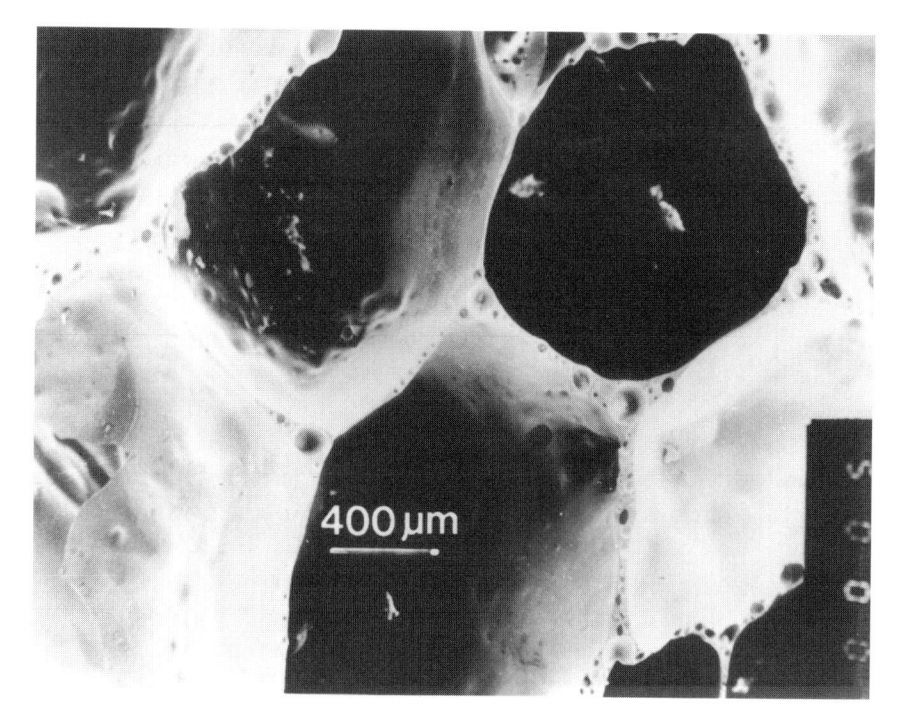

Fig. II.2. Structure of FOAMGLAS® cellular glass.

Water absorption...............	zero except for some temporary water retention on the surface
Hygroscopicity	zero
Permeability.....................	zero (μ = practically infinite)
Capillarity........................	zero
Combustibility..................	non combustible
Glass specific heat............	0.84 kJ/(kg.K)
Coefficient of thermal expansion at 10°C............	$9 \times 10^{-6} K^{-1}$

II.1.3 DIMENSIONS OF THE SLABS

The slabs are 0.45 × 0.6 m. These dimensions are fixed by the size of the moulds, which are limited to about 0.45 × 0.6 m. Larger size moulds would be technically possible but would require more time consumed annealing the slab. This would mean reduced production capacity or a longer annealing furnace. Since 60 cm is a very well accepted dimension on jobsite, increasing the size of the slabs with the corresponding cost increase is not a very challenging proposition.

II.1.4 FIRE BEHAVIOUR

This subject is discussed in detail in V.1 when comparing cellular glass and other insulants from a fire standpoint. Consequently, it will be sufficient here to summarise the fire behaviour of FOAMGLAS® cellular glass and insulation systems incorporating it.

Regarding fire reaction, FOAMGLAS® cellular glass is strictly incombustible according to all the existing test methods, including the most severe such as ISO 1182, ASTM E-136, BS 476 Part 4 and NEN 3881. These tests have been successfully carried out a number of times. The results have been confirmed by many certificates of approval issued by internationally acknowledged authorities [2,3,4,5,6,7,8]. Moreover FOAMGLAS® cellular glass has been classified M0 in France and A0 in Belgium according to the epiradiator test. This is the best possible classification [9,10]. Finally, the gross calorific value has been found at 0 kJ/kg which is also the best possible result [11].

---

[2] "Combustibility Test of FOAMGLAS® Insulation for Pittsburgh Corning Corporation", *Factory Mutual Research*, 14 October (1986) p. 1.
[3] "Test Report on Heated Vertical Tube Test for Nippon Pittsburgh Corning K.K.", *Japan Ship Machinery Quality Control*, 81-116E, 30 May (1981), p. 4.
[4] "Lloyd's Register of Shipping Certificate", *Certificate ICD/F85, 207*, 2 September (1985), p. 1.
[5] "Fire Test Report", University of Hong Kong, 29 September, (1986), p. 1.
[6] "Noncombustibility Test on FOAMGLAS® Material", *Singapore Institute of Standards and Industrial Research*, June (1980), p. 3.
[7] "Report Concerning an Examination of FOAMGLAS® cellular glass Type T4 on Non-combustibility According to IMO Resolution A 472", *Instituut TNO voor Bouwmaterialen en Bouwconstructies, B-86-302 (E)*, August (1986), p. 2.
[8] "Rapport Betreffende een Onderzoek van FOAMGLAS® cellular glass Type T4 op Onbrandbaarheid volgens NEN 3881", *Instituut TNO voor Bouwmaterialen en Bouwconstructies, B-86-302 (E)*, 6 August (1986), pp. 1-2.
[9] "Essai de Réaction au Feu d'un Matériau", *Centre Scientifique et Technique du Bâtiment, P.V. 78, 13991*, November (1978), p. 1.
[10] "Verslag van Proeven Nr. 5503 Isolatiemateriaal FOAMGLAS® cellular glass type T2", *Rijksuniversiteit Gent*, 7 July (1986) p. 2.
[11] "Verslag van Proeven Nr. 5486 Isolatiemateriaal FOAMGLAS® cellular glass type T2", *Rijksuniversiteit Gent*, 26 September (1986) p. 2.

As far as smoke and toxic gases are concerned, the glassy nature of cellular glass prevents any emission of smoke and toxic gases. This fact has been confirmed by tests carried out by the TNO according to NEN 3883. This is indicated in the report [12]:

> "The measurement of smoke development during fire of FOAMGLAS® cellular glass type T2 was zero".

Concerning fire resistance of insulation systems incorporating cellular glass, a number of major tests have been run on various installations. They called for a general conclusion: provided they are adequately designed, FOAMGLAS® cellular glass based systems, even when using organic adhesives, which normally burn, give a high degree of fire resistance. This is explained by the fact that these organic adhesives are applied between FOAMGLAS® cellular glass slabs providing slow temperature increase and little oxygen penetration. Consequently, organic adhesives in well-designed cellular glass systems do not cause substantial deterioration of the fire resistance of the system. Regarding coatings and finishes, they generally do not affect considerably the fire resistance of the system but their choice may affect flame propagation on the surface of the insulation. When aiming at optimal performances, stainless steel jackets should be preferred to galvanised steel and to aluminium. Metallic jackets should also be preferred to organic coatings. A description of the fire resistance tests on flat systems, pipe insulation, penetrations and panels is given in V.1.6. It is complemented by the review of a number of significant fires, which confirm the test conclusions.

Together with the fire behaviour, it may be worth to point out that cellular glass does not absorb combustible liquid and consequently does not present fire hazard by this cause.

II.1.5 WATER AND WATER VAPOUR

As discussed in V, FOAMGLAS® cellular glass does not absorb a significant amount of water at room temperature, even after accidental contact during prolonged periods of time.

---

[12] "Rapport Betreffende een Onderzoek Volgens NEN 3883 naar de Mate van Rookontwikkeling bij Brand van FOAMGLAS® cellular glass Type T2", *Instituut TNO voor Bouwmaterialen en Bouwconstructies, B-81-60*, 3 February (1981), p. 4.

The only indication that may mislead the user consists of the surface water retention observed when running an immersion water absorption test. Due to the open cells on the surface of the material caused when sawing it to the required dimensions, cellular glass may give the impression of water absorption during the test. Actually, it is surface water retention. This can easily be proven by carefully drying the surface.

At high temperatures, chemical reaction between water (or steam) and the glass may take place. It is nevertheless rather slow and will only affect the performance of the system if the minimum design caution has not been taken. Many successful applications prove that this potential problem is easily overcome by adequate design and installation. It is also worth pointing out the durability of FOAMGLAS® cellular glass which is in hydrolytic class III according to DIN 12 111.

At low temperatures, when cycling around 0°C, some water could fill the outside surface cells of cellular glass, particularly when the slab is in a horizontal position. When this water expands to turn to ice, it can exert mechanical pressure high enough to break some cells of the material. This phenomenon that can be repeated during many temperature cycles, is called freeze and thaw in wet conditions. Basically it only takes place on horizontal, unprotected surfaces, the amount of water remaining on the outside surface of vertical slabs is not sufficient to enable the mechanical pressure to build up. Again, correct design can easily overcome this remote possibility.

As far as the resistance of cellular glass to water vapour is concerned, it is practically absolute at ambient temperature. This is proven by many test reports concluded by a "practical infinite µ coefficient". The reaction to steam at high temperatures is discussed in V.3.4.

II.1.6 CONSTANT THERMAL EFFICIENCY

As explained later, the thermal conductivity of FOAMGLAS® insulation remains constant over time since the product does not absorb either water or water vapour. Several tests have been carried out taking back samples of FOAMGLAS® cellular glass from already old building or industrial applications and measuring their density and thermal conductivity: on 26, 32 and even

38 years old references, these properties were the same as at installation time [13] [14] [15]. These results justify that FOAMGLAS® cellular glass service life can be evaluated at 50 years or even longer and presents an economic solution. This can be demonstrated by applying discounted cash flow methods to the total life of the system.

II.1.7 CORROSION

As explained in V.6.4, the use of cellular glass alone does not constitute an absolute guarantee against corrosion but any well designed and correctly installed insulation system based on cellular glass will minimise corrosion danger. The reasons for this are as follows:

1. Cellular glass does not absorb water and is water vapour tight
2. Given adequate care, it is possible to design systems that remain water and water vapour tight
3. Chemically, cellular glass does not contribute to corrosion, neither carbon steel corrosion, nor stainless steel stress corrosion. This last point has been proven by many tests and certificates [16] [17] [18].

Many existing industrial references also prove this by long term practical experience.

II.1.8 CHEMICAL DURABILITY

FOAMGLAS® cellular glass is composed exclusively of glass, with the exception of a very small amount of carbon, burnt to produce CO_2 or encapsulated in the cell walls.

---

[13] "FOAMGLAS® Insulation: 32 Years Performance", Case Study of Petro-Wax PA, Emlenton, Pennsylvania; Pittsburgh Corning Corporation, *FI-165*, 10 M Rev. 3/92, pp. 1-2.
[14] "FOAMGLAS® Insulation: 26 Years Performance", Case Study of Eliza Coffee Memorial Hospital, Florence, Alabama; Pittsburgh Corning Corporation, *FI-193*, December (1984).
[15] FOAMGLAS® Insulation: 38 Years Performance. The roof of Equitable Life Assurance Society Office Building, Pittsburgh, Pennsylvania; Holometrix report, *PIC-48*.
[16] "Report from Professional Service Industries, Inc, Pittsburgh Testing Laboratory Division, Pittsburgh, Pennsylvania; *831-73080*, 7 December (1987), pp. 1-2.
[17] Test Reports from the University de Liège – Institut du Génie Civil, Liège, Belgium; 10 avril (1986), *P/S.5 No.39.512*, pp. 1-3, and *P/S.3 No.40.385/1*, pp. 1-3.
[18] "Cellular Glass as Insulation Material for Industrial Equipment Applications", AGI Work sheet; Working Association Industrial Constructions e.V., Q137, August (1985), pp. 1-12.

PART II. PROPERTIES OF FOAMGLAS® CELLULAR GLASS AND ITS ACCESSORIES

The glass selected to manufacture FOAMGLAS® cellular glass corresponds to hydrolytic class III according to DIN 12 111 or class 3 according to an ISO test. To qualify the durability of this type of class, it can be mentioned that when tested according to the USP (United States Pharmacopœia) procedure, the FOAMGLAS® cellular glass is as chemically durable as the glass used for packing pharmaceutical preparations.

To summarise its behaviour, its resistance towards acids – except hydrofluoric acid and orthophosphoric acid – is excellent, its resistance against bases (alkali) fair with a few exceptions and its resistance against organics excellent.

The following table provides data on chemical resistance in specific conditions. It is based on a 100% immersion for a period of almost two years at 23°C.

	Liquids tested		Rating
(1)	5% Hydrochloric acid	(HCl)	E
(2)	5% Sulphuric acid	(H_2SO_4)	E
(3)	50% Sulphuric acid	(H_2SO_4)	E
(4)	Conc. sulphuric acid	(H_2SO_4)	E
(5)	5% Nitric acid	(HNO_3)	E
(6)	5% Phosphoric acid	(H_3PO_4)	G-P
(7)	Acetic acid	(CH_3COOH)	E
(8)	Hydrofluoric acid	(HF)	S
(9)	Water	(H_2O)	E
(10)	Ethyl alcohol	(C_2H_5OH)	E
(11)	Benzene	(C_6H_6)	E
(12)	Carbon tetra chloride	($C Cl_4$)	E
(13)	5% Potassium chromate	($K_2CR_2O_7$)	G
(14)	5% Ammonium hydroxide	(NH_4OH)	G
(15)	5% Sodium chloride	(NaCl)	G
(16)	5% Sodium sulfite	(Na_2SO_2)	G
(17)	5% Sodium carbonate	(Na_2CO_3)	G
(18)	5% Potassium hydroxide	(KOH)	P
(19)	5% Sodium hydroxide	(NaOH)	P
(20)	50% Sodium hydroxide	(NaOH)	P
(21)	Mersol		G

E = Excellent. G = Good. P = Poor. S = Severe attack.

More information is given in V.7.

Nevertheless, it is not recommended – although sometimes done – to systematically use unprotected cellular glass as a chemical barrier. The prime function of cellular glass remains insulation. This standpoint may be explained by the number of parameters which influence chemical durability in a real case, like the temperature and the concentration of the reagent, the duration of the contact, the exposed surface, etc. In case of accidental contact, for instance caused by spillage, FOAMGLAS® cellular glass offers a very valid protection in the vast majority of cases.

II.1.9 DIMENSIONAL STABILITY

As a consequence of its glassy composition, FOAMGLAS® cellular glass is dimensionally stable. It does not experience any change of dimension or shape due to water, humidity, sun rays or high loads. Slabs and fabricated elements do not warp or bow. Joint openings, damages to the coatings or waterproofings are avoided. The only dimensional changes are those corresponding to the coefficient of thermal expansion/contraction. They are reversible and predictable, with accuracy.

II.1.10 COEFFICIENT OF THERMAL EXPANSION

The coefficient of thermal expansion of FOAMGLAS® cellular glass is given in Fig. II.3. As can be seen, it slightly increases with the higher temperature and decreases more sharply at low temperature.

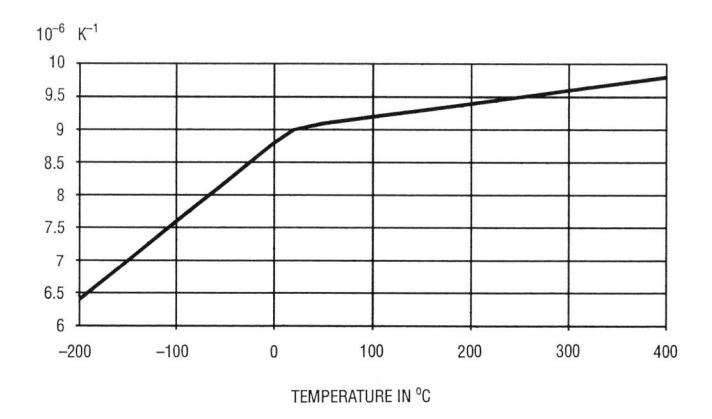

Fig. II.3. Thermal expansion coefficient of FOAMGLAS® cellular glass.

II.1.11 ECOLOGICAL STANDPOINT: PAY BACK IN ENERGY

In liaison with the manufacturing method, the pay back time in energy is more and more often addressed today. We can quickly give an order of magnitude to see how efficient FOAMGLAS® cellular glass is in energy saving.

To produce one cubic meter of FOAMGLAS® cellular glass type T4, one needs an energy of about 800 kWh.

We can easily calculate the pay back time for energy in two cases.

We shall consider one square metre of a liquefied ammonia tank wall operating at −33°C when the outside temperature is 10°C with a typical insulation thickness of 120 mm FOAMGLAS® cellular glass T4. The U value is given by:

$$\frac{1}{U} = 0.04 + \frac{d}{k}$$

$$= 0.04 + \frac{0.12}{0.039}$$

$$= 3.12$$

$$\text{or} \quad U = 0.32 \text{ W}/(\text{m}^2 \cdot \text{K})$$

and the heat loss per m² is

$$q = U \times S \times \Delta T$$
$$= 0.32 \times 1 \times (10 + 33)$$
$$= 13.8 \text{ W}$$

Since the energy to produce 1 m² of FOAMGLAS® cellular glass type T4 in 12 cm thickness amounts to

$$800 \times 0.12 = 96 \text{ kWh},$$

we would need

$$\frac{96{,}000}{13.8} = 6{,}957 \text{ hours}$$

i.e. 290 days or less than 10 months to recover the energy needed for manufacturing the product.

We shall now consider one square metre of a tank wall operating at 150°C, with an insulation of 60 mm FOAMGLAS® cellular glass type T4 covered by a thin coating having an emissivity of 0.3 placed in an ambience at 25°C, with a wind speed of 5 km/h. The heat loss is 105.88 W/m², the outside surface temperature 36°C, which is quite acceptable.

Since the energy needed to produce 60 mm FOAMGLAS® cellular glass type T4 amounts to 800 × 0.06 = 48 kWh, the time to recover it reads:

$$\frac{48{,}000}{105.88} = 453 \text{ hours or } 19 \text{ days !}$$

It is easy to see how fast the energy is recovered, especially when the temperature difference between the two faces of the insulation is large.

II.1.12 ECOLOGICAL STANDPOINT: CO_2 IN THE CELLS

As a second "ecological aspect" of FOAMGLAS® insulation, one can raise the question of the CO_2 in the cell of the product.

It is now well known that excessive CO_2 concentration in the atmosphere should be avoided since it causes the "green house effect" which is one of the reasons for the earth temperature increase. What is the situation with FOAMGLAS® cellular glass?

First of all, it should be repeated that cellular glass cells are tight to CO_2 as well as to water vapour. Consequently the CO_2 that is produced with the added carbon and the oxygen during the manufacturing process does not have any possibility of escaping from the material during its service life.

When, after many years, the plant or building will be demolished, the CO_2 will escape into the atmosphere from the cells that may be broken (not necessarily from all cells of the material).

Keeping in mind that the pressure of the gas in the cell is about 0.3 atmosphere at ambient temperature, it is easy to calculate that the amount of CO_2 under normal conditions in one cubic meter of FOAMGLAS® cellular glass type T4 is only:

$$1000 \times 0.95 \times 0.99 \times 0.3 = 282 \text{ litres}$$

(0.95 because 5% of the volume are occupied by glass, 0.99 because CO_2 represents 99% of the gas in the cells).

Since the density of CO_2 is about 1.96 g/l, crushing completely 1 m^3 of cellular glass would free

$$282 \times 1.96 = 553 \text{ g of } CO_2$$

To get a feeling for this amount, one could compare it with the amount of CO_2 given by a man breathing during one day in normal conditions, i.e. about 1 kg or about twice as much.

A fortiori, one could compare the amount of CO_2 liberated when crushing cellular glass at the end of its service life with the amount of CO_2 generated to produce the energy that can be saved by 1 m^3 cellular glass properly installed.

As a summary, the energy pay back of FOAMGLAS® cellular glass is excellent and its use permits great saving of CO_2 compared to the very small amount of CO_2 that can be released in the atmosphere at the end of its service life.

Finally, as proven by experience and several scientific reports, the disposal of FOAMGLAS® cellular glass after its life does not create any ecological problem. And the life of cellular glass can almost always match the life of the construction in which it is incorporated.

II.1.13 DIFFERENT TYPES OF FOAMGLAS® CELLULAR GLASS

Some mechanical properties of cellular glass – compressive strength, flexural strength, tensile strength, shear strength and modulus of elasticity – generally increase with the density.

This dependence explains why several types of cellular glass have been developed to fit as perfectly as possible various applications. Without being too specific, FOAMGLAS® cellular glass for industrial applications exists in the following types for which typical applications are mentioned (in order of increasing density/compressive strength):

T4............... pipe covering, vessel and tank wall insulation
T2............... pipe covering, vessel and tank wall insulation
S3 .... idem but more specifically when high stresses/loads are expected
HLB 115.... tank base insulation and highly stressed supporting parts
HLB 135.... tank base insulation and highly stressed supporting parts
HLB 155.... tank base insulation and highly stressed supporting parts
HLB 175.... tank base insulation and highly stressed supporting parts

As example, the properties of T4, T2 and S3 were as follows, at the time of edition of this book [19] when measured according to the indicated test methods:

FOAMGLAS® cellular glass type	T4	T2	S3
Nominal density (kg/m^3) (+/–10%)	120	125	135
Thermal conductivity at 10°C (maximum average measured in W/(m.K))	0.042	0.044	0.046
Compressive strength (minimum average in N/mm^2)	0.7	0.8	0.9
Flexural strength (minimum average in N/mm^2)	0.4	0.45	0.5
Flexural modulus of elasticity (minimum average in N/mm^2)	800	1000	1200
Thermal diffusivity in m^2/s	4.2×10^{-7}	4.2×10^{-7}	4.2×10^{-7}

and the properties of FOAMGLAS® cellular glass HLB types were [20]:

FOAMGLAS® cellular glass HLB type	115	135	155	175
Nominal density (kg/m^3) (+/– 10%)	136	140	148	152
Thermal conductivity at 10°C (maximum average measured in W/(m.K))	0.0495	0.0515	0.0520	0.0525
Compressive strength (minimum average in N/mm^2)	0.8	0.93	1.06	1.2

---

[19] 1992.
[20] FOAMGLAS® cellular glass properties are regularly compared to application requirements and customer requests. Therefore they are sometimes modified or new types are defined. It is suggested to check the FOAMGLAS® cellular glass properties with the up-to-date data sheet when needed.

PART II. PROPERTIES OF FOAMGLAS® CELLULAR GLASS AND ITS ACCESSORIES

It should be added that FOAMGLAS® cellular glass HLB types are specially produced for tank base insulation and that a very stringent quality control is enforced, corresponding to the risk associated with this application.

II.1.14 THERMAL CONDUCTIVITY

Due to technical progress, the gas in the cells is now almost exclusively composed of CO_2, less than 1% of the volume consisting of other gases [21]. One should also note that the pressure in the cells is about 0.3 atmosphere at ambient air temperature, the value being explained by the fact that the cells are closed at high temperature and cooled.

From what has been said of the production process and the product, one can easily understand that the heat transfer through cellular glass is caused by

- Conduction in the glass matrix
- Convection of the gas enclosed in the cells
- Radiation.

At ambient temperature, the radiation term remains small compared to the conduction and the convection. Its importance increases very fast with the temperature.

Lowering the density tends to lower the thermal conductivity provided that other factors remain unchanged, for instance that the cell size is kept low enough to avoid increase of convection. As a general rule, the thermal conductivity of cellular glass increases with the density although several other factors can influence it.

II.1.15 COMPRESSIVE STRENGTH

Complementary to the values previously indicated, one should note several important points regarding cellular glass compressive strength.

---

[21] The typical odour which is smelt when breaking one cell of FOAMGLAS® cellular glass is caused by a very small amount of H_2S which is part of the remaining 1 %. It may be worth pointing out that less than 1ppm (Part Per Million) of H_2S is sufficient to be able to smell the characteristic odour.

In general terms, compressive strength increases with density, although other factors influence the value. This is reflected in the table of the different types of FOAMGLAS® cellular glass.

The numerical values correspond to a precise testing method that basically calls for the application of a layer of bitumen of about 1.25 kg/m² and a thin roofing felt on the two bearing faces of the test specimens. The purpose of this layer is to simulate many applications in which bitumen or a mechanically similar coating/adhesive is used. With this layer, the surface open cells are filled and the material can provide its full compressive strength instead of being progressively damaged by local rupture of unfilled cell walls.

The test procedure for compressive strength measurement is described in detail in several norms: ASTM C 240-91, DIN 18 174, and ÖNorm B 6041. Give or take some minor details, these three procedures are equivalent and provide equivalent results.

Some specific, adequately chosen, other cappings give compressive strengths comparable to the "ASTM conditions". For instance, a well-defined roofing felt, PLUVEX No.1 DAMP PROOF COURSE CLASS A manufactured to BS 6398 CLASS A, with a Hessian base, gave compressive strength comparable to "ASTM conditions". Of course, the equivalence to the reference conditions has to be checked on a regular basis. In the other direction, several different although more modern roofing felts, with glass or polyester reinforcement provide lower compressive strength than the reference conditions. A fortiori, dry cellular glass compressive strength was found substantially lower than the reference conditions. Figure II.4 illustrates this point.

The deformation measured on the stress-strain diagram is very limited in the case of cellular glass and virtually corresponds only to the settlement of the capping in the surface open cells of the product and a small creep of bitumen at the test specimen edges. It is worth pointing out that this bitumen edge creeping is possible on small test specimens but limited to the periphery of the large surface applied in a real jobsite. It is easily proven that the cellular glass deformation is very limited. For a material having a compressive modulus of elasticity of 800 N/mm² and a compressive strength of 0.7 N/mm², the relative deformation, just before rupture, would be given by

$$\sigma = \varepsilon \times E \quad \text{where } \varepsilon = \frac{\Delta L}{L}$$

$$\text{or} \quad \varepsilon = \frac{\sigma}{E} = \frac{0.7}{800} = 0.875 \times 10^{-3}$$

which corresponds to 0.0875 mm for a 100 mm thick slab.

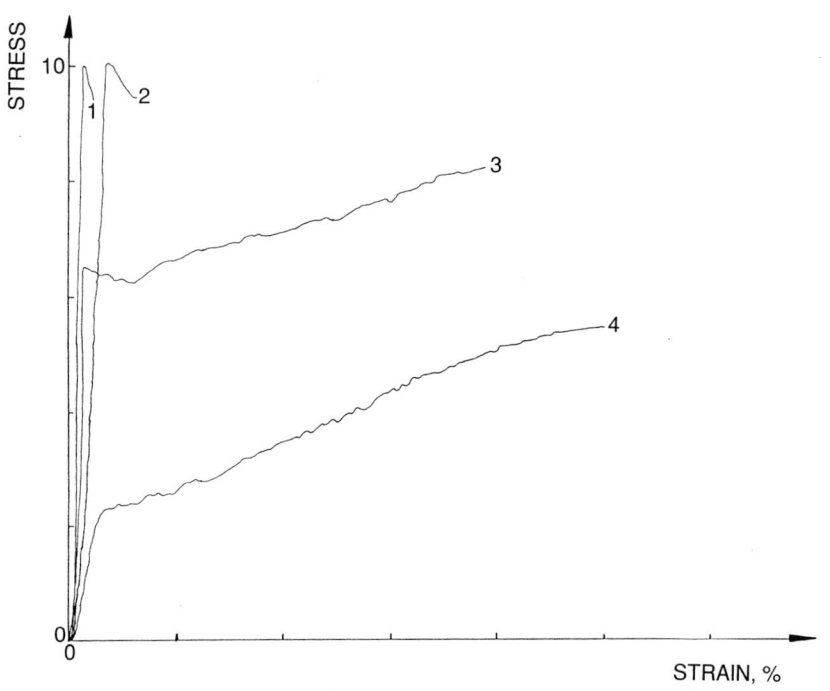

Fig. II.4. Typical stress-strain response of a FOAMGLAS® HLB insulation as a function of capping type.

1. ASTM C 240-91.
2. Approved bituminous felt.
3. Typical low organic or inorganic felt.
4. Bare, no capping.

Note: The above is just to represent typical behaviour and should not be used for design purposes.

It is clear that the usual deformation at failure of 1 to 2 mm is mainly caused by bitumen/roofing felt influence on the surface, the cellular glass internal deformation being close to negligible.

As indicated by several test reports, the compressive strength of FOAMGLAS® cellular glass is only very moderately reduced by elevated temperatures, at least up to its given temperature limit of 430°C.

Similarly, its compressive strength is only very moderately reduced by long term loading [22] [23].

The usually published compressive strength corresponds to tests run with the force applied perpendicular to the main face of the slab (thus parallel to the foaming direction). Should the force be applied in a direction parallel to the main faces (in length or width), the measured compressive strength will be increased by a factor of about 1.2. This phenomenon can be explained by the structure of the glass cells.

The published compressive strength is a minimum average value in ASTM conditions. This means that it represents the minimum which should be found, when testing to failure in ASTM conditions, a statistically significant number of test specimens and calculating the average value.

Without putting too much emphasis on the numerical value, it can be mentioned that the standard deviation of cellular glass compressive strength generally ranges from 8% to 15% for a well defined and controlled production process.

Since the published value is an ultimate value (= at failure), a safety factor is needed. For industrial applications, a safety factor of three based on the maximum working stress and the average compressive strength has been generally adopted [24].

II.1.16 FLEXURAL STRENGTH

FOAMGLAS® cellular glass flexural strength depends on the type and, as a general rule, increases with density.

---

[22] "Cellular Glass as Insulation Material for Industrial Equipment Applications", AGI work sheet, *Q137*, August (1985), p. 4.
[23] "Staatliches Materialprüfungsamt Nordrhein Westfalen", Test Report PCE Technical Sales Letter, *E 188*, 14 October (1975), p. 2.
[24] For some less risky applications in the architectural field, this value of three has sometimes been slightly reduced.

The measurement is carried out by the flexural beam method. See for instance ASTM C 203.

The values given in II.1.13 for the various types of FOAMGLAS® cellular glass being average ultimate values, a safety factor is needed. A value of three based on the maximum working stress and the average flexural strength is generally adopted.

II.1.17 SHEAR STRENGTH

FOAMGLAS® cellular glass shear strength also increases with density.

A distinction is made between internal shear that is a true property of the material and surface shear that also involves adhesive and is more an application related property.

The internal shear of cellular glass is usually measured by determining a test area limited by two incisions in the test specimen that will be subject to shear when the test specimen is subjected to compressive stress, in given conditions.

It is important to follow exactly the prescribed test method since the quality and the validity of the results strongly depend on the test method.

Typical average values for FOAMGLAS® cellular glass internal shear strength read as follows, when tested according to the described method:

FOAMGLAS® Type	Shear strength in N/mm²
T4	0.30
T2	0.35
S3	0.40
HLB 115	0.35
HLB 135	0.40
HLB 155	0.45
HLB 175	0.50

The surface shear strength is measured by adhering a disk on the surface of the FOAMGLAS® cellular glass slab with a chosen adhesive, measuring the rupture torque and calculating the corresponding shear stress.

As it can be easily understood, the measured value depends not only on the FOAMGLAS® cellular glass but also on the selected adhesive and the results may differ substantially in relation to the adhesive. For the FOAMGLAS® cellular glass/bitumen couple, typical average values would read 0.30 N/mm² for type T4, 0.35 N/mm² for type T2 and 0.40 N/mm² for type S3.

II.1.18 TENSILE STRENGTH

Tensile strength also tends to increase with density. Typical average values for the various types of FOAMGLAS® cellular glass would be:

FOAMGLAS® Type	Tensile strength in N/mm²
T4	0.24
T2	0.26
S3	0.28

For tensile strength, as for shear strength, flexural strength and compressive strength, the given values are average ultimate values, which means that they are obtained by calculating the average of the results at rupture measured on a meaningful number of test specimens.

For application, a safety factor is needed. It should be chosen by the design engineer who masters the data for a given case. A value of three calculated by dividing the average ultimate value by the working stress is often chosen.

II.1.19 TEMPERATURE INFLUENCE ON THE PROPERTIES OF CELLULAR GLASS

As an example, the Fig. II.5 shows how the thermal conductivity rises with temperature.

As far as the mechanical properties are concerned, like compressive strength, flexural strength, tensile strength and shear strength, they slightly improve at lower temperatures as it is generally the case for glass.

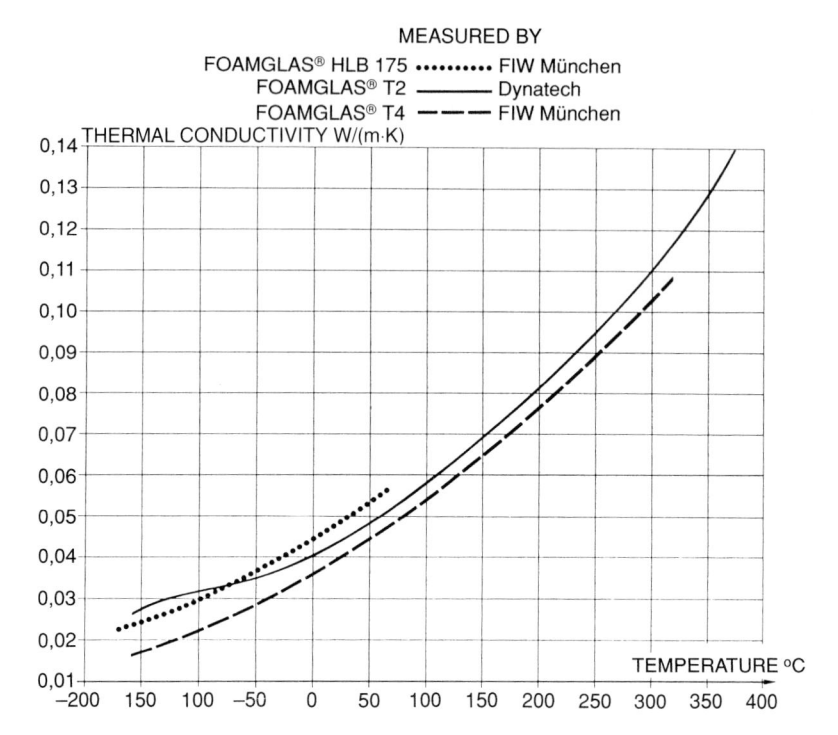

Fig. II.5. Thermal conductivity curve in function of temperature.

At high temperature, without exceeding the theoretical limit of 430°C, the previously mentioned mechanical properties are slightly reduced, the temperature influence being marginal from an application standpoint. It should be remembered that the softening point is at 730°C.

II.1.20 LIMITATIONS TO THE USE OF CELLULAR GLASS

FOAMGLAS® cellular glass is only subjected to a few limitations which can be overcome by adequate installation methods. As the main causes of limitations we can quote:

- Thermoshock that leaves hairline cracks. These very minute cracks do not reduce significantly the insulation value of FOAMGLAS® cellular glass.

And it can be installed in such a way that they do not have negative consequences
- Surface state and choice of adhesive and coating. Very rigid adhesives or coatings having a coefficient of thermal expansion different from cellular glass will create stresses when the temperature varies considerably. These stresses may result in shearing near the surface of the FOAMGLAS® cellular glass. A good choice of adhesive and coating will eliminate this problem. A number of adhesives, coatings and sealers, etc. have been tested and reports are available.
- Permanent contact with hot water or, even more, with high temperature steam
- Freeze and thaw cycle in wet conditions..

These two last situations, which are unusual, will be analysed in V.3.4

II.2
ACCESSORY PRODUCTS

The title "accessory products" usually means a series of products such as adhesives, sealers, coatings, anti-abrasives and mechanical fixings used to install insulation products, more specifically FOAMGLAS® cellular glass in our case. Contrary to what the adjective "accessory" suggests, the accessory products have a strong influence on the insulation system performance and constitute a substantial part of its installed cost.

If the distinction between the various categories of accessory products is useful from an explanation standpoint, it should be pointed out that one single product frequently can fulfil several functions, for instance, being an adhesive and a sealer.

We will review the main types of accessory products in general, mention some important requirements and characteristics and finally present the main products compatible with FOAMGLAS® insulation.

II.2.1 ADHESIVES

Adhesives are used to fix insulation material to a surface which is generally metallic in the case of industrial applications, sometimes concrete. The quality of their bond to the substrate surface strongly depends on the substrate, which should be clean, dry and not frozen. Traces of rust, oil and foreign matter should be eliminated before the use of adhesives. When an anti-corrosive layer is applied on the substrate before the adhesive, the limiting factor can be the adherence of the anti-corrosive layer to the substrate or the adherence of the adhesive to the anti-corrosive layer. The same should be noted if a primer is applied on the substrate before the adhesive. The designer has to check the compatibility of all products included in the system. Should the substrate surface, for instance in the case of an existing application to be reinsulated, be already painted, compatibility tests are needed (Fig. II.6).

Fig. II.6. Three-foot adhesion tester.

Regarding the service temperature, a distinction should be made between the permanent service temperature and the temporary service temperature (lower or higher than the normal service temperature) since adhesives are generally organic products that can experience ageing problem when subjected to excessive temperature. It should also be pointed out that the adherence strength or tensile strength generally decreases sharply when a given temperature is exceeded.

Since cellular glass does not absorb water or solvent and since the same applies to steel and more generally metallic support, adhesives that are used between FOAMGLAS® cellular glass layers or to adhere FOAMGLAS® cellular glass to metal cannot set when the hardening is based on water or solvent evaporation.

This physical point restricts the choice of adhesives to those which set on the basis of other mechanisms. Typically, these are temperature change, two component chemical reaction, contact adhesive or hydraulic setting cement adhesive.

II.2.1.1 Bitumen

Bitumen is heated to a liquid state at a temperature of about 180°C, depending on the type of bitumen, without exceeding about 220°C. FOAMGLAS® cellular glass slabs or elements are dipped into the hot bitumen in such a way that all the surfaces to be adhered are well covered. If necessary the dipping is done twice. Immediately, the surfaces to be adhered are firmly pressed together. The slabs are kept in place until the temperature decreases and the bitumen hardens and develops its tensile strength. This usually takes slightly less than one minute. Obviously this adherence based on temperature modification limits the use of bitumen to temperatures at which it does not lose its strength.

II.2.1.2 Two Component Adhesives

In the case of two component adhesives, usually a liquid and a powder, sometimes two liquids, their mixing initiates a chemical reaction that develops the strength of the product.

The mixing ratio of the two components, the mixing time and temperature and the type of mixing equipment strongly influence the final performance. The mixer has to be suitable for the characteristics of the products to mix, particularly their viscosity. Mixers are specified or have been developed for PC® 56 ADHESIVE and PC® 88 ADHESIVE.

II.2.1.3 Contact Adhesives

Contrary to the two previous adhesives and to hydraulic setting cements, the contact adhesives are applied with such a low coverage that they do not fill completely the surface cells of FOAMGLAS® cellular glass but more or less bridge them and adhere to the exterior part of the cell walls. Usually they are single component adhesives that are applied to the two surfaces to be adhered. After a given drying time, when they start to set, the two surfaces are applied to each other and a pressure is applied, eventually with a moderate temperature increase. In the case of cellular glass, contact adhesives are mainly used for sandwich panel production but are not used on industrial sites.

II.2.1.4 Hydraulic Setting Cement Adhesives

Hydraulic setting cement adhesives, in the case of FOAMGLAS® cellular glass, are basically limited to the product used for manufacturing large size pipe coverings and more generally fabricated elements aimed at high service temperature. It should be mentioned that the use of cement based adhesives is not recommended with FOAMGLAS® insulation, specially when the presence of water is expected.

II.2.2 SEALERS

Sealers are applied in the joints between the slabs, as a sliding layer in expansion joints between two layers of FOAMGLAS® cellular glass and at singular points such as around nozzles, etc. They are also applied at metal jacket overlaps to obtain water vapour tightness (Fig. II.7 and Fig. II.8).

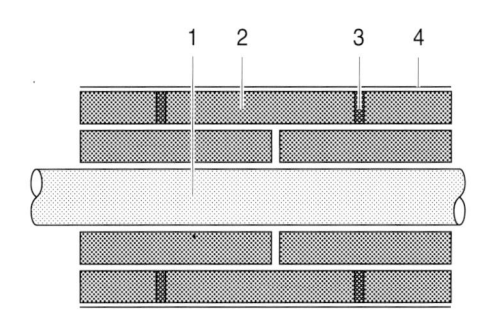

Fig. II.7. Joints between shells with sealers.
1. Pipe.
2. FOAMGLAS® insulation.
3. PITTSEAL® 444.
4. Reinforced mastic finish.

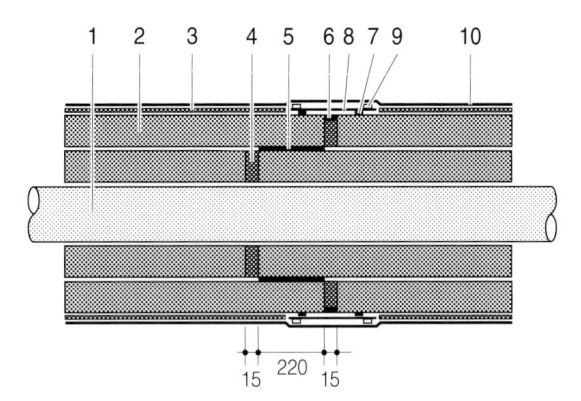

Fig. II.8. Expansion joint with sealers.
1. Pipe.
2. FOAMGLAS® insulation.
3. Cell filler.
4. Joint filled with resilient insulation material.
5. Sliding compound compatible with temperature, for instance PITTSEAL® 444.
6. PITTSEAL® 444.
7. PITTSEAL® 444.
8. Polyisobutylene local vapour barrier.
9. Fixing straps.
10. Metal jacket allowing contraction.

For lower temperature applications, joint sealers must be water vapour tight. For high temperature applications, water tightness is generally sufficient.

Sealers are typically made of resins, fillers and a low percentage of solvents. A recent trend has been the choice of butyl based sealants that have very low permeability and good ageing.

The small amount of solvent should remain in the sealers without escaping with time to maintain the flexibility of the product over many years of service. FOAMGLAS® cellular glass favours this permanence of properties since solvents cannot pass through it and consequently the surface through which they can evaporate is very limited. The flexibility, characterised by a low modulus of elasticity is needed at the service temperature of the sealer. Sealers should also adhere firmly to the faces of the joints to avoid water vapour passage between them and the insulation. They should also have low shrinkage to avoid crack formation and enable the insulation system to follow movements of thermal origin or due to the support.

Their resistance to water vapour migration has to be very high to achieve an actual water vapour barrier when sealing the joints between the FOAMGLAS® cellular glass elements that are perfectly vapour proof themselves. This can be achieved since, contrary to the exterior water vapour barriers that are generally about 2 mm thick and which are exposed to all sorts of abuses, the sealers present a "length" for the passage of water vapour equal to the thickness of the insulation layer, i.e. from 40 to 100 mm and even more. Finally the sealers are applied between two FOAMGLAS® cellular glass slabs and thus protected from weather or mechanical abuses.

II.2.3 COATINGS AND JACKETS

The function of these products is mainly to provide a protection against water or water vapour that can penetrate in the insulation system.

As explained in V.3 and V.4, water and water vapour absorption are very detrimental to an insulation system.

For low temperature application, the presence of water vapour coming from the exterior easily explains the problem. Although less obvious and somewhat less severe the problem also exists for high temperature applications, for instance when they are idling.

A weather barrier is installed on the outside of an insulation system to protect it from the weather actions, such as rain, snow, hail-stones, wind, solar radiation and from human actions such as mechanical abuse, chemical spillage, deluge testing, etc.

As an example of mechanical abuse, workers tend to use insulated lines as scaffolding and walk or place tools and other instruments on them. Chemical spillages can contain products, which may deteriorate the weather or vapour barriers. Fire is another source of degradation. Although less obvious, movements of equipment, piping, insulation products and vibration can create forces that develop stresses in the mastic. For instance, thermal contraction and expansion of the vessels, pipings and insulations, can result in stresses in the coating.
Fortunately, FOAMGLAS® insulation has a coefficient of thermal expansion slightly lower than steel and consequently this type of stress is limited.
Therefore the classical problem of joint opening between the insulation slabs with corresponding stress development in the weather barrier in front of the joints, is reduced with FOAMGLAS® cellular glass (Fig. II.9).

A vapour barrier theoretically prevents the passage of water vapour from the exterior into the insulation system.

In fact, as explained later, a perfect vapour barrier is virtually impossible to achieve and to maintain. In the majority of the cases the word vapour brake or vapour retarder, which only suggests that the passage of water vapour is reduced and not prevented, is more appropriate to the real situation.

On permeable insulation, specially when the service conditions foresee a temperature lower than the ambient air such as for instance chilled water, liquefied gas etc. a good vapour barrier is needed.

Fortunately on FOAMGLAS® insulation, a weather barrier is generally sufficient provided that the joints between the slabs of the outside layer are well filled with an adequate sealer. Cellular glass as such is a vapour barrier and this characteristic applies to the whole product and not only to a fragile film applied on the outside.

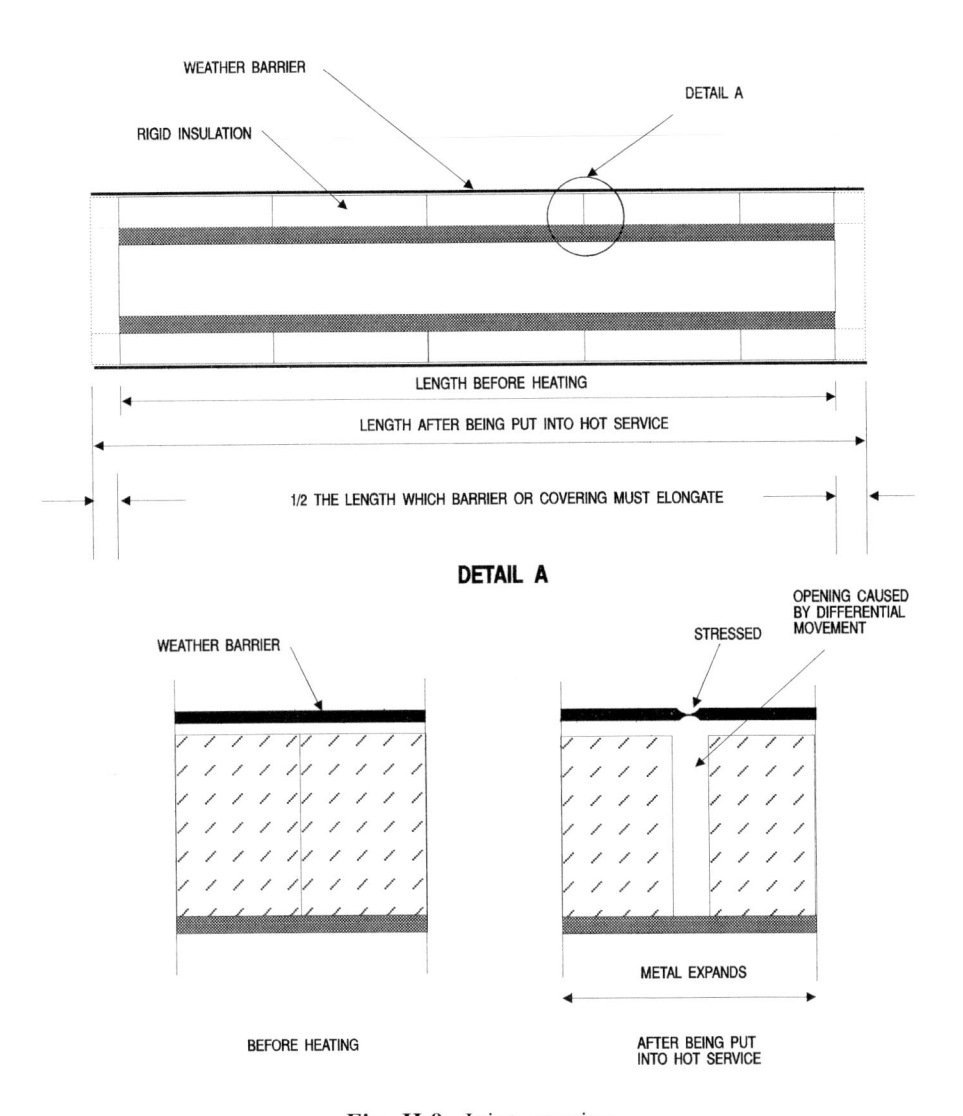

Fig. II.9. Joint opening.

A number of products are excellent weather barriers, few are good vapour barriers. Physically the functions of a weather barrier or vapour barrier can be obtained with a variety of products: they are classified in two main families: reinforced mastics and metal jackets.

II.2.4 MASTICS

Reinforced mastics are made of bituminous components or are resin based. The liquid component necessary to make them suitable for application is either water in the case of emulsion or solvent in the case of cutback. They incorporate also a large amount of fillers to give them body and often some fibre to improve the mechanical properties. The main family of resin based products are acrylic, butyl and polyvinyl acetate. Acrylic and butyl based mastics tend to age better than the others.

II.2.5 REINFORCEMENT FABRICS

When mastics set by evaporation of water or solvents, they shrink. Cracks should not result from this shrinkage. Being applied directly on an insulating material, they experience considerable temperature changes. Their coefficient of thermal expansion and their modulus of elasticity should be such that further stress is limited and cracking does not occur. Regarding service temperature, it is necessary that they do not become too hard at low temperature or lose their strength at high temperature. They should not age badly when subjected to these conditions. It should be noted that on some insulated cold applications, the surface temperature can decrease to values around $-20°C$ during a cold night. On a hot insulated line, the surface temperature can easily reach $60°C$ and even more, for instance $90°C$ if the solar radiation is intense. These extreme temperatures can cause mastic brittleness or fast ageing if the product does not have adequate properties.

Weather proofing mastics or vapour barriers are usually reinforced with glass or polyester fabric, perhaps another type of fabric. These reinforcements give a much higher tensile strength to the mastic layer to enable it to resist the stresses previously mentioned; it is particularly important where the mastic layer bridges joint openings, voids or cracks and when the reinforcements are applied at particular points such as nozzles, tees, etc. Moreover the reinforcement imposes a minimum thickness of the mastic layer and would appear visible if this minimum coverage is not correctly applied. It also practically imposes an application of the mastic in two layers that is favourable as far as the drying time is increased. From this standpoint, the choice of the mesh size has to be adapted to the required mastic layer thickness.

Regarding the choice of reinforcement material it should be adapted to the type of mastic. Glass fibre reinforcements that have low elongation before they break should be used with rather hard mastic. Polyester based reinforcements that give higher elongation before they break are well fitted for more flexible mastics such as the acrylic ones. As a rule, polyester fabrics presenting high elongation are often preferred with the more flexible mastics although it should be recognised that glass fabrics present a marginal advantage from a fire standpoint. *Pittsburgh Corning* markets PC® FABRIC 79P – a polyester fabric, PC® FABRIC 79G – a glass fabric and PC® FABRIC 79 HD – a heavy duty glass fabric.

Before ending this paragraph on mastics, it is appropriate to mention a few points concerning these applications:

1. Sufficient time has to be allowed for the application of insulation and coatings. This is specially important in the case of a shut down.
2. For outdoor applications, weather is a critical factor in the final quality of these systems. Saving on weather protection, like scaffolding with tarpaulins may be regretted during many years.
3. Outside conditions, for a given accessory product, such as application temperatures, have to be taken into account.
4. Equipment, such as a spraying machine, has to be adapted to the product being applied.
5. When the manufacturer recommends mastics or sealers to be applied in two layers it is wise to follow the suggestion. Too thick layers of these materials often get blistered and damaged because of the long drying time. It is always important to apply mastics, adhesives and sealants according to the manufacturers' instructions.

II.2.6 METAL JACKETS

For outdoor applications, metal jackets are usually made of galvanised, pre-painted, aluminised, stainless steel or aluminium. On a galvanised steel jacket, impregnated Kraft paper or equivalent is sometimes applied on the inside face to protect the steel from corrosion favoured by condensation coming from the inside insulation on a hot line. Needless to say, this extra-protection is not necessary in the case of FOAMGLAS® cellular glass in which neither water nor steam can be present.

Stainless steel jackets offer a better chemical resistance and improve the fire resistance of insulation systems but they are more expensive.

The thickness of metal jackets should not be forgotten: it has direct influence on the resistance against mechanical abuse. Specially at points where traffic could be expected, saving on thickness may prove a false economy particularly when compressible insulation is used.

If the metal jackets as such are obviously water and water vapour tight, their joints constitute a real problem to obtain system tightness. Although it is conceivable to seal all of them with adequate sealer, it should be recognised that it is practically impossible to obtain and maintain perfect integrity. For this reason, metal jackets should be considered as a weather barrier rather than as a water vapour barrier. And even to be a weather barrier, tightness to liquid water passage has to be achieved. This requires a great care at all the singular points such as for instance overlaps which have to be done in the correct manner and sealed in many cases. For instance, unsealed overlaps are sufficient on a vertical line but sealed joints are required on a horizontal line (Fig. II.10).

If this is not done, water will penetrate the system and the metal jacket can only be evaluated as a mechanical covering but not as a weather barrier.

Of course the system can be improved by more sophisticated joint designs that prevents water passage. In many cases, metal jacket systems can be designed with drain holes at the lowest part. These remarks explained why metal jackets are more suitable for straight lines than for complicated equipment or valves, elbows etc.

For these reasons, some design engineers who want the advantages of the metal jacket, i.e. fire resistance and good mechanical protection, provided it is thick enough, specify a reinforced mastic vapour barrier on the insulation and a metal jacket on top of it. On FOAMGLAS® insulation that is itself a vapour barrier provided the joints are correctly sealed, such a solution should be applied only in case of extreme requirements.

Some designers also specify a cell filler (generally PITTCOTE® 300) on the FOAMGLAS® cellular glass surface before the metal jacket is applied. This solution is advisable only in high wind areas or when many significant temperature variations are expected.

CHAPTER II.2 ACCESSORY PRODUCTS

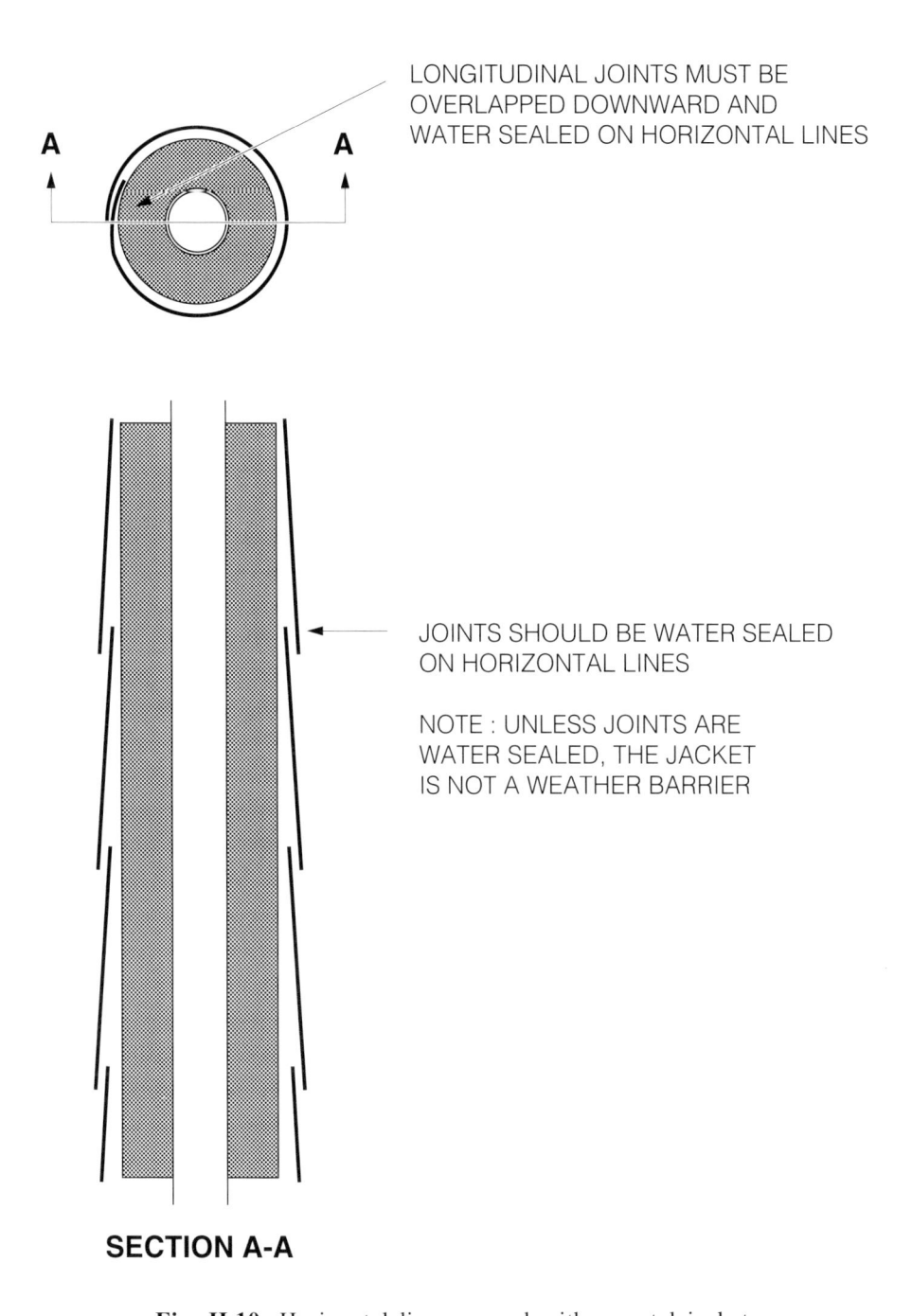

LONGITUDINAL JOINTS MUST BE OVERLAPPED DOWNWARD AND WATER SEALED ON HORIZONTAL LINES

JOINTS SHOULD BE WATER SEALED ON HORIZONTAL LINES

NOTE : UNLESS JOINTS ARE WATER SEALED, THE JACKET IS NOT A WEATHER BARRIER

SECTION A-A

Fig. II.10. Horizontal line covered with a metal jacket.

A last consideration on the choice between a reinforced mastic coating or a metal jacket should be made depending on the emissivity. This decision that governs the emissivity should be made at the design stage since it may have a strong influence on the thickness of insulation required. On low temperature lines, the thickness needed to prevent condensation will be different depending whether an emissivity of 0.9 for a dark reinforced mastic or an emissivity of 0.1 to 0.2 for an aluminium jacket is in the calculations. It may substantially increase thickness in the second case. As an example, if we consider a 6 inch pipe operating in the following conditions:

Service temperature ..... −104°C (liquefied ethylene)
Ambient temperature ... +10°C
Wind speed .... 0 km/h
Relative humidity 70%

the surface condensation could be avoided with an 80 mm thickness of FOAMGLAS® cellular glass when the mastic emissivity is 0.9 but a FOAMGLAS® cellular glass thickness of 130 mm will be needed to avoid it, if the jacket has an emissivity of 0.2.

In a similar way the surface temperature on a hot line can be much higher, everything else being equal, if a dark reinforced mastic is replaced by a white finish or a low emissivity aluminium jacket. It can result in an increased danger of personnel hazard.

As another example, if we consider a vessel of 3 m diameter operating in the following conditions:

Service temperature .... 300°C
Ambient temperature .... +25°C
Wind speed .... 0 km/h

a 50 mm FOAMGLAS® cellular glass thickness will be required to limit the surface temperature to 59°C if the emissivity is 0.9, as in the case of a dark mastic. But with an emissivity of 0.2 and the same 50 mm FOAMGLAS® cellular glass thickness, the surface temperature will reach 77°C. A 90 mm FOAMGLAS® cellular glass thickness will be necessary to limit the surface temperature to 59°C.

II.2.7 ANTI-ABRASIVE COATINGS

Anti-abrasive is applied on the bearing surfaces of insulation products to avoid deterioration due to abrasion caused by repeated thermal movements or mechanical vibrations. In the case of FOAMGLAS® cellular glass, they can also be recommended to avoid the surface open cell damaging an eventual anti-corrosion layer applied on the metal surface of the pipework or equipment. It should be pointed out that an anti-abrasive is only necessary on FOAMGLAS® cellular glass when important temperature variations are frequent since the cellular glass coefficient of expansion is very close to that of steel. Another instance where anti-abrasive is required is when large amplitude vibrations are expected.

Typical anti-abrasive products are based on polyurethane resin, on other solvent resin or on modified calcium sulphate mixed with water.

The resin based products are applied from low temperature to about 100°C, depending on the type of resin used.
The inorganic product is used for high temperature applications.

For dual temperature and cryogenic applications, operating at a temperature lower than –183°C at which oxygen condensation requires the use of inorganic products, the modified calcium sulphate is generally chosen, another reason more to avoid water penetration in the system.
The thickness of the anti-abrasive coating shall be taken into consideration when cutting or preparing fabricated elements.

As a last point, two classical mistakes to avoid should be mentioned: applying the insulation with its anti-abrasive on the support before it has completely dried or applying the anti-abrasive on the support instead of to the insulation material.

II.2.8 CHOICE OF ACCESSORY PRODUCTS

As already mentioned, the selection of accessory products, constitutes a vital decision in the design of industrial insulation systems and should be made at an early stage.

The following list outlines points that should be considered, at least from a qualitative standpoint, when making the selection.

II.2.8.1 Points to be Considered when Selecting Accessory Products

- Service temperature limits (low and high)
- Chemical nature of the accessory product:
  - Emulsion
  - Cutback
  - Two components (mix ratio)
  - Moisture curing
- Viscosity (workability)
- Solid percentage at 105°C (for quality control purpose)
- Ashes percentage at 800°C (for quality control purpose)
- Adherence to the substrate (or bonding strength)
- Cohesive tensile strength
- Chemical compatibility with existing substrate and other products used in the insulation system
- Coefficient of thermal expansion
- Shrinkage
- Modulus of elasticity (or flexibility)
- Resistance to water penetration
- Resistance to water vapour diffusion
- Flash point
- Combustibility
- Chemical compatibility to products which potentially could spill on them
- Resistance to sun radiation
- Resistance to abrasion and mechanical abuse
- Ageing
- Application temperature limits (low and high)
- Drying time in standard conditions (temperature and relative humidity) and in foreseen conditions
- Number of layers needed for the application
- Pot-life and packing size
- Shelf life
- Storage conditions
- Coverage

- Mixing facilities, if needed
- Method of application:
  - by trowel
  - by brush
  - by spray

  and necessary equipment
- Health hazard and safety (material safety data sheet).

II.2.8.2 Properties for Particular Applications

- Suitable for use on stainless steel without causing stress corrosion
- Oxygen compatibility index
- Approval for the food industry
- Approval for the nuclear industry
- Other specific cases.

This check list is general by nature and should be enlarged for specific applications, particularly when the application is unusual.

All the points mentioned in the above list are not necessarily relevant to each application but many deserve some consideration.

It is often difficult to obtain quantitative information on one brand of accessory products but even qualitative information is safer than no consideration at all.

II.2.9 ACCESSORY PRODUCTS MARKETED BY *PITTSBURGH CORNING* FOR INDUSTRIAL APPLICATIONS

Pittsburgh Corning markets a number of accessory products that have been designed or modified to specially fit the requirements of FOAMGLAS® cellular glass based insulation systems. We will quote the main ones with a short description of the application possibilities and some comments when appropriate. It is recommended to read carefully the corresponding data sheet before specifying and applying them [1].

---

[1] Since *Pittsburgh Corning* is permanently trying to improve its line of accessory products, the detailed properties are not presented in this Handbook. It is suggested to check them with the up-to-date data sheet at the appropriate time.

II.2.9.1 Adhesives

A. PC® 56 ADHESIVE

PC® 56 ADHESIVE is a two-component adhesive based on a bitumen emulsion improved by synthetic material for component one and cement with filler for component two, these to be mixed in the ratio of 3 part component one with 1 part component two by weight.

A 10 times diluted solution of PC® 56 ADHESIVE is used as its own primer on absorbent surfaces. PC® 56 ADHESIVE should not be applied with other primers.

PC® 56 ADHESIVE is used to adhere FOAMGLAS® cellular glass to concrete or FOAMGLAS® cellular glass to itself.

Its service temperature ranges from –15 to +45°C, when used as an adhesive [2]. It should be noted that this temperature range is conservatively fixed for applications in which the maximum tensile strength is needed in steady state. It can be reasonably increased when other conditions prevail. *Pittsburgh Corning* can provide assistance in specific cases.

A description of mixers suitable for PC® 56 ADHESIVE is available.

B. PC® 88 ADHESIVE

PC® 88 ADHESIVE is a two-component adhesive based on bitumen and polyol for component one and isocyanate for component two, to be mixed in a ratio of 47 parts of component one and 1 part of component two by weight. No primer is needed.

PC® 88 ADHESIVE is used to adhere FOAMGLAS® cellular glass to itself, concrete, steel or aluminium. After curing, it forms a bond that remains flexible and absorbs mechanical and thermal stresses even at a temperature as low as –40 to –55°C depending on the support.

Its wettability and its flexibility at low temperature make it unique as a FOAMGLAS® cellular glass adhesive.

---

[2] PC® 56 ADHESIVE is also frequently used in the building field as an adhesive and for other functions.

Its service temperature ranges from –40/-55°C to +80°C, when used as an adhesive. Here also it should be noted that this temperature range is conservatively fixed for PC® 88 ADHESIVE used as an adhesive in applications for which the maximum tensile strength is required in steady state. Some applications have been carried out successfully at higher temperatures. Its service temperature can be reasonably increased when other conditions prevail. *Pittsburgh Corning* can provide assistance in specific cases.

A description of a suitable mixer for PC® 88 ADHESIVE is available. For larger jobsites *Pittsburgh Corning* can loan a suitable mixer.

C. PC® 86T

PC® 86T is a one-component bitumen based product, with fillers, offering a good resistance to high temperature.

It is used as an adhesive to adhere FOAMGLAS® cellular glass to concrete and as a bedding compound between FOAMGLAS® cellular glass layers for hot tank base insulation. No primer is needed. Its service temperature ranges from –10 to +90°C.

D. PC® 80M MORTAR

PC® 80M MORTAR is a two-component inorganic mortar based on specially formulated glass powder and fillers for component one and modified silica dispersion for component two to be mixed in the ratio of 100 part component one with 23 part component two by weight.

PC® 80M MORTAR is used to adhere FOAMGLAS® cellular glass to itself when the temperatures are very low, very high, or when they cycle rapidly, using the laminated technique that consists of embedding a specific reinforcement in the adhesive layer. No primer is needed. PC® 80M MORTAR service temperature (as an adhesive between two FOAMGLAS® slabs) has been tested from –196 to +320°C. It is incombustible and its coefficient of thermal expansion is almost identical to FOAMGLAS® cellular glass.

Special equipment for mixing and spraying it has been identified. Details are available.

II.2.9.2 Sealer

A. PITTSEAL® 444

PITTSEAL® 444 is a single component, non hardening butyl based sealer that does not dry but forms a thin skin after 1 to 3 hours.

It is used to seal joints between FOAMGLAS® cellular glass slabs or pipe coverings. It is also used to seal protrusions, metal jacket laps and as a sliding layer in expansion/contraction joints.

Its resistance to water vapour diffusion is excellent. Although its published service temperature only ranges from –50 to +80°C, it has been successfully tested at cryogenic temperatures at which it still gives an acceptable elasticity.

Its adherence to FOAMGLAS® cellular glass and many other surfaces such as steel, concrete, wood etc., is so remarkable that it can practically only be applied with a gun. With this tool, it can be placed exactly where it should be without having the problem of cleaning it back from other surfaces.

Several gun types have been selected, suitable for any application. PITTSEAL® 444 can be delivered in drums or in cartridges (Fig. II.11).

Fig. II.11. PITTSEAL® 444 Gun.

II.2.9.3 Coatings and Jackets

A. PITTCOTE® 300

PITTCOTE® 300 is a single component solvent cutback of a selected bitumen base containing a high percentage of fibre and other mineral fillers blended to give a resilient, resistant and watertight coating with a low water vapour permeability.

It is used as a coating on FOAMGLAS® insulation with a glass fabric or a polyester fabric reinforcement.

Its service temperature ranges from –40 to +80°C. It certainly qualifies as a weather barrier and its properties allow to classify it as a vapour barrier.

Should it be necessary to paint it, for instance for aesthetical reasons, or to improve its ageing, it should be allowed to dry at least one month in normal weather conditions before the paint application to enable the solvent to completely dry out.

It can be applied by trowel or sprayed with airless equipment.

B. PITTCOTE® 404

PITTCOTE® 404 is a highly flexible, acrylic latex coating containing about 60% of solids by volume. It is used as a weather barrier on FOAMGLAS® insulation with a polyester fabric reinforcement.

Its service temperature ranges from –35 to +80°C. It should be noted that the drying time of PITTCOTE® 404 should be respected and that the recommended layer thickness should not be exceeded. If the insulation coating is exposed to severe conditions like sand wind, a protection layer should be applied on PITTCOTE® 404 to reinforce its surface. Chemglaze II A 276 aliphatic urethane from *Lord Chemical* has been successfully used for this application. PITTCOTE® 404 has to be fully dry before applying the Chemglaze finish. An alternative solution is to apply the Chemglaze during the first maintenance shut down. PITTCOTE® 404 may be applied with glove, trowel, brush or spray. Specific airless equipment has been identified for the spray application.

C. PC® 74A

PC® 74A is a two-component coating for indoor application based on a dispersion of polyvinyl propionate polymer with aluminium oxide and silicium oxide fillers for component one and modified Portland cement for component 2 to mix in the ratio of 4 part component one and one part component two by weight.

Due to its composition, PC® 74A is classified A2, non combustible in Germany according to DIN 4102.

PC® 74A is mainly used to coat FOAMGLAS® insulation on chilled water lines where its incombustibility is appreciated. It has also been used, although less frequently, as an adhesive to adhere FOAMGLAS® cellular glass to metal decks, here also mainly due to its incombustibility.

On FOAMGLAS® insulation its service temperature ranges from –30 to +80°C. As a coating, it is reinforced with a glass fabric. The coverage of 4 kg/m^2 should not be exceeded.

D. PC® 85 POWDER

The totally inorganic mineral based PC® 85 POWDER has been specially designed to fill the open surface cells when FOAMGLAS® cellular glass becomes load bearing insulation of liquid oxygen or hot tank bottoms. Together with appropriate interleaving material (for instance some specific mineral sheets and fabrics), PC® 85 POWDER can achieve the required load distribution.

Its temperature ranges from –200 to +250°C. Coverage: About 1.2 kg of dry powder per m^2.

E. PC® 18

PC® 18 mastic contains a small quantity of a rapidly evaporating non combustible solvent. Its initial adhesiveness is very high. Because of its pasty consistence, it is possible to compensate moderate unevenness of the surface.

PC® 18 mastic is used as a cold applied bituminous adhesive, or as a joint sealer. Its main application is the adhesion of FOAMGLAS® cellular glass pipe segments onto chilled water lines.

PC® 18 mastic may be used on various grounds, such as concrete, bricks, metal, numerous plastics, bitumen based coatings, wood, etc.

On FOAMGLAS® cellular glass its service temperature ranges from –30 to +90°C. Coverage: about 3 kg/m^2.

F. PITTWRAP®

PITTWRAP® is a jacket specially designed to cover FOAMGLAS® insulation applied on underground pipes.

It consists of a prefabricated laminate containing a bitumen impregnated glass fabric and an integral aluminium foil, sandwiched between three layers of modified bitumen mastic. The exposed surface of PITTWRAP® jacket is coated with a protective plastic film and the inner surface with a special release paper.

PITTWRAP® is applied around FOAMGLAS® cellular glass in a cigarette type wrap and is heat-sealed at the overlaps of the longitudinal direction. Butt-joints are covered with a 10 cm wide strip that is also heat sealed with a propane torch at the edges around the circumference. Its service temperature ranges from −5 to +90°C (Fig. II.12).

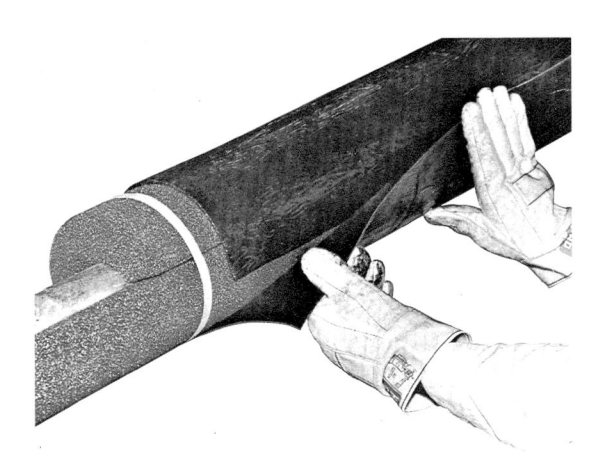

Fig. II.12. PITTWRAP® "cigarette" application.

II.2.9.4 Reinforcement Fabrics

Pittsburgh Corning markets several reinforcement fabrics:

- PC® FABRIC 79 P is an open mesh polyester fabric for reinforcing mastic finishes having square openings of about 3 mm. It is characterised by substantial deformation before breaking.

- PC® FABRIC 79 G is an open mesh glass fabric for reinforcing mastic finishes having square openings of about 3 mm. Its tensile resistance is greater than the polyester fabric but the deformation is lower.
- Finally PC® FABRIC 79 HD is an open mesh glass fabric for reinforcing mastic finishes where impact is expected. The mesh openings are 4 × 5 mm and its weight reaches 720 g/m². Its tensile resistance is very high and its deformation before breaking very low. It is more often used in the building field than for industrial applications. The selection of PC® FABRIC 79 HD requires an increase of the mastic coverage.

II.2.9.5 Anti-abrasives

A. PC® ANTI-ABRASIVE COMPOUND 2A

PC® ANTI-ABRASIVE COMPOUND 2A is a one-component oil modified urethane cut with mineral spirits. It is applied on FOAMGLAS® cellular glass to reinforce its surface for low temperature applications. Due to its nature, typical safety measures are needed.

Its service temperature ranges from –180 to +120°C.

It is applied by brush or by spray.

B. PC® HIGH TEMPERATURE ANTI-ABRASIVE

PC® HIGH TEMPERATURE ANTI-ABRASIVE is a dry one-component mix of modified calcium sulphate with some fillers.

It is applied on FOAMGLAS® cellular glass to reinforce its surface for high temperature applications and when needed for dual temperature applications. Before its application, it is poured into water at the rate of 10 kg anti-abrasive powder into about 7 litre of water. PC® HIGH TEMPERATURE ANTI-ABRASIVE is incombustible. Although its published service temperature ranges from +20 to +250°C, it has been successfully applied at low temperatures, including temperature lower than –180°C. In this temperature, avoiding moisture penetration takes a special importance. It should be noted that PC® HIGH TEMPERATURE ANTI-ABRASIVE is also used as an adhesive to fix together the various parts of some fabricated ware of large dimensions where the service temperature is too high to use bitumen.

Important note:

The present short description of the main accessory products marketed by *Pittsburgh Corning* is aimed at providing general guidance for their use. Before specifying these products, the designer and the contractor should consult the corresponding product data sheet and the corresponding material safety data sheet.

II.2.9.6 Chemical Resistance Table for Coatings (pp. 118-119)

II.2.10 SOME OTHER ACCESSORY PRODUCTS COMPATIBLE WITH FOAMGLAS® INSULATION

The number of accessory products, adhesives, sealers, coatings, anti-abrasives that are claimed to be compatible with FOAMGLAS® cellular glass is so considerable that we will comment only on some of those which have been used in substantial quantities.

II.2.10.1 Adhesives

A. Foster® 81-80 and 81-82

These two products are two component urethane based adhesives designed for all insulating materials. They can replace PC® 88 ADHESIVE with FOAMGLAS® cellular glass in applications for which no extreme temperatures are expected but it should be noted that their modulus of elasticity increase to very high value at low temperatures and preclude their use with FOAMGLAS® cellular glass when the temperature is lower than about −30°C. When the temperature decreases to less than −15°C, a stress analysis based on the modulus of elasticity and all the other parameters should be carried out before specifying them.

It should also be noted that *Foster* recommends the application of oxide chromate primer 51-11 on metal when using Foster® 81-80 or Foster® 81-82.

Exposure	PITTCOTE® 404		PITTCOTE® 300	
	24°C	66°C**	24°C	66°C**
Acids:				
Acetic 5%	R*	LR*	R	NR
Acetic 50%	NR	NR	LR	NR
Acetic glacial........................	NR	NR	NR	NR
Citric 10%	R*	LR*	R	LR
Hydrochloric 10%	R*	NR	LR	NR
Hydrochloric 35%	NR	NR	LR	NR
Nitric 10%	LR*	NR	R	NR
Nitric conc.	NR	NR	NR	NR
Phosphoric 10%	LR*	LR*	R	LR
Phosphoric 85%	NR	NR	R	R
Sulphuric 10%	NR	NR	R	LR
Sulphuric 50%	NR	NR	R	LR
Sulphuric 60% conc.	NR	NR	NR	NR
Bases:				
Ammonium 10%	R*	–	R	R
Hydroxide 28%	NR	–	R	R
Potassium 10%	NR	NR	R	R
Hydroxide 50%	NR	NR	R	LR
Sodium 10%	NR	NR	R	R
Hydroxide 50%	NR	NR	R	LR
Salts:				
Acetates Ammonium.............	R*	R*	R	R
Acetates Potassium	R*	R*	R	R
Acetates Sodium	R*	R*	R	R
Carbonates Ammonium	R*	R*	R	R
Carbonates Potassium	R*	R*	R	R
Carbonates Sodium	R*	R*	R	R
Nitrates Ammonium	R*	R*	R	R
Nitrates Potassium................	R*	R*	R	R
Nitrates Sodium....................	R*	R*	R	R
Sulphates Ammonium	R*	R*	R	R
Sulphates Potassium.............	R*	R*	R	R
Sulphates Sodium.................	R*	R*	R	R
Sulphites Ammonium...........	R*	R*	R	R

Exposure	PITTCOTE® 404 24°C	PITTCOTE® 404 66°C**	PITTCOTE® 300 24°C	PITTCOTE® 300 66°C**
Sulphites Potassium	R*	R*	R	R
Sulphites Sodium	R*	R*	R	R
Chlorides Ammonium	R*	R*	R	R
Chlorides Potassium...........	R*	R*	R	R
Chlorides Sodium................	R*	R*	R	R
Solvents:				
Acetone................................	NR	NR	R	NR
Butyl Alcohol	LR	–	R	R
Ethyl Alcohol	R*	–	R	R
Isopropyl Alcohol................	R	LR	R	R
Methyl Alcohol	R	R	R	R
MEK	NR	NR	NR	NR
MIBK	NR	NR	NR	NR
Mineral spirits	R	R	NR	NR
Chlorinated..........................	NR	NR	NR	NR
Aromatic..............................	NR	NR	NR	NR
Gasoline...............................	NR	NR	NR	NR
Turpentine	NR	NR	NR	NR
Misc.:				
Ammonia Wet	R*	R*	R	NR
Ammonia Dry......................	R	R	R	R
Chlorine Wet	–	–	R	NR
Chlorine Dry.......................	–	–	R	NR
Animal Oils	NR	NR	NR	NR
Vegetable Oils	NR	NR	NR	NR
Mineral Oils	LR	LR	NR	NR
Sewage	R*	R*	R	R
Brine	R*	R*	R	R
Water	R*	R*	R	R
SO_2	LR	–	R	R
SO_3	LR	–	R	R

R = Recommended.
LR = Limited Recommendation, may need wash down. Test.
NR = Not Recommended.
* = Not recommended for continuous or occasional submersion (Unless dried out between the occasional submersions).
** = Surface temperature could be above 66°C in practice (Non-volatile materials can concentrate on drying).

B. Foster® 82-10 and 81-33

These two adhesives although very satisfactory with some other insulating products should not be used with FOAMGLAS® cellular glass.

II.2.10.2 Sealers

A. Foster® FOAMSEAL® 30-45

This product is, to our knowledge and at the time of writing this book, rather similar to PITTSEAL® 111 that has now been replaced by PITTSEAL® 444. As PITTSEAL® 111, it may become hard and eventually crack with age.

B. Foster® 95-50

It is a one component sealer which can be used with FOAMGLAS® cellular glass instead of FOAMSEAL® 30-45.

It is sometimes presented as a replacement for PITTSEAL® 444.

Compared to PITTSEAL® 444 which in cold weather conditions can practically only be applied with a special gun, Foster® 95-50 has the advantage that it can also be applied with a trowel, but it does not present, in service, the same exceptional properties of elongation at low temperatures.

II.2.10.3 Coatings

A. Foster® 60-25

Foster® 60-25 is, to our knowledge and at the time of writing this book, rather similar to PITTCOTE® 300.

B. Foster® 60-75

Foster® 60-75 is a solvent borne vinylic polymer that can be used as a coating on FOAMGLAS® insulation. It gives a better resistance to fire than PITTCOTE® 300 and Foster® 60-25. It should always be reinforced by glass, polyester or nylon fabric. It should be noted that this product is not compatible with PC® 88 ADHESIVE.

C. Foster® Lagtone® 30-70

Foster® Lagtone® 30-70 is a polyvinyl acetate resin emulsion based finish applicable on FOAMGLAS® insulation. It does offer some anticryptogamic protection. It does not have a direct equivalent in the PC® accessory product series.

D. Foster® Monolar®

Foster® Monolar®, a hypalon based product, has been used successfully on FOAMGLAS® insulation. It should be noted that this product is not compatible with PC® 88 ADHESIVE.

E. Encacel T

Encacel T is an elastomeric polymer based product that can be used as a finish on FOAMGLAS® insulation. It should be noted that Encacel T is not compatible with PC® 88 ADHESIVE.

II.2.10.4 Anti-abrasive

A. Foster® Anti-abrasive 30-16

Foster® Anti-abrasive 30-16 is a solvent borne vinylic resin product applicable on FOAMGLAS® insulation as an anti-abrasive for low and medium temperatures from −180 to +120°C. To our knowledge it is rather similar to PC® ANTI-ABRASIVE COMPOUND 1A which has been replaced by PC® ANTI-ABRASIVE COMPOUND 2A. It is not recommended for use over stainless steel piping and equipment.

Important note:

This short description of some accessory products compatible with FOAMGLAS® insulation is given for general guidance only. Before specifying these products, the designer and the contractor should consult the corresponding product data sheet and the material safety data sheet.

III

HOW TO DESIGN AND INSTALL FOAMGLAS® INSULATION FOR INDUSTRIAL APPLICATIONS

INTRODUCTION

We have seen for which applications cellular glass insulation should be chosen. We have also reviewed the main properties of FOAMGLAS® insulation and the accessory products that are suitable for its installation. We can now develop the principles that should be followed to design and install FOAMGLAS® insulation. Obviously we cannot provide all the detailed specifications for every specific application but we will concentrate on the main concepts and the general rules, asking the reader to request, or to develop, the detailed specifications when he has decided the type of specification needed for the application he is designing.

It is necessary to point out the importance of design to any insulation system. High quality insulating products, like FOAMGLAS® cellular glass, applied with high quality accessory products are not enough to guarantee excellent service over many years. A correct design is essential to achieve that goal.

Labour for installation often being an important part of the total cost, if not the main one, the choice of high quality insulating material and accessory products will only moderately affect the total installed cost.

As a general remark, insulating materials are by nature mechanically rather weak, since their manufacturers must work with a minimum of solid material and a maximum of air or gas to achieve the desired thermal performance. Their mechanical performance is generally one order of magnitude lower than a typical construction product for the strongest and two or more orders of magnitude lower for the weakest.

Moreover, due to their thermal characteristics, insulating products are generally subjected to extreme temperatures and temperature variations much more than classic building products.

III.1

APPLICATION REQUIREMENTS

The first thing to do before writing any application specification consists in establishing as accurately as possible the application requirements to be fulfilled in the particular project. From the list of application requirements presented, in this chapter, the reader should in each case decide which application requirements apply to his project and, if necessary, add whatever other requirements are suitable. The given list, without being absolutely exhaustive, nevertheless covers the main requirements needed for general FOAMGLAS® cellular glass applications.

Since omitting a requirement may often make the difference between a successful application and a failure, it is suggested that a choice should be made carefully of the requirements for any specific application before starting to design.

A list of requirements is given below with some comments.

III.1.1 THERMAL REQUIREMENTS

III.1.1.1 Heat Transfer Limitation

Heat transfer limitation for economic reason.

This requirement is of prime importance and often the major reason for which insulation is installed.

To determine the economic thickness which meets the heat transfer limitation for economic reasons, one can either adopt a value that is generally chosen for a typical application, or one can use a detailed economic analysis.

The first method is easier and can provide valid results provided none of the parameters of the application substantially differ from those generally experienced.

To carry out a detailed economic analysis, one has to know a considerable number of parameters among which the most obvious are:

- The cost of energy
- The installed cost of the insulating system
- The expected life of the insulation system
- The maintenance cost of the insulation system, specially the coating, sealer...
- The expected rate of interest
- The expected rate of inflation
- The cost of a shut down of the installation in the event of a failure of the insulation
- The probability of a shut down caused by a given insulation system
- The utilisation factor (percentage of the time during which the installation will operate).

Various papers can be consulted on the question of economic insulation thickness.

III.1.1.2 Surface Condensation Limitation

Due to the physical nature of cellular glass, interstitial condensation does not occur in it. Provided they are correctly designed, the systems based on cellular glass also do not risk interstitial condensation. But surface condensation on the outside face may occur.

With the exception of some special applications such as kilns or steam rooms surface condensation should only be considered for low temperature applications. For hot applications, the surface temperature is generally high enough to prevent condensation.

An important question to be answered is why to limit surface condensation and how stringent should one be in this respect. Since cellular glass does not absorb neither water nor water vapour, it is not necessary to avoid surface condensation provided that the joints between the slabs or pipe covering elements are especially tight. But the coating or even the metallic finish applied on FOAMGLAS® cellular glass can deteriorate substantially – especially some coatings – if they are subjected to permanent contact with water. On the other hand, one has to recognise that, particularly for outside exposed insulation, rain water will stay on the insulation system, at least during some part of the year.

It would not be economical to select a surface condensation limitation requirement so high that it would imply excessive insulation thickness when designing an installation that would anyhow be exposed to rain during a substantial part of the year.

To avoid this mistake, it is advisable to express the surface condensation requirement in term of percentage of the time during which surface condensation should not take place, preferably to the classic method based on imposing arbitrarily the relative humidity under which condensation has to be avoided in given climatic conditions.

The selection of the level of this requirement becomes specially important when designing an insulation system to be installed in extreme climatic conditions, such as a terminal based on a small island or a platform situated at sea in a cold area.

Whenever possible, calculations should be based on representative statistical climatic data.

III.1.1.3 Protection from Burns

This requirement that is rather obvious leads the designer to limit the surface temperature on hot installations at temperatures that do not cause personnel burns in case of accidental contact. This temperature is generally evaluated at about 60°C.

For installations on which personnel have no access, a somewhat higher temperature can be acceptable but the maximum service temperature of the mastic, sealer etc. should be taken into consideration. A careful analysis is needed before accepting surface temperatures higher than 60°C.

Although physical skin damage can also happen following contact with a low temperature surface, generally no specific requirement is needed since other limitations for heat transfer or to restrict surface condensation almost always give a surface temperature that is not dangerous.

III.1.1.4 Surface Temperature Limitation

For some applications, the limitations of surface temperature to avoid deterioration of coating (or eventually sealer in the joints close to the outside surface) constitute an essential requirement. Insulated underground lines with a bitumen based coating or wrapping constitute a typical example of this case. Due to the insulating value of the ground, especially during a dry period, the surface temperature of the coating tends to increase and eventually exceeds the material limit. The insulation thickness placed between the pipe and the coating should be calculated in such a way that the surface temperature of the coating remains within the acceptable range.

III.1.1.5 Temperature Process Control

Many processes are fairly flexible regarding the temperature at which they take place and the temperature variations they can withstand without being affected, but other processes require strict control of their temperature and temperature variation. The designer will have to know the acceptable temperature variation of the plant he is working on to avoid mistakes that can be catastrophic.

As an example of a delicate process, one can quote the installation in which a change of phase – liquid to solid for instance – can occur. Water turning to ice, heavy oil to solid, liquid sulphur to solid,etc. can illustrate the potential problems due to lack of process temperature control or, of an insulation system not performing as expected. Although it only applies to certain processes, this point should be very seriously considered. As an extreme case, poor temperature control will generate excessive boil off in a storage tank containing a liquefied medium that can vaporise, creating excessive pressure with the corresponding risk if the safety valves do not react in due time.

III.1.1.6 Rapid Temperature Variations

This requirement concerns production in which rapid temperature changes are needed to obtain some physical modification in the processed material. The choice of insulation system will focus more on low thermal inertia than on the classic requirements like heat transfer limitations. To a large extent, this requirement is related to process control but with the aim of accomodating rapid temperature variations instead of temperature constancy.

III.1.1.7 Protection of Structural Material against Excessive Temperature

At supporting points or for their bases, installations operating at high or low temperatures are in contact with the structural materials that carry their weight, generally through some insulation products such as FOAMGLAS® cellular glass, very high density polyurethane foam or even specially selected hard wood.

Since the mechanical properties of the usual structural materials such as carbon steel or concrete generally decrease when the temperature deviates from ambience, the thermal limitation becomes a structural requirement.

For instance, when cryogenic installations are supported by non cryogenic structural materials through insulation, the achievement of the expected performance of this insulation cannot be too much emphasised.

Freezing of the ground under large low temperature installations like storage tanks can result in movement and very detrimental frost heaving.

III.1.1.8 Fire Protection Requirements

Although fire protection is related to thermal requirements, its importance justifies to single it out as a specific requirement.

In spite of the fact that it concerns all types of construction products and installations, it takes on some special importance in the case of insulation systems.

Insulating materials being light materials, with very large internal surfaces that become exposed to air when they are damaged by fire, generally burn very easily with the notable exception of inorganic materials, such as cellular glass. V.1 provides details of a number of fires in which insulating products were involved.

Moreover industrial insulation includes applications – as for instance in the chemical and particularly the petrochemical field – for which fire hazard represents a cause for concern. Consequently, fire requirements should be established in a very responsible way, preferably after due consideration of human safety, the shut down cost and the time needed to react, should a fire occur.

The fire requirements can be expressed in a fire resistance rating – i.e. the time after which some criteria on the protected face of the insulated system will no longer be met, in given conditions.

It should be pointed out that the testing conditions influence very strongly the fire resistance of a given system and apparent minor modifications of the testing conditions can drastically change the results.

The fire requirements can also be expressed in terms of reaction to fire of the specified products for instance in imposing incombustible insulating products when tested according to a given norm. The two methods of expressing fire requirements can complement each other and should not be regarded as contradictory.

As general remarks when fixing fire protection requirements, the following three points should be kept in mind:

1. Insulating products having a good reaction to fire, for instance being non combustible, generally improve the fire resistance rating of the systems in which they are involved.
2. More than thermal conductivity, thermal diffusivity governs fire resistance of the system. This observation corresponds to the fact that a fire is an unsteady state thermal situation.
   From the formula of thermal diffusivity

$$a = \frac{k}{c\rho}$$

where
   a = thermal diffusivity
   k = thermal conductivity
   c = specific heat
   ρ = density

it can be deduced that a high density helps to reduce the thermal diffusivity since the thermal conductivity increases generally less than the density.
3. Insulating products having a wide service temperature range on the high side usually perform better in a fire since they keep their mechanical properties for a longer time when the temperature increases. They consequently offer a better protection to the structure on which they are installed.

For the determination of fire protection requirements, the reader should consult the national fire regulations applicable to the specific project. Should they not exist, industry professional codes or recommendations are generally available.

It is certainly not the purpose of this book to establish fire protection requirements but it can be said that in the absence of other recommendations, a fire resistance of half an hour or preferably one hour should be a reasonable level for ordinary industrial installations where fire fighting equipment is readily available. The testing conditions should correspond for instance to ASTM E 119 (ISO 834) in the general case and the fire curve should be the hydrocarbon one for petrochemical applications.

This standpoint corresponds to the fact that half an hour represents the absolute minimum to detect a fire, give the alarm, enable the firemen to arrive, install their equipment and fight the fire. In many cases, one hour is a more realistic assessment of the time needed for these operations.

Some petrochemical companies have established internal rules requiring two hour fire resistance, especially for installations that are critical from the personal safety or the process standpoint.

In our view relaxing the half an hour fire resistance requirement should only be allowed exceptionally for isolated equipment of little importance to the general safety or to the production.

III.1.1.9 Non Absorption of Flammable Liquids

Although this requirement is certainly related to fire, its specificity justifies a special heading. The requirement of non absorption of flammable liquids by insulating systems, (primarily by the insulating products), generally applies to petrochemical, chemical and pharmaceutical plants in which a risk of contact with such liquids, for instance due to leakage, exists. The special danger related to this situation is explained in detail in V.2. Needless to say, when this requirement has to be considered, it should be of predominant importance.

Among the flammable liquids presenting particularly high risk, we can quote, although not exclusively, the following: heat transfer fluid, ethylene oxide, methanol, MTBE and many other petrochemical fluids.

III.1.1.10 Compatibility with Liquid Oxygen

This requirement which also relates to the fire / explosion issue is an essential one, but it only concerns a few very specific applications.

When the temperature becomes lower than $-183°C$, the oxygen contained in the air liquefies and presents a very high danger of explosive reaction if it comes into contact with organic substances. A normal requirement for installation operating at a temperature lower than the liquefaction temperature of oxygen consists of imposing the use of only non combustible, inorganic materials.

III.1.1.11 Nuclear Applications

For these very specific applications, some particular requirements have been developed. Since they are related to nuclear safety, they are mandatory. They should be found in national codes and nuclear regulations and/or in more or less international regulations such as the American NRC 1.36 (Nuclear Regulation Commission). These requirements are generally associated with the stringent limitation of radioactive dust and other typical nuclear properties. Nuclear piping and equipment often being in stainless steel, compatibility with this steel can also be a requirement for nuclear applications.

III.1.2 MOISTURE RESISTANCE REQUIREMENTS

Since, as detailed in V.3, V.4, and V.5 moisture can have a very detrimental influence on thermal insulation, the moisture requirements are logically mentioned immediately after the thermal and associated requirements.

Moisture penetration into insulation systems should be avoided, both in the liquid water phase and in particular in the water vapour phase. This requirement is generally understood for low temperature insulation from which water does not escape and can turn to ice with the problems of thermal and mechanical damage. But it is also important for high temperature insulation from which water vapour at temperatures exceeding 100°C, steam in other words, tends to escape but in which water vapour at temperatures lower than 100°C tends to remain.

The requirement of moisture resistance has to be considered in various conditions with different levels of pressure driving water and water vapour into the insulation system.

Climatic conditions, including the expected relative humidity, the period of time during which it rains, the maximum force that pushes the rain water, the importance of snow falls which melt into water, play an obvious role.

When establishing the requirements for high temperature insulation, the number and the length of idling periods should be considered to take into account that during these periods, the installation cannot eliminate the water and water vapour that could have penetrated the system.

A special requirement concerns the resistance against water running or sprayed under pressure on the insulation during test of the eventual deluge system. In some cases – when fire hazards are considerable – such tests of the deluge systems can be carried out on a monthly basis.

Finally, some processes call for temperature and relative humidity much more severe than any exterior climatic conditions: for instance, drying rooms or kilns.

Adequate requirements have to be established for these applications.

III.1.3 CORROSION RESISTANCE REQUIREMENTS

Insulation materials are normally not applied on installations to protect them from corrosion for which more suitable systems, such as anti-corrosion products have been developed but at least insulation materials should not contribute to corrosion and insulation systems should offer some protection against corrosion of the installation.

The possibility of corrosion caused by insulation materials or insulation systems is very important and corrosion can lead to the total failure of the installation with possible accidents and loss of production. Although corrosion requirements are less mentioned than heat transfer limitations and usually play a somewhat reduced role, they can be vital in specific cases.

A relation exists between moisture and corrosion resistance requirements: if moisture is totally eliminated from an insulation system, the principal condition favouring corrosion is also eliminated. But since completely avoiding moisture in an insulation system is not always achieved and maintained during the expected life of the insulation system, it is reasonable to set up corrosion resistance requirements for insulation materials and for insulation systems.

Depending on whether the installation to be protected is built of carbon steel or stainless steel (or possibly other metals), the requirements for the insulation and the accessory products will differ. In the first case, a requirement imposing a pH value that does not favour corrosion is generally sufficient. In the second case, the corrosion resistance criteria should include a limitation of the chloride content and other detrimental components. Guidance can be found in ASTM C 795 and in *AGI* documents.

Regarding insulation systems, their design and application methods should be decided to meet the corrosion resistance requirements, for instance in eliminating moisture penetration or if needed, in imposing anti-corrosion layers particularly when the steel structure has to wait before insulation is applied.

The level of the corrosion resistance requirements should also be related to the service temperature: for instance, they can be less severe for an equipment operating at −40°C than for another one operating at 80°C.

A final system requirement concerns the eventual imposition of a complementary protection, such as anti-corrosion coating applied on the equipment, electrical protection etc.

III.1.4 CHEMICAL RESISTANCE REQUIREMENTS

Here again, chemical resistance is not a basic requirement of insulation materials and insulation systems but it can be of considerable importance for some well-identified applications.

Chemical attack can come from a leakage affecting the insulated equipment. If leakage from straight pipes or tanks is rather rare except in case of severe corrosion, the same cannot be said from special equipments such as flanges and valves, etc. A part of the danger comes from the fact that the leaking liquid can be absorbed into the insulation on the inside and cause severe damage before it is noticed on the outside.

Another source of chemical attack may come from spillage from the insulated equipment or from another installation in the neighbourhood.

A third source is related to the atmosphere that can contain chemical or petrochemical products reacting with the insulation material and/or the sealer and coating. Typically a hydrocarbon loaded atmosphere will pose problems with a bitumen cutback coating.

Another source of concern would be a marine atmosphere containing salt that can attack the metal jacket and degrade it within a short period of time.

All these considerations and any others related to this question should be taken into account by the designer when he makes his decision on the level of chemical resistance required.

For some equipment containing neutral liquids installed within a building, the requirements can be moderate. For installations containing aggressive liquids situated in petrochemical plants near the sea, stringent requirements are generally needed.

The possibility of easy replacements should also be evaluated. Obviously, changing a metal jacket is less difficult than renewing a complete insulation system, specially if it has to be carried out in service.

Finally in some specific cases, chemical resistance requirements are of predominant importance: for instance, in kilns or furnaces insulated on the inside and in which no protection can be applied. Some applications impose so severe chemical conditions that the concept of periodic insulation replacement has to be accepted in the same way as refractory bricks for glass tanks or metallurgical blast furnaces.

The same applies to a large extent for chimneys in which insulation becomes the chemical barrier against aggressive fumes which may attack the structural part of the construction if the refractory bricks fail. In this case, the insulation plays the role of a secondary protection barrier (Fig. III.1).

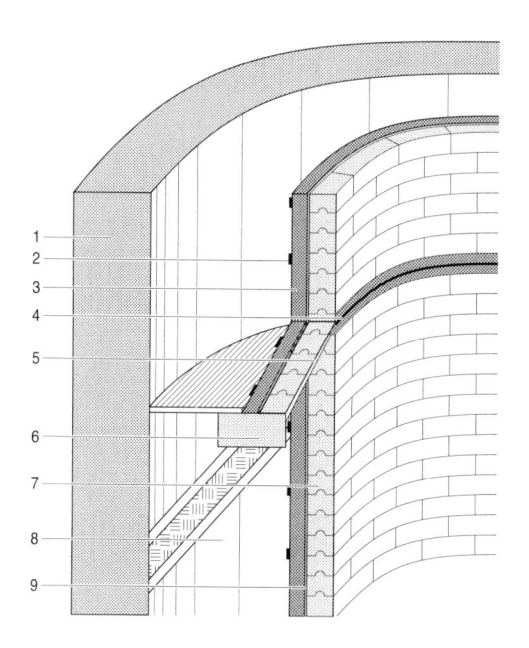

Fig. III.1. Typical simplified cross section of a high rise industrial chimney.

1. Outside wall of stack.
2. Glass fabric strip.
3. FOAMGLAS® insulation.
4. Sheet lead.
5. Packing cord.
6. Supporting ring.
7. Acid resistant brick lining.
8. Ventilated space.
9. Distance layer.

III.1.5 MECHANICAL REQUIREMENTS

Dead loads are generally seen as the first ones to be considered when establishing the mechanical requirements: equipment weight, wind, snow and ice load, etc. The local building codes can generally be taken as a first guidance, checking that they correspond to the real situation for an industrial plant: for instance, the special shape of an industrial equipment may result in higher than expected wind loads; this affects not only the steel or concrete structure but also the method of fixing insulation to it; in the same way, lower surface temperature of a cold installation may lead to thicker than expected ice layers with the corresponding increase of weight.

Although not exactly a dead load, hail stones can impose requirements on the mechanical protection of insulation systems in some areas.

Finally, living loads, such as people walking on insulated equipment or installing scaffolding on them, constitute severe problems and can result in coating or jacket perforations leading to water intrusion if the system relies on these elements to remain waterproof.

If a long insulated pipeline connecting one plant to another one is not designed for use as a gangway, some local parts of a plant may be used as passage for personnel. These points should be taken into consideration when establishing the mechanical requirements. It is often necessary to fit them to the local situation, for instance to specify stronger jackets in places where more mechanical abuse is expected.

Here again the designer should use judgement and experience to establish the mechanical requirements leading to satisfactory performance during the expected life of the plant at an acceptable cost.

III.1.6 IMPORTANT NOTES

Although they do not constitute specific requirements, the following notes should be kept in mind when establishing the level of other requirements.

1. The service temperature and particularly its variations and their gradient should be carefully considered before deciding the requirement levels. In the same way, possible idling periods should be considered.

For example, rapid temperature variations can impose substantially higher stress to the insulation system than slow temperature changes.

Should equipment be frequently idling, the moisture resistance requirements should be verified in both conditions, in service and when idling. These double evaluations may also be necessary for other requirements.

2. The shut down costs including repair costs, loss of production and customers, etc. should be carefully calculated before setting up the other requirement levels.

3. Finally, the time needed for the application of the insulation systems may become an important limitation specially in the case of repair work. But, in spite of its stringent nature, this limitation should not be a justification for poor application leading to major problems later.

III.2
THICKNESS DETERMINATION

The determination of insulation thickness is generally considered as the first step to design an insulation system, since it has some influence on other parameters, such as for instance the number of layers.

The requirements being different for cold, hot or dual temperature insulation, these cases will be analysed separately.

III.2.1 COLD APPLICATIONS

The requirements to take into consideration are the limitation of heat transfer for economic reasons, the limitation of surface condensation and the control of the process. In some conditions, the fire protection requirement may also impose a certain insulation thickness. Each requirement has to be satisfied, the most stringent one dictates the insulation thickness. Usually in chemical and petrochemical process, the control of process is sufficient even with heat transfer as high as 100 W/m². Except in unusual cases, the thickness determination can thus be only based on two requirements: the limitation of the heat transfer and of the surface condensation.

In the case of a FOAMGLAS® cellular glass based system – and provided the joints are correctly sealed – avoiding internal condensation within the insulating material is not necessary since cellular glass is impermeable to water and water vapour. For the same reasons, the limitation of surface condensation can be less severe than for insulation relying heavily on a vapour retarder.

It may be useful to mention that surface condensation is influenced by the ambient air temperature, relative humidity, wind speed and exterior surface emissivity.

The heat transfer limitation is more influenced by economic than technical factors. The principle consists in calculating the thickness that, in a given case, will minimise the sum of the installation cost and the operating cost. Many formulas and computer programs have been published for this. They are generally based on the factors previously listed. This economical study is fairly complicated and assumptions have to be made for the numerical values of many parameters. Fortunately, the economic optima are often represented by rather flat curve near their minimum, and a small error on the numerical value of an input does not seriously affect the results.

If it is advisable to run a detailed economic analysis particularly for applications having unusual parameters, it can nevertheless be said that, in the present economic situation and for installation in permanent service, without special characteristics, the optimum heat transfer often ranges from 25 to 30 W/m^2 for low temperature insulation.

It should be noticed that this value is 3 to 4 times smaller than the usual value needed for control of the process.

Table III.1 provides the FOAMGLAS® cellular glass T4 thickness needed to limit heat transfer to 30 W/m^2 in the indicated conditions. See annex regarding calculation method at the end of this part.

Regarding the limitation of surface condensation requirement, it should be pointed out that this has more importance for insulation systems installed within a building, such as chilled water lines, than for insulation installed on the outside of a building that will be subjected to rain, snow and fog. As already suggested, the limitation of surface condensation on FOAMGLAS® cellular glass based insulation systems is needed more for the coating or the jacket than for the insulation itself. It is thus judicious to avoid too stringent requirements that will lead to uneconomic thickness.

The following example shows that an increase of a few percent in the relative humidity can impose a very substantial thickness increase.

In some conditions, for instance on a terminal based on a small island, the relative humidity and sometimes the temperature could be so severe that limitation of surface condensation could lead to thicknesses exceeding the

TABLE III.1
Recommended FOAMGLAS® cellular glass T4 insulation thicknesses
in mm in order to limit heat gain to a maximum of 30 W/m²
Outdoor application

Conditions: Ambient temperature = 20°C   Wind speed = 10 km/h
Emissivity = 0.2

Pipe diameter		Service temperature (°C)								
(inch)	(mm)	0	−20	−40	−60	−80	−100	−120	−140	−160	−180
1/2	21.3	*20	30	40	50	60	70	70	80	80	90
1	33.7	*20	40	50	60	60	70	80	90	90	100
1¹ᐟ²	48.3	*20	40	50	60	70	80	90	90	100	100
2	60.3	*20	40	50	60	70	80	90	100	100	110
3	88.9	*20	40	60	70	80	90	100	110	110	120
4	114.3	*20	40	60	70	80	90	100	110	120	130
6	168.3	30	50	60	80	90	100	110	120	130	140
8	219.1	*30	50	60	80	90	100	120	130	130	140
10	273.0	*30	50	70	80	100	110	120	130	140	150
12	323.9	*30	50	70	80	100	110	120	130	140	150
14	355.6	*30	50	70	80	100	110	130	140	150	160
16	406.4	*30	50	70	90	100	110	130	140	150	160
20	508.0	*30	50	70	90	100	120	130	140	150	160
24	609.6	*30	50	70	90	110	120	130	150	160	170
30	762.0	*30	50	70	90	110	120	140	150	160	170
36	914.4	*30	50	70	90	110	130	140	150	170	180
	1000.0	*30	50	70	90	110	130	140	150	170	180
	3000.0	*30	50	80	100	120	130	150	170	180	190
	6000.0	*30	50	80	100	120	140	150	170	190	200
	FLAT	*30	50	80	100	120	140	160	180	190	210

On the basis of the thicknesses calculated with the heat gain requirement, the relative humidity should not exceed 81% to avoid condensation.
* Thicknesses should be rounded off to the nearest higher manufacturer's thickness.

economic thickness by a considerable margin. In such a case, it is more realistic to foresee frequent maintenance of the coatings than to choose excessive insulation thicknesses.

Statistical Study Based on Weather Data

Moreover, in the case of a detailed study, one should favour the method limiting the percentage of time during which surface condensation should be avoided and run calculations based on the climatic conditions of the specific application, when they are available.

Energy analysis studies can show the percentages of time when condensation occurs and also the heat gain using actual weather conditions.

As an example, the following calculation was made for an LNG loading line of 12" NPS (pipe diameter 323.9 mm) operating at −160°C, covered with a coating having an emissivity of 0.8, using the weather data for 1986 in Antwerp (Belgium).

Twelve monthly tables have been prepared in which the reader can find for each hour of the day (from 0 to 23 horizontally) and for each day of the month (1 to 31 vertically) a symbol going from A to X indicating the minimum thickness needed to prevent condensation (Table III.2).

Symbol A means that 100 mm FOAMGLAS® insulation are needed to avoid condensation in the given conditions, with the weather data prevailing during the concerned hour.

Symbol B means that the required insulation thickness ranges from 100 to 110 mm.

Symbol C means that the required insulation thickness ranges from 110 to 120 mm, and so on.

A blank position means that the calculation has not been carried out for that hour since weather data indicate rain during this period.

As an example the symbol for January 25 during the 15th hour is B, which means that 110 mm FOAMGLAS® insulation thickness is needed.

The twelve monthly tables do not mention the weather data corresponding to each one hour period since they would become excessively complicated.

But it may be interesting to mention that the average weather data on which the calculation are based read as follows:

Temperature .... Mean hourly ambient air temperature over the year: 9.7°C
minimum hourly ambient air temperature: −11.3°C
maximum hourly ambient air temperature: +31.4°C
Relative humidity.. Mean hourly value over the year: 79.9%
minimum hourly value over the year: 24%
maximum hourly value over the year: 100%
Wind speed.......... Mean hourly value over the year: 12.23 km/h
minimum hourly value over the year: 0.0 km/h
maximum hourly value over the year: 74.2 km/h

TABLE III.2

JANUARY

```
                       HOURS
DAY              1 1 1 1 1 1 1 1 1 1 2 2 2 2
     0 1 2 3 4 5 6 7 8 9 0 1 2 3 4 5 6 7 8 9 0 1 2 3
 1  E E F A A A C B B     X         X X X X X
 2  X     X   X X X X J   I D E D A B B   A A A   A
 3  B B B A B B F             X X X X   B B H H I
 4  X X X X X X F G E J I I C B B B B E E E E E E E
 5    H B X   X X X       X       X     X X
 6  X         X X X X X       I           X X X X
 7  X X X X X X X           X         X X H H I
 8  X X X H H H H                 X
 9      X X X X J           X H D   F
10                                          X X X
11  I I I X X X F F F X X X I H H A A A A     H H
12  H I H H H H D D D B A B A A A A A A A A A A A
13  A A A A A     K C C C C E C A A A D C C
14  X X X X A A   A A A A A A A A A A A A A A A A
15  A       A   A         A       F A B B X X X X X X
16    X X X X X D B C A A A B B B A A A A X X X
17  G G H H H H B B B E E E   X           X X     X
18      I I                       X X X X
19        X X F G       X     X       A A A   A A
20  A A A A A A A A A B B B A A A A A A A A A A A
21  A A A A A A A A A A   A A   A           B E E E
22  A B A E     E A B B D B D A A A A
23        X X D           E X X X A A A A A A A A
24  A A A A B A D       X X A A A A   A     X   X X
25  A A A C D D I X J     A A H I X B C A D C C D E D
26  X X X X X X F I J X X X X X X G H H X X X X X
27  X X         X   X         X X X X X X X X X X
28    X X X X         X I I J B B B A A A C B B A A A
29  C B C C C C E F G D C D A A A A A B B B A A A
30  A A A E E E G F H       A A A A A A A A D D E
31  C D C E   E F F F       D A A A A A A       C C
```

FEBRUARY

```
                       HOURS
DAY              1 1 1 1 1 1 1 1 1 1 2 2 2 2
     0 1 2 3 4 5 6 7 8 9 0 1 2 3 4 5 6 7 8 9 0 1 2 3
 1                B           J         F E B B B
 2  B B B D D E F J I                   X X   K X
 3                                                F
 4  G H G A B A B B B A B B E D E E D E A   B G H H
 5  C     X X X E E       C B B B B       X X X
 6  F                     X X             X
 7                          A A A D D E C E C H H H
 8  E F F C D D K K I   D E A   A A A A A A   G F
 9  I I G A B B E E E A A B B C A B B E F F J J J
10  A A A A A C D     K K   A A H           F
11  E E E C C C C C C     X X A B A A A A   B D D D
12  F           F E F F       A B A A A A C C H H H
13  I H I B B B E E H G E E B A A A A A B D D B B B
14  A A A A A A E E E A A A A A A A A A A A A
15  A A A A A A B     F   D A A A A A H
16                          C B B A A A A B       E
17  B B B C                   A A A A A A A A A A
18  A A A A A A D D E C C D C C B A A A D B C A A A
19  B A B C E E G H       H G F F   E   I I I G
20  G G G D D D F F E     I A A A A A A A C H F H
21  X X X X X X           G A       C C A A A E F G
22  X X                   K I A A A A A A H   K F F I
23  J X X X X X G G E G G E A A A A A A A A A A A A
24  B B C A A A D D E J J I A A A A A A A A A A A
25  D D D J J J C C C A   A A A A A A A A A H G G
26  D D D C C C F F F A A A A A A A A A A A A A A
27  A A A A A A A A A A A A A A A A A A A A A A A A
28  A A A A A A A A   A A A A A A A A A A A A A
```

MARCH

```
                       HOURS
DAY              1 1 1 1 1 1 1 1 1 1 2 2 2 2
     0 1 2 3 4 5 6 7 8 9 0 1 2 3 4 5 6 7 8 9 0 1 2 3
 1                C C C A A A A A A A A A A A A J H I
 2  A A A A A A A A A A A A A A A A A A A A I I I
 3  X X X F F F H J I C C D A A A A A A A C C C
 4
 5  X         X     X         X               X
 6  X X X J J K           X D C C A A A A A A B A
 7  G F H X X X X X X                           X X
 8  X                         A A A A A A A A C
 9                        X X A A A A A A A A K K X
10  X X X X         D A A A A A A A A A A X X X
11  X                     G A A A A A A A A A B B
12              X
13                        C A A A B A A A A C
14  C C B C C F A G F C B B A A A A A A A A
15                    E D A A A A A A A A A A A A
16  A A B                 X A A A A A A A A C D D
17  X             X             X               X
18  X X X   X             X   A A A A A A A A C B E
19  C C B B B C J J J A A A A A A A A A A
20                A A A A                 X X X J K J
21  A A B A A A C E C B B B A A A A A A A A B B B
22  X X X X J X X           X A A A           H C C C
23  C C C F   F             X B C E A A A A A A A A
24  A A A A                 X X   K A A A A A A A A
25  A A A A A A A A A A A A A A A A A A A A A A
26  D E E D E E E G E I I A A A A A A A A A A A
27                    X X X       X       X X F
28        X X   E C C B B A A A A A A A A B B
29  A A       H H     X A   A A A A A A G
30  J     D   A A A A A A A                 X G G
31  K K J X X X         X   D C C A A A A A   A A A
```

APRIL

```
                       HOURS
DAY              1 1 1 1 1 1 1 1 1 1 2 2 2 2
     0 1 2 3 4 5 6 7 8 9 0 1 2 3 4 5 6 7 8 9 0 1 2 3
 1  I G G A A A D D C D D D A A A A   A A A A I G G
 2  X X X F F G X X X X E E E A A A A A A A A F G F
 3  D E E H J I X X X G H G A A A A A A A A C D D
 4  C E E F F J I   H X X X A A A C B B A A A A A A
 5  H G G A A A J H J A A A A A A A A A A A C D D
 6  D F G J H I I     H H H F A A A A A A A A A A A
 7                I           J                   X
 8  X       X                 X X                 J
 9      X                         J J I B   C A A A
10  A A A A A A A A A A A A A A A A A A A A A A A
11  A A A A A A A B A C C C X X X A A A A A A A A A
12  A A A C D F K J F A A A A A A A A A A A I I X
13  X X X X I       C     C A A A A A A A I H H
14  D C                 I J           X         X X X X X
15  X X X X X X X X X B B A A A A G F G A A A G G G
16  H F F G H G X X X A A A A A A G F E D B C   G H
17  X X X J X X X X X B B B A B A A   A A A A B B D
18  J I J X X X             H A   A A A A   A A G G I
19  H J H H G   X         B B E F E A A A A A A A X
20  X X X G G E                         X X   X G F F
21  F F G E E E C D D C B A     D A A B B B   F F E
22  F F G K J H A A A A A A A A A A A A A A A
23  A A A A A A E D B A A A A A A A A A A A A A A
24  A A A C C C C C B       E A A A A A A A A F I E
25  X X X X X       X X C A A A A A A   A A A B A A
26  E G G B A             X           X       A A     B C
27  X       X   X X J A A B A A B   C A A A X J J
28  X J J X X               C F F A A A A A A G H J
29  F D D A A A X K K B F C C         E C B C D A A B
30  E G G X X X     X X A A A A A A A A A A A A A A
```

CODE:
A = 100 mm   E = 140 mm   I = 180 mm   BLANK = RAIN
B = 110 mm   F = 150 mm   G = 190 mm
C = 120 mm   G = 160 mm   K = 200 mm
D = 130 mm   H = 170 mm   X = OVER 200 mm

145

TABLE III.2 (continued)

```
                        MAY                                              JULY
                       HOURS                                            HOURS
DAY            1 1 1 1 1 1 1 1 1 1 2 2 2 2       DAY            1 1 1 1 1 1 1 1 1 1 2 2 2 2
       0 1 2 3 4 5 6 7 8 9 0 1 2 3 4 5 6 7 8 9 0 1 2 3         0 1 2 3 4 5 6 7 8 9 0 1 2 3 4 5 6 7 8 9 0 1 2 3
 1   A A A K K K G F G A A A A A A A A A A A A A A    1   I I         A A A A A A A A A A A   A
 2   A A A I           A A A A A A A A A A A A A A    2   J J J X         C D C A A A A A A A A A A A A
 3   A A A B B B C B B A A A A A A A A A A A A E      3   G I I X     A A A A A A A A A A A A A A A A A
 4       X X X X X X X A A A A A A A A A A A A A      4   J X X G G H H F G A A A A A A A A A A A A A
 5   A               X X X X X A A A B B B A A A E E E   5   A A A A A A B B B A A A A A A A A A F E F
 6   X X X X X X     X X K K G C B A X X F   E G F   I   6         X X X X X F F G B C C A A A A A A A A
 7   X X X E E E E E D A A A A A A A A A G G I X X X     7   C         X X X X D D D A A A F F F A A A A A A
 8   A A A G G F C B B A A A A A     A A A D C D D C C   8   K K I X X X K X X E D D A A A A A A A A A A A
 9   F F G X K J X X X C C C A A           X     X X     9   H H F   X X K X K A A A A A A A A A A A A A A
10     K K       J   X X     X X E D E   E   X X X D E  10     E D K K K A A A C     C H H H A A A A A A A A
11   F F F A A A B B B A A A A A A A A A A A A A A A    11   B C E           X J I A A B A A A A A A A A A
12   G G G C C D F F D A A A A A A A A A A A D H F      12   A A E X X X X X A A A A A A A A A A A A A A
13   X X X E E E E E     X J A A A A A A A A B B A      13   A     I G       X X H H J A A A A A A A A A A B
14   B A A B A A     A A A A A A A A A A A A A A A A    14   X X X X                 B A A A A A A A     I I
15   A A A B B A         E       I H F A A A A A A A A  15   X X X             A A A A A A A A A A B B
16   B B B H H J X X X F E E A A A A A A A A A A A A    16   X X X X             E D D A A A A A A A A A A A
17   A A A E E D J G G A A A A A A A A A A           H H   17   X X X               B A A A A A A A A A A A A
18   X X X X X X X     X E E E A A A A A A A A A B B    18   A A B X X X X X A A A A A A A A A A A A A A A
19   G G X X X X X X X A A A A A A A A A A A A A A      19   A B B K X X G F G A A B A A A A A A A A A A A
20   A A B X K K E F D A A A A A A A A A A A A          20   C C C X X X G F D A A A A A A A A A A A A H H H
21   C C C C C C C C D A A A A A A A A A A A A          21   G G G X X     X X A A A B D C B B B A A A A A B
22   A A A A A G F E A A A A A A A A A A A A A A        22   X X X J X     F G F E E E A A A A A A A A A A A
23   A B C K G X X X X F C C     X   F H H B C E E H H  23   E E F E F H H I A A B A A A A A A A A A K J K
24   X K             X X X X B B   A A A A A A A A A A  24   F F G K X K X X X A A A A A A B A A A A A A A A
25   C C B D D D D D B A A A A A A A A A A A A A A A    25   A A A A A A A       B B       I   H H I X X X F F F
26   A A             X X X A A A A A A A A A A A A A    26   E E H I I K X K       X E E E A A A A A A A A A
27   A A A X X           X     E E A A A A A A A A A A  27   A A A C D E X I I A A A A A A A A A A A A A A
28   A A A B B C X X X A A A A A A A A A A     A X X X  28   X X X X X X X X         A A A A A A A A A A A B A A
29   X X X X X X X             G H G     A A A A A A A A A A   29   A A A A A A A A A A A A A A A A A A A A A A A
30   B C D F F F X X K A A A A A A A A A A A A A       30   A A A X X K G G H D D C A A A A A A A A A A A A
31   H K K X X X X X X X A A A A A A A A A A A A A A   31   D D C I I K H G F A A A A A A A A A A A A A A A
```

```
                        JUNE                                           AUGUST
                       HOURS                                            HOURS
DAY            1 1 1 1 1 1 1 1 1 1 2 2 2 2       DAY            1 1 1 1 1 1 1 1 1 1 2 2 2 2
       0 1 2 3 4 5 6 7 8 9 0 1 2 3 4 5 6 7 8 9 0 1 2 3         0 1 2 3 4 5 6 7 8 9 0 1 2 3 4 5 6 7 8 9 0 1 2 3
 1   B C C I X G K X                 X     X X X         1   A A A C C D X X X A A A A A A A A A A A A A A A
 2     X             E F K A A C A A A A A A A C         2   A A A C C F F E C A A A A A A A A A A A A A A
 3   I E F X             X             A             X   3   F F F X X X X X X A A A A A A A A A A A G     D
 4       H I A A A A A     A A G E F A     A E F G F G G   4   X X X X         X     X J J I A A A A A A A A G     X
 5   D C D G F F F F E A A A A A A         E E D         5   X X X X J J X X X A A A A A A A A A A A A A A
 6       X X X X       X A     A E D D A A               6   X K X X X X X X X A A A A A A A A A A A A A A
 7                                     G B A A A A A A A A   7   A A A D C D A A A A A A A A A     A A A A A A A
 8   A A A F F         C B B A A A A A A A A B E C       8   B A A X X X X X X C B A A A A A A A A A A A A
 9   D C C B B B B I I I A A A A A A A A A A A A A       9   X X X X X X X X     X J H H A A A A A A A A A A
10   E E F H K I H F E A A A A A A A A A         A A A   A  10   C C C X X X         X A A A A A A A A A A A A A A
11   A A A A A A G G G C D C A A A A                 X X X X  11   A A A E E E X K K B B C A A A A A A A A A A B A A
12   X X X X X         X X X A A A A A A A A A A A A     12   X X X X X X X X X G E E A A A A A A A A A B B B
13   E E X X X X X X C A A A A A A A A A A A A           13   X X X X X X                 X X     G G A A A A A A
14   E E F I I I E E G A A A A A A A A A A A A A         14                             B A A A A A A A A A A A A A
15   A A A A A A A A A A A A A A A A A A A A A A         15   A A A A A A E E D A A A A A A A A A A A A A A
16   A A A D E D I H E A A A A A A A A A A A A A         16   A A A J J K K X I A A A A A A A A A A A A A A
17   D D E                 H A A A A A A A A A A         17   A A A E E E X X     X X     A A A A A A A A A A A
18   X X X X X             X A A A A A A A A A A A A A   18   A B D F G F G H I D B     A A     A A A A A A A A A
19   B B B F F           A A A A A A A A A A A A A A     19   A A A A A A H           X X X A A     A A A A A A
20   A A A B B             A A A A A A A A A A A A A A   20   X X X X X X             B B A A A A A A A A A B C E
21   D D D       X B B B A A A A A A A A A A             21   X X X           X       A A A A A A A A A A A B D
22   D D E X X X X X X X X C C C A A A A A A A A A A     22   X X X     X     J G K X             A   E             X X X X
23   A A A A A A F D D A B B A A A A A A                 X    23       X X X X X X             X X     E E A A A A A A B B C
24   X X X C B B E D D C B B A A A A A A A A A A A       24   X X X X X X X X X A A A A A A A B A A A A A A A A
25   X X X X X                 X X A A A A A A A A A A A A   25   X X X X X X X X X A B A A A A A A A A A
26   A A                       A A A A A A A A A A A A A A     26                         I H E E C A A A A A A A A A A A
27   A A A D D D I H H A A A A A A A A A A A A A A       27         X X X     X X X A A A A A A A A A A A A A
28   A A A F             K K B A A A A A A A A A A A A   28   A A B C D D X X X C C D A     A A       A H H G
29   A A A A A A E D D A A A A A A A A A A A A A A       29   X X X X X X             J C C C A A A A A A A A     A A
30   A A A A A B G E E A A A A A A A A A A A A A A       30   I       J                                             X X                             X
                                                         31         X X X X X X     X X X A A A A A A A A C D A A A
```

CODE:
A = 100 mm   E = 140 mm   I = 180 mm   BLANK = RAIN
B = 110 mm   F = 150 mm   G = 190 mm
C = 120 mm   G = 160 mm   K = 200 mm
D = 130 mm   H = 170 mm   X = OVER 200 mm

TABLE III.2 (*continued and the end*)

```
                  SEPTEMBER                                          NOVEMBER
                    HOURS                                              HOURS
DAY              1 1 1 1 1 1 1 1 1 1 2 2 2 2        DAY              1 1 1 1 1 1 1 1 1 1 2 2 2 2
     0 1 2 3 4 5 6 7 8 9 0 1 2 3 4 5 6 7 8 9 0 1 2 3      0 1 2 3 4 5 6 7 8 9 0 1 2 3 4 5 6 7 8 9 0 1 2 3
 1   X X X X X X G F F F F E A A A A A A A A A F     1   X X X X X X   X     X X X   B B G H K B D
 2         X X X I H H I H H A A A A A A A A A D C C  2  G E F A A A A A B B C A A A A A A A A A A A
 3   A A A C C B B       X B B     D E   D E A A A   3   H K F X X X X X X X X A A A A A A A A A A A A
 4   A A A A A A E E E A A A A A A A A A A X X K K   4     X     X     X   X X X A A A D D D E E F
 5   X X X X X X X   X H F F A A A A A A A A X X X   5   X X X K K K G F F B B B A B   A   A E E
 6   X X X X                 A A A A A A     G G G   6       X X X X D C C G F G A A A A A A G H I X X X
 7   C D D X X X           X A A A A A A A A A A F E D 7  X X X X X X X X X     A A A J J J D C C
 8   D D E X X             X A A A A A A A A A C B C  8  C C C X X X A A A A A A A     H H A A A F G G
 9   G K K X X             X X X X A A A A A A A A   9   C C D D C D G F F C C C A A A A A A A A A A A
10   B C C X K X           X B B B A A A A A A A C C E 10 A A A A A A A A A D C D           J H   E E   F E
11   X X X X X X             X X A A A A A A A A A A  11  A A A A A A A A A A A A A A A A
12   A A A A A C B C D D D A A A A A A A A A A       12  X X X X X X X X           H
13   E D D E E F I I I D D C B B A             X X   13     X   X X X X X X X J A A A A A A A A A A A
14      X X X X X H H   H H H         X         X   X X X 14 A A     J K X X X K J I A A A A A A K K K B B B
15   X X X X X             X               X     X   15  B C B A A A C B A B   B C   C X X   X X X X X X
16   X X X X X X X X X X H G I B A B A A A A A A C B C 16  X X X X X X       X             X A A B K X X G G G
17   D E E D E D B B B   G F A A A A A A A A A A A C A 17  G G G D D C       J X X X C C B A A A C B B A A B
18   K J E C C E K J I B A A A A A A A A A A A A    18  F F F F E E A A A A A     F F H H H A           X X
19   H H I X X X X X X D E D A A A A A A A A A B B   19  X X X
20   X X X X X X             X I A A A A A A A A G X X 20             X E E D A A A A B A X K X X X X
21          X               X   J A A A A A A A A   C D  21 X X K K K J A         X X X   E     X   X C B B
22                X                 J A A A A        22  F F G E E       X X I I H A A A   A   E       X X X
23   X X                           C H A A A A       23  E E E F E E X X X D D D A A           X X E E E
24                               X A A A A A A A A E D C 24  A A B A A A B B A         X I                   X
25   D D E E E D X X X C A A A A A A A A A A A C C C 25     X X D D E H         H H D E E A A A I E E B B B
26   D F E X X X X X X J I G A A A A A A A A A X X   26  A A A B B B A A A           J J J B B B F F F E F
27   X X X X X X                     F F A A A A A A G I X 27 H F F X X X X X X X X X X X X G H H I J I X X
28   X X                             A A A A A A A A X X X 28 X
29                                         X           I X J I 29      X
30                         X                 B B       E X X X 30     X                             A
```

```
                   OCTOBER                                           DECEMBER
                    HOURS                                              HOURS
DAY              1 1 1 1 1 1 1 1 1 1 2 2 2 2        DAY              1 1 1 1 1 1 1 1 1 1 2 2 2 2
     0 1 2 3 4 5 6 7 8 9 0 1 2 3 4 5 6 7 8 9 0 1 2 3      0 1 2 3 4 5 6 7 8 9 0 1 2 3 4 5 6 7 8 9 0 1 2 3
 1                         X   D H A A A C            1        X X     D E X         E D F E
 2                         X A A A A A A C C E K X    2            X                 X     X X
 3   X                       X X X X A A A A A A G F G 3         J J I C B B A A A A A A A A A A
 4   X X X X X X X X   H G F A A A A A A A A A A A   4   A A A A A F E E I H H A B B A A A A A A A A
 5   G H H X X X X           A A A A       A A A     5   A A A A A A A A A A A A A A A A A A A A A
 6   C D D F F F F         D     B B A A E X X X     6                         X X X X X X F K K X X X
 7   X X X X X X X           E D D A A A A A A A     7                               X   X X C C C G   F J K J
 8   B B A X     X         X X B A A A A C C B X X X 8   I J J C D C X X X X X J A A A A A D D C A A A
 9         X       X         X   C D A A A     X X X 9   A A A A A A A A A A A A B A A A C D D G G H
10   X                             A A A A A A A A B B B 10 D D F X X X F G F X X   X   X G H   X           X
11   A A A X X X             X   X H I H F E   X X K 11    X X X X X X X X X K K F F G A A A B B B I I J
12   I K H E F I X           X   B B A A A A A A E F 12  F         X                 X     X X X X   X
13                            X B A B A A A          13  K X K F G G E D D A                 X X J K J X X X
14      X                 X       A A A A A A A A H H G 14 F F G C C D X X X X X           X X X X X
15   X X X X X X X X         X A A A A A G F E X X X 15                       X   X X J I H                 X X
16                                     K B C C B B B B C 16 X X     X                       A A A A A A     A A
17   E F F J I I X X X F F D A A A A A A C E E X X   17  C C       G G H K X J X X X H H I B B B B B A A A
18                 X                     G   X X     18           X X D D C D E E D E D B B         X X X X X
19   X X X H H H E D D X   X A A A A A A A A A A     19  A A A A A             A A A A A   X X A A A B C A A A
20   F F F B B B                 E E A   A A A A A A 20  J I I D D           D D   J H A A     A A B B B     B C
21   A A A A A A A A A A A   D                        21 D C D C C C C C D F F E E D E A A A F   I X X X
22                       X X                 X X X X X J J I 22      X X X   X X X     X X H F G A A A X X X X X
23   A B B A B B G F E A A A A A A A     A A A A A A A 23 X X X X X X X X X C D D A A A X X     J X J E D D
24   A A A A B B F F F F E A A A A A A A A A A A A A 24  A A A B B F X X   X X X X X X E H I           X X
25   A A A     I I I H H G A A A A A I         X X   25                       X X X                           X
26   A     A A A A A     A A B B A A A A A A B B B D C D 26 X X X X X X X J K K K K D D E F F G F F F       F D
27   D D C X K K X X X X X X A A A B A A A           27  I H H A A A A A B C B A A A A A A E                 X
28                                   X   X       X X X X X 28 I I J B B B B B A H H H B A A     X X X X X
29   I H I I           X X F F E A A A A A A A A H H H 29 A A A A A A A A A A                   C A A A D D D H G G
30   X X X X K K X K J A A A A A A A A A A A A     B B 30                                       X           X A A A
31   H H I C C D X X X X X X A A A A A A A A A     K X 31 A A       D C   A A A A A A A A A A A A A       A B
```

CODE:

A	=	100 mm	E	=	140 mm	I	=	180 mm
B	=	110 mm	F	=	150 mm	G	=	190 mm
C	=	120 mm	G	=	160 mm	K	=	200 mm
D	=	130 mm	H	=	170 mm	X	=	OVER 200 mm

BLANK = RAIN

The Table III.3 presents a summary indicating the number of hours of rain (2073 hours) and the number of hours during which each FOAMGLAS® insulation thickness will prevent surface condensation. For instance condensation will be avoided during 3271 hours with 100 mm thickness. An additional 378 hours will be free from surface condensation if the FOAMGLAS® insulation thickness is increased to 110 mm, etc.

TABLE III.3

Codes	FOAMGLAS® thickness (mm)	Number of times referenced in Table — for each code (n)	Number of times referenced in Table — cumulated number of times	Percentage for each code (%)	Percentage of time insulation jacket is dry — including rain (%)	Percentage of time insulation jacket is dry — excluding (rain) (%)
A	100	3271	3271	37.34	37.34	48.92
B	110	378	3649	4.32	41.66	54.57
C	120	294	3943	3.35	45.01	58.97
D	130	255	4198	2.91	47.92	62.78
E	140	281	4479	3.21	51.13	66.98
F	150	233	4712	2.66	53.79	70.45
G	160	176	4888	2.01	55.80	73.10
H	170	185	5073	2.11	57.91	75.86
I	180	144	5217	1.64	59.55	78.02
J	190	115	5332	1.32	60.87	79.74
K	200	112	5444	1.28	62.15	81.41
X	over 200	1243	6687	14.19		
Rain (blank)		2073	8760	23.66		
Total		8760		100.00		

For instance for code E, which means that the FOAMGLAS® thickness is increased from 130 to 140 mm, the additional number of one hour periods during which surface condensation is avoided is 281. It brings the total number of hours free from condensation from 4198 to 4479.

In percentage of the total year, it represents $\frac{281}{8760} = 3.21\%$ and the total percentage of time during which the insulation jacket is free of condensation increases from 47.92% to 51.13% (from $\frac{4198}{8760}$ to $\frac{4479}{8760}$).

Should one exclude the period during which it rained, the possibly dry total number of periods is reduced from 8760 to 6687 and the total percentage of periods free from surface condensation becomes $\frac{4479}{6687} = 66.98\,\%$. This second approach seems more logically since surface condensation calculation is irrelevant when it rains, the insulation jacket being wet, whatever the insulation thickness is.

From the table it can be seen that increasing the insulation thickness from 180 mm (code I) to 200 mm (code K) only increases the percentage of time during which the jacket is dry from 59.55% to 62.15%. A 10% thickness increase is necessary to obtain a 4.17% "dry period" increase.

In Table III.4 the maximum heat gain in W/m^2 is also given, for each thickness.

TABLE III.4
Heat gain calculation using mean weather data values:
Ambient temperature = +9.7°C Wind speed = 12.23 km/h
Relative humidity = 79.9%

FOAMGLAS® thickness (mm)	Maximum heat gain (W/m^2)	Maximum relative humidity (%)
100	42.02	79.7
110	37.62	81.3
120	34.00	82.8
130	30.93	84.0
140	28.35	85.1
150	26.10	86.1
160	24.15	86.9
170	22.46	87.7
180	20.95	88.4
190	19.61	89.0
200	18.41	89.5

The results of many calculations have shown that the limitation of heat transfer to 25 to 30 W/m^2 corresponds rather often to an acceptable control of the surface condensation in usual conditions.

Influence of the Type of Finish

It may be useful to explain here what was said when describing coating and jacket: the choice of the type of finish – or at least its emissivity – has to be made before the thickness determination since emissivity influences strongly the result of the calculation.

Each finish has an emissivity coefficient (ε) which compares the product with a black body for which $\varepsilon = 1$.

Higher ε increases the heat absorbed by radiation, according to Stefan-Boltzmann's law

$$Q_r = 5.67 \times \varepsilon \left[\left(\frac{T_s}{100} \right)^4 - \left(\frac{T_a}{100} \right)^4 \right]$$

in metric units where

Q_r = heat transfer by radiation in W/m²
ε = surface emissivity
T_s = absolute temperature of finish surface in K
T_a = absolute temperature of air and bodies around the pipe in K.

Higher ε means higher heat gain, higher surface temperature and so higher acceptable relative humidity before dew point is obtained.

The higher surface heat exchange has a negligible influence on the total heat gain of the cold pipe or equipment because this total resistance to heat transfer is very high thanks to the insulation material. Changing the finishing from a low emissivity ($\varepsilon = 0.2$ for stainless steel) to a high one ($\varepsilon = 0.9$ for mastic), changes the total heat gain only very moderately.

Influence of Wind

For similar reasons, when the wind speed increases, the heat exchange on the surface increases according to Langmuir's law

$$Q_c = 1.94 \, (T_a - T_s)^{1.25} \sqrt{\frac{W + 1.26}{1.26}}$$

in the given units, where

Q_c = heat transferred by convection in W/m²
T_s = temperature of surface in °C or K
T_a = temperature of air in °C or K
W = wind speed in km/h.

The acceptable relative humidity increases with the wind speed.

Recommended Thickness Tables

The following tables can be used as a general guide when a detailed study cannot be made or when climatic data is not available. They provide the FOAMGLAS® cellular glass T4 thickness needed to avoid surface condensation in the indicated conditions. See annex regarding calculation method at the end of this part.

For indoor applications, the Tables III.5a and III.5b have been calculated in the given conditions, the difference between the two tables being the emissivity. Its very strong influence should be noticed.

TABLE III.5a
Recommended FOAMGLAS® T4 insulation thicknesses in mm in order to avoid condensation on the outside surface of the insulation system
Indoor application

Conditions: Ambient temperature = 20°C   Wind speed = 0 km/h
Emissivity = 0.2   Relative humidity = 75.0%

Pipe diameter		Service temperature (°C)									
(inch)	(mm)	0	−20	−40	−60	−80	−100	−120	−140	−160	−180
1/2	21.3	30	40	60	70	80	90	100	110	120	130
1	33.7	30	50	70	80	90	100	110	120	130	140
1¹ᐟ²	48.3	30	50	70	90	100	110	120	130	140	150
2	60.3	30	50	70	90	110	120	130	140	150	160
3	88.9	30	60	80	100	110	130	140	150	160	170
4	114.3	30	60	80	100	120	140	150	160	170	180
6	168.3	40	70	90	110	130	150	160	180	190	200
8	219.1	40	70	90	120	140	150	170	190	200	210
10	273.0	40	70	100	120	140	160	180	190	210	220
12	323.9	40	70	100	120	150	170	180	200	210	230
14	355.6	40	70	100	130	150	170	190	200	220	230
16	406.4	40	70	100	130	150	170	190	210	220	240
20	508.0	40	70	110	130	160	180	200	220	230	250
24	609.6	40	80	110	140	160	180	200	220	240	250
30	762.0	40	80	110	140	170	190	210	230	250	260
36	914.4	40	80	110	140	170	190	220	240	250	270
	1000.0	40	80	110	140	170	200	220	240	260	270
	3000.0	40	80	120	150	190	210	240	270	290	310
	6000.0	40	80	120	160	190	220	250	280	300	320
	FLAT	40	80	120	160	200	230	260	290	310	340

On the basis of the thicknesses calculated with the anti-condensation requirement, the heat gain does not exceed 18.0 W/m².

TABLE III.5b
Recommended FOAMGLAS® T4 insulation thicknesses in mm in order to avoid condensation on the outside surface of the insulation system
Indoor application

Conditions: Ambient temperature = 20°C Wind speed = 0 km/h
Emissivity = 0.9 Relative humidity = 75.0%

Pipe diameter		Service temperature (°C)									
(inch)	(mm)	0	−20	−40	−60	−80	−100	−120	−140	−160	−180
1/2	21.3	*20	30	40	40	50	60	60	70	70	80
1	33.7	*20	30	40	50	60	60	70	70	80	80
1 1/2	48.3	*20	30	40	50	60	70	70	80	90	90
2	60.3	*20	30	50	50	60	70	80	80	90	100
3	88.9	*20	40	50	60	70	80	80	90	100	100
4	114.3	*20	40	50	60	70	80	90	100	100	110
6	168.3	*20	40	50	70	80	90	100	100	110	120
8	219.1	*20	40	50	70	80	90	100	110	120	120
10	273.0	*20	40	60	70	80	90	100	110	120	130
12	323.9	*20	40	60	70	80	100	110	120	120	130
14	355.6	*20	40	60	70	90	100	110	120	130	130
16	406.4	*20	40	60	70	90	100	110	120	130	140
20	508.0	*20	40	60	70	90	100	110	120	130	140
24	609.6	*20	40	60	80	90	100	110	130	140	140
30	762.0	*20	40	60	80	90	110	120	130	140	150
36	914.4	*20	40	60	80	90	110	120	130	140	150
	1000.0	*20	40	60	80	90	110	120	130	140	150
	3000.0	*20	40	60	80	100	110	130	140	150	160
	6000.0	*20	40	60	80	100	120	130	140	160	170
	FLAT	*20	40	60	80	100	120	130	150	160	170

On the basis of the thicknesses calculated with the anti-condensation requirement, the heat gain does not exceed 35.9 W/m².
* Thicknesses should be rounded off to the nearest higher manufacturer's thickness.

The comparison of Tables III.1, III.5a and III.5b shows that the results of the calculation made on the base of a 30 W/m² heat gain (Table III.1) and in view of avoiding surface condensation at 75% relative humidity and 20°C ambient air temperature (Tables III.5a and III.5b) are not too different except in the case of low emissivity (0.2 – Table III.5a) which requires considerable thickness to limit condensation. The corresponding heat gains are low and probably not economic in many circumstances. When considering the ageing of jackets, this last case is rather theoretical.

For outdoor applications, the Tables III.6a, III.6b and III.6c have been calculated in the given conditions, the difference between the three tables being again the emissivity. Here also, its influence is strong.

The comparison of Tables III.1, III.6a, III.6b and III.6c shows rather similar results when calculating on the base of a heat gain of 30 W/m^2 (Table III.1) and when limiting the surface condensation to 80% relative humidity except in the case of an emissivity of 0.9 where low thicknesses are found.

TABLE III.6a
Recommended FOAMGLAS® T4 insulation thicknesses in mm in order to avoid condensation on the outside surface of the insulation system
Outdoor application

Conditions: Ambient temperature = 20°C   Wind speed = 10 km/h
             Emissivity = 0.2   Relative humidity = 80.0%

Pipe diameter		Service temperature (°C)								
(inch)	(mm)	0	–20	–40	–60	–80	–100	–120	–140	–160	–180
1/2	21.3	*20	30	40	50	60	60	70	70	80	80
1	33.7	*20	30	40	50	60	70	80	80	90	90
1$^{1/2}$	48.3	*20	40	50	60	70	70	80	90	90	100
2	60.3	*20	40	50	60	70	80	80	90	100	100
3	88.9	*20	40	50	60	70	80	90	100	110	110
4	114.3	*20	40	50	70	80	90	100	100	110	120
6	168.3	*20	40	60	70	80	90	100	110	120	130
8	219.1	*20	40	60	70	90	100	110	120	130	130
10	273.0	*20	40	60	80	90	100	110	120	130	140
12	323.9	*20	50	60	80	90	100	120	130	140	140
14	355.6	*20	50	60	80	90	110	120	130	140	150
16	406.4	*20	50	60	80	100	110	120	130	140	150
20	508.0	*30	50	70	80	100	110	120	130	150	150
24	609.6	*30	50	70	80	100	110	130	140	150	160
30	762.0	*30	50	70	90	100	120	130	140	150	160
36	914.4	*30	50	70	90	100	120	130	140	160	170
	1000.0	*30	50	70	90	100	120	130	150	160	170
	3000.0	*30	50	70	90	110	130	140	160	170	180
	6000.0	*30	50	70	90	110	130	140	160	170	190
	FLAT	*30	50	70	90	110	130	150	160	180	190

On the basis of the thicknesses calculated with the anti-condensation requirement, the heat gain does not exceed 32.3 W/m^2.
* Thicknesses should be rounded off to the nearest higher manufacturer's thickness.

TABLE III.6b
Recommended FOAMGLAS® T4 insulation thicknesses in mm in order to avoid condensation on the outside surface of the insulation system
Outdoor application

Conditions: Ambient temperature = 20°C Wind speed = 10 km/h
Emissivity = 0.4 Relative humidity = 80.0%

Pipe diameter		Service temperature (°C)									
(inch)	(mm)	0	–20	–40	–60	–80	–100	–120	–140	–160	–180
1/2	21.3	*20	30	40	40	50	60	60	70	70	80
1	33.7	*20	30	40	50	60	60	70	70	80	80
1 1/2	48.3	*20	30	40	50	60	70	70	80	90	90
2	60.3	*20	30	50	50	60	70	80	80	90	90
3	88.9	*20	40	50	60	70	80	80	90	100	100
4	114.3	*20	40	50	60	70	80	90	100	100	110
6	168.3	*20	40	50	70	80	90	100	100	110	120
8	219.1	*20	40	50	70	80	90	100	110	120	120
10	273.0	*20	40	60	70	80	90	100	110	120	130
12	323.9	*20	40	60	70	80	100	110	120	120	130
14	355.6	*20	40	60	70	80	100	110	120	130	130
16	406.4	*20	40	60	70	90	100	110	120	130	140
20	508.0	*20	40	60	70	90	100	110	120	130	140
24	609.6	*20	40	60	80	90	100	110	120	130	140
30	762.0	*20	40	60	80	90	100	120	130	140	150
36	914.4	*20	40	60	80	90	110	120	130	140	150
	1000.0	*20	40	60	80	90	110	120	130	140	150
	3000.0	*20	40	60	80	100	110	130	140	150	160
	6000.0	*20	40	60	80	100	110	130	140	160	170
	FLAT	*20	40	60	80	100	120	130	150	160	170

On the basis of the thicknesses calculated with the anti-condensation requirement, the heat gain does not exceed 36.3 W/m².
* Thicknesses should be rounded off to the nearest higher manufacturer's thickness.

III.2.2 HOT APPLICATIONS

The requirements to take into consideration are the limitation of surface temperature to avoid burns, the heat transfer for economic reasons and the control of the process. Here again, in particular conditions, a fire protection requirement may also impose a certain insulation thickness. Generally the process can support considerable heat transfer without being upset. Consequently the requirements are often reduced to limit the heat transfer and the surface temperature to avoid burns to personnel.

TABLE III.6c
Recommended FOAMGLAS® T4 insulation thicknesses in mm in order to avoid condensation on the outside surface of the insulation system
Outdoor application

Conditions: Ambient temperature = 20°C   Wind speed = 10 km/h
Emissivity = 0.9   Relative humidity = 80.0%

Pipe diameter || Service temperature (°C)										
(inch)	(mm)	0	−20	−40	−60	−80	−100	−120	−140	−160	−180
1/2	21.3	*20	*20	30	40	40	50	50	60	60	60
1	33.7	*20	30	30	40	50	50	60	60	70	70
1 1/2	48.3	*20	30	40	40	50	60	60	70	70	80
2	60.3	*20	30	40	50	50	60	70	70	80	80
3	88.9	*20	30	40	50	60	60	70	80	80	90
4	114.3	*20	30	40	50	60	70	70	80	90	90
6	168.3	*20	30	40	50	60	70	80	90	90	100
8	219.1	*20	*30	50	60	70	70	80	90	100	100
10	273.0	*20	*30	50	60	70	80	80	90	100	110
12	323.9	*20	*30	50	60	70	80	90	90	100	110
14	355.6	*20	*30	50	60	70	80	90	100	100	110
16	406.4	*20	*30	50	60	70	80	90	100	100	110
20	508.0	*20	*30	50	60	70	80	90	100	110	110
24	609.6	*20	*30	50	60	70	80	90	100	110	120
30	762.0	*20	40	50	60	70	80	90	100	110	120
36	914.4	*20	40	50	60	70	90	100	110	110	120
	1000.0	*20	40	50	60	80	90	100	110	110	120
	3000.0	*20	40	50	70	80	90	100	110	120	130
	6000.0	*20	40	50	70	80	90	100	110	120	130
	FLAT	*20	40	50	70	80	90	100	120	130	140

On the basis of the thicknesses calculated with the anti-condensation requirement, the heat gain does not exceed 46.3 W/m².
* Thicknesses should be rounded off to the nearest higher manufacturer's thickness.

The calculation to determine the economic thickness can be carried out using the same approach as for cold insulation. The thicknesses compared to the temperatures are smaller than for cold insulation because the cost of producing heat is lower than the cost of producing cold. Depending of the conditions, heat transfer can range from 50 W/m² to very high values.

To limit the danger of burns, the surface temperature should not exceed about 60°C for outdoor applications, possibly less if substantial sun loads are expected. For indoor applications, a limit of 40°C is often imposed to avoid excessive heating in closed spaces. Here again surface emissivity plays a considerable role and can change substantially the required thickness.

For the final decision, the thickness calculated to limit surface temperature should be increased if it is economically justified.

The following tables can be used as guidance to determine the FOAMGLAS® cellular glass T4 thickness in the indicated conditions.

The Tables III.7a and III.7b provide the thickness needed to limit the surface temperatures to 60°C in the given conditions.

TABLE III.7a
Recommended FOAMGLAS® T4 insulation thicknesses in mm for personnel protection
Indoor application

Conditions: Ambient temperature = 20°C   Wind speed = 0 km/h
Emissivity = 0.2   Surface temperature = 60°C

Pipe diameter		Service temperature (°C)						
(inch)	(mm)	100	150	200	250	300	350	400
1/2	21.3	*10	*20	*20	30	40	50	50
1	33.7	*10	*20	30	30	40	50	60
1 1/2	48.3	*10	*20	30	40	50	50	60
2	60.3	*10	*20	30	40	50	60	70
3	88.9	*10	*20	30	40	50	60	70
4	114.3	*10	*20	30	40	50	60	80
6	168.3	*10	*20	30	40	60	70	80
8	219.1	*10	*20	*30	40	60	70	90
10	273.0	*10	*20	*30	50	60	70	90
12	323.9	*10	*20	*30	50	60	70	90
14	355.6	*10	*20	*30	50	60	80	90
16	406.4	*10	*20	*30	50	60	80	90
20	508.0	*10	*20	*30	50	60	80	100
24	609.6	*10	*20	*30	50	60	80	100
30	762.0	*10	*20	40	50	60	80	100
36	914.4	*10	*20	40	50	60	80	100
	1000.0	*10	*20	40	50	60	80	100
	3000.0	*10	*20	40	50	70	90	110
	6000.0	*10	*20	40	50	70	90	110
	FLAT	*10	*20	40	50	70	90	110

The minimum and maximum calculated heat losses in this table are respectively: 140.4 W/m² and 251.1 W/m².
* Thicknesses should be rounded off to the nearest higher manufacturer's thickness.

TABLE III.7b
Recommended FOAMGLAS® T4 insulation thicknesses in mm for
personnel protection
Indoor application

Conditions: Ambient temperature = 20°C   Wind speed = 0 km/h
Emissivity = 0.9   Surface temperature = 60°C

Pipe diameter		Service temperature (°C)						
(inch)	(mm)	100	150	200	250	300	350	400
1/2	21.3	*10	*10	*20	*20	30	30	40
1	33.7	*10	*10	*20	*20	30	30	40
1¹/²	48.3	*10	*10	*20	*20	30	40	40
2	60.3	*10	*10	*20	30	30	40	40
3	88.9	*10	*10	*20	30	30	40	50
4	114.3	*10	*10	*20	30	30	40	50
6	168.3	*10	*10	*20	30	40	40	50
8	219.1	*10	*20	*20	*30	40	40	50
10	273.0	*10	*20	*20	*30	40	50	60
12	323.9	*10	*20	*20	*30	40	50	60
14	355.6	*10	*20	*20	*30	40	50	60
16	406.4	*10	*20	*20	*30	40	50	60
20	508.0	*10	*20	*20	*30	40	50	60
24	609.6	*10	*20	*20	*30	40	50	60
30	762.0	*10	*20	*20	*30	40	50	60
36	914.4	*10	*20	*20	*30	40	50	60
	1000.0	*10	*20	*20	*30	40	50	60
	3000.0	*10	*20	*20	*30	40	50	60
	6000.0	*10	*20	*20	*30	40	50	60
	FLAT	*10	*20	*20	*30	40	50	60

The minimum and maximum calculated heat losses in this table are respectively: 140.4 W/m²
and 446.9 W/m².
* Thicknesses should be rounded off to the nearest higher manufacturer's thickness.

It can be noted, that Table III.7b based on an emissivity of 0.9 corresponds to fairly high heat loss and would generally not be acceptable for indoor applications. Moreover these heat losses exceed what is economically justified.

The Tables III.7c and III.7d provide the thickness needed to limit the surface temperature to 40°C in the given conditions.

TABLE III.7c
Recommended FOAMGLAS® T4 insulation thicknesses in mm for personnel protection
Indoor application

Conditions: Ambient temperature = 20°C   Wind speed = 0 km/h
Emissivity = 0.2   Surface temperature = 40°C

Pipe diameter		Service temperature (°C)						
(inch)	(mm)	100	150	200	250	300	350	400
1/2	21.3	*20	30	50	60	70	90	110
1	33.7	*20	40	50	70	80	100	120
1 1/2	48.3	30	40	60	70	90	110	130
2	60.3	30	40	60	70	90	110	130
3	88.9	30	40	60	80	100	120	140
4	114.3	30	50	60	80	110	130	150
6	168.3	30	50	70	90	110	140	160
8	219.1	*30	50	70	90	120	150	170
10	273.0	*30	50	70	100	120	150	180
12	323.9	*30	50	80	100	130	160	190
14	355.6	*30	50	80	100	130	160	190
16	406.4	*30	50	80	100	130	160	190
20	508.0	*30	50	80	110	140	170	200
24	609.6	*30	50	80	110	140	170	210
30	762.0	*30	60	80	110	140	180	210
36	914.4	*30	60	80	110	150	180	220
	1000.0	*30	60	80	110	150	180	220
	3000.0	*30	60	90	120	160	200	250
	6000.0	*30	60	90	120	160	200	250
	FLAT	*30	60	90	120	160	210	260

The minimum and maximum calculated heat losses in this table are respectively: 65.8 W/m^2 and 107.4 W/m^2.
* Thicknesses should be rounded off to the nearest higher manufacturer's thickness.

TABLE III.7d
Recommended FOAMGLAS® T4 insulation thicknesses in mm for personnel protection
Indoor application

Conditions: Ambient temperature = 20°C　　Wind speed = 0 km/h
　　　　　　Emissivity = 0.9　　　　　　　Surface temperature = 40°C

Pipe diameter		Service temperature (°C)						
(inch)	(mm)	100	150	200	250	300	350	400
1/2	21.3	*20	*20	30	40	50	60	70
1	33.7	*20	30	30	40	50	60	70
$1^{1/2}$	48.3	*20	30	40	50	60	70	80
2	60.3	*20	30	40	50	60	70	80
3	88.9	*20	30	40	50	60	80	90
4	114.3	*20	30	40	50	70	80	100
6	168.3	*20	30	40	60	70	90	100
8	219.1	*20	*30	40	60	70	90	110
10	273.0	*20	*30	50	60	80	90	110
12	323.9	*20	*30	50	60	80	100	120
14	355.6	*20	*30	50	60	80	100	120
16	406.4	*20	*30	50	60	80	100	120
20	508.0	*20	*30	50	60	80	100	120
24	609.6	*20	*30	50	60	80	100	130
30	762.0	*20	*30	50	70	80	110	130
36	914.4	*20	*30	50	70	90	110	130
	1000.0	*20	*30	50	70	90	110	130
	3000.0	*20	*30	50	70	90	110	140
	6000.0	*20	*30	50	70	90	120	140
	FLAT	*20	*30	50	70	90	120	150

The minimum and maximum calculated heat losses in this table are respectively: 88.9 W/m^2 and 95.9 W/m^2.
* Thicknesses should be rounded off to the nearest higher manufacturer's thickness.

The comparison of Tables II.7c and III.7d shows once again the strong influence of emissivity, the 0.2 emissivity being rather theoretical for an aged jacket.

The Tables III.8a and III.8b provide the FOAMGLAS® cellular glass T4 thickness needed to limit the heat loss to respectively 150 W/m² and 200 W/m² in the given conditions.

TABLE III.8a
Recommended FOAMGLAS® T4 insulation thicknesses in mm in order to control heat losses
Indoor application

Conditions: Ambient temperature = 20°C   Wind speed = 0 km/h
Emissivity = 0.9   Maximum heat loss = 150 W/m²

Pipe diameter		Service temperature (°C)						
(inch)	(mm)	100	150	200	250	300	350	400
1/2	21.3	*20	30	40	50	60	70	80
1	33.7	*20	30	40	50	60	80	90
1 1/2	48.3	*20	30	40	60	70	80	100
2	60.3	*20	30	50	60	70	90	100
3	88.9	*20	40	50	60	80	100	110
4	114.3	*20	40	50	70	80	100	120
6	168.3	*20	40	50	70	90	110	130
8	219.1	*20	40	60	70	90	110	140
10	273.0	*20	40	60	80	100	120	140
12	323.9	*20	40	60	80	100	120	140
14	355.6	*20	40	60	80	100	120	150
16	406.4	*20	40	60	80	100	120	150
20	508.0	*20	40	60	80	100	130	160
24	609.6	*20	40	60	80	110	130	160
30	762.0	*30	40	60	80	110	130	160
36	914.4	*30	40	60	90	110	140	170
	1000.0	*30	40	60	90	110	140	170
	3000.0	*30	40	70	90	120	150	180
	6000.0	*30	40	70	90	120	150	190
	FLAT	*30	40	70	90	120	150	190

The maximum calculated surface temperature in this table in 35.9°C.
* Thicknesses should be rounded off to the nearest higher manufactuer's thickness.

Although tables have not been recalculated for various wind speeds in this book, a general remark could be formulated: higher wind speeds decrease the surface temperature of hot applications and increase the heat loss. Influences exceeding 10°C surface temperature difference and 20 W/m² are not unusual.

TABLE III.8b
Recommended FOAMGLAS® T4 insulation thicknesses in mm in order to control heat losses
Indoor application

Conditions: Ambient temperature = 20°C   Wind speed = 0 km/h
Emissivity = 0.2   Maximum heat loss = 200 W/m²

Pipe diameter		Service temperature (°C)						
(inch)	(mm)	100	150	200	250	300	350	400
1/2	21.3	*10	*20	30	40	50	60	70
1	33.7	*10	*20	30	40	50	60	70
1¹ᐟ²	48.3	*10	*20	30	40	50	70	80
2	60.3	*10	*20	30	50	60	70	80
3	88.9	*20	30	40	50	60	70	90
4	114.3	*20	30	40	50	60	80	90
6	168.3	*20	30	40	50	70	80	100
8	219.1	*20	*30	40	50	70	90	100
10	273.0	*20	*30	40	60	70	90	110
12	323.9	*20	*30	40	60	70	90	110
14	355.6	*20	*30	40	60	70	90	110
16	406.4	*20	*30	40	60	80	90	110
20	508.0	*20	*30	40	60	80	100	120
24	609.6	*20	*30	40	60	80	100	120
30	762.0	*20	*30	40	60	80	100	120
36	914.4	*20	*30	40	60	80	100	120
	1000.0	*20	*30	40	60	80	100	130
	3000.0	*20	*30	50	60	80	110	130
	6000.0	*20	*30	50	60	90	110	140
	FLAT	*20	*30	50	60	90	110	140

The maximum calculated surface temperature in this table is 53.3°C.
* Thicknesses should be rounded off to the nearest higher manufacturer's thickness.

III.2.3 DUAL TEMPERATURE APPLICATIONS

These installations are operating at temperatures varying between a positive and a negative temperature.

In this case, the limitation of the surface condensation applies for the low temperature and the limitation of the surface temperature for the high temperature.

The heat transfer limitation for economic reasons applies in both cases, in function of the time periods during which the installation is running at the given temperatures.

The process control requirement and the eventual fire protection requirement are often less stringent requirements but they can become predominant in some particular cases.

Although more complicated than for cold or hot applications, the calculation of the insulation thickness is basically carried out by the same methods. The final thickness is selected by comparison between the results found with the different requirements.

The tables previously given for guidance can be used in the appropriate context.

III.2.4 OVERFIT/RETROFIT ON HIGH TEMPERATURE APPLICATIONS

This system consists in applying an outside layer of FOAMGLAS® cellular glass on a high temperature permeable insulation. It gives a reduction of heat transfer, a diminution of surface temperature and if needed, a drying out of the moisture that may have penetrated into the permeable insulation. As explained in some details in V.4.4 and V.4.5, the drying out of the permeable insulation requires that the temperature of the interface between the complementary cellular glass layer and the existing layer exceeds 100°C.

Consequently, three requirements have to be taken into consideration: the limitation of the heat transfer, the limitation of the outside surface temperature and the minimum of 100°C at the FOAMGLAS® cellular glass layer existing insulation interface.

Usually the last requirement is the predominant one, the previous ones being generally met when the last one is fulfilled.

To be on the safe side, the determination of the interface temperature should be done with the dry thermal conductivity of the permeable insulation. This choice gives also the advantage of reflecting the final state when the permeable material has been dried out by the Overfit application.

The calculation is based on the distribution of the temperature differences in the two insulation materials in proportion to their thermal resistances.

Although a precise determination is always advisable, many examples in normal conditions have shown that the required thickness to obtain the drying out phenomena amounts to about 50 mm.

III.2.5 INSULATED UNDERGROUND PIPINGS

Because of cellular glass properties and particularly its mechanical characteristics such as compressive strength, the direct burial of FOAMGLAS® cellular glass insulated lines is possible in spite of the pressure developed by the ground and possible loads on it. Using FOAMGLAS® cellular glass eliminates the need for tunnels but will involve direct contact between the coating applied on the insulation and the ground. In these conditions, organic coatings, often bitumen based, are generally specified. Although it is conceivable – and sometimes done – to use the system for very low temperature liquid transport, it is rarely done due to the problems associated with potential freezing of the ground, and heaving of it, especially in the presence of water.

By far the majority of the applications carry high temperature fluids. The thermal requirements for their applications are: limit coating temperature to an acceptable value, limit the heat transfer for economic reasons, limit the temperature variations of the fluid between the beginning and the end of the line for process reasons.

Without going into calculation details, it is useful to quote the classic formula giving the heat transfer and the formula giving the surface temperature.

The heat transfer Q reads:

$$Q = \frac{2\pi k_i k_g (t_s - t_g)}{k_g \ln \frac{r_2}{r_1} + k_i \ln \frac{r_3}{r_2}}$$

where:

Q = heat transfer in W/ linear meters
k_i = thermal conductivity of the insulating material in W/(m.K)
k_g = thermal conductivity of the ground in W/(m.K)
t_s = service temperature of the pipe in °C
t_g = average temperature of the ground at a point where it is not affected by the heat of the pipe in °C
r_1 = outside radius of the pipe in m

PART III. HOW TO DESIGN AND INSTALL FOAMGLAS ® INSULATION FOR INDUSTRIAL APPLICATIONS

r_2 = outside radius of the insulation in m
r_3 = distance in m for the centre of the pipe to the point at which the temperature of the ground becomes t_g
ln = symbol for natural logarithm.

Usually r_3 is estimated at 12 m and t_g is estimated at 5°C.

The insulation surface temperature is given by:

$$t = t_s - \frac{Q}{2 \pi k_i} \ln \frac{r_2}{r_1}$$

TABLE III.9a

Conditions: Ground temperature = 5°C
Insulation surface temperature = 90°C
Ground thermal conductivity = 0.9 W/(m.K) (dry soil)

Pipe diameter		Service temperature (°C)								
		150		175		200		225		250	
(inch)	(mm)	(mm)	(W/m)	(mm)	(W/m)	(mm)	(W/m)	(mm)	(W/m)	(mm)	(W/m)
1/2	21.3	*30	29.9	*30	36.5	*30	43.8	*30	50.3	*30	58.4
1	33.7	*30	37.5	*30	45.8	*30	53.5	*30	62.9	30	73.1
1 1/2	48.3	*30	45.3	*30	53.9	*30	64.5	30	75.8	30	88.0
2	60.3	*30	51.0	*30	60.8	30	72.6	30	85.3	40	85.6
3	88.9	*30	61.7	*30	75.2	30	89.6	40	91.0	50	94.2
4	114.3	*30	70.9	30	86.2	40	89.1	50	93.4	60	98.6
6	168.3	30	87.4	40	92.8	50	99.0	60	105.8	80	105.7
8	219.1	*40	100.4	40	107.3	60	105.0	70	113.8	90	115.7
10	273.0	40	99.8	50	109.3	70	110.4	90	113.6	110	118.4
12	323.9	40	109.3	60	110.5	80	114.0	100	119.0	120	125.1
14	355.6	40	114.9	60	116.6	80	120.5	100	125.9	130	126.8
16	406.4	50	112.5	70	117.0	90	122.9	110	129.8	140	132.1
20	508.0	50	126.9	80	125.1	100	133.2	130	136.1	160	141.0
24	609.6	60	130.2	90	131.7	120	135.6	150	141.2	190	143.7

* Reduced thickness would be sufficient to fulfil the 90°C surface temperature requirement. The heat transfer calculated is expressed in Watt per meter (W/m).

As it can be seen, these formulas reflect the influence of the ground thermal conductivity in the calculation of heat transfer and of the insulation outside surface (or coating) temperature. This can be physically explained by the fact that the insulation of the pipe is provided both by the insulation material and by the ground around it.

It should be noted that a low ground thermal conductivity will increase the insulation surface temperature and will reduce the heat transfer. On the contrary, a high ground thermal conductivity will decrease the insulation surface temperature and will increase the heat transfer.

As a matter of fact, the ground thermal conductivity varies with its moisture content, during the years, being higher during a wet period than during a dry period. Moreover the value cannot be accurately measured, particularly when considering a long pipeline going through various zones.

Consequently, many specialists calculate the insulation outside surface temperature with a relatively low thermal conductivity of the ground and the heat transfer with relatively high thermal conductivity of the ground to be on the safe side.

The Tables III.9a and III.9b have been calculated to limit the insulation outside surface temperature (the coating) at 90°C. They give the required thickness and the corresponding heat transfer.

TABLE III.9b

Conditions:	Ground temperature = 5°C
Insulation surface temperature = 90°C
Ground thermal conductivity = 1.3 W/(m.K) (wet soil)

Pipe diameter		Service temperature (°C)								
		150		175		200		225		250	
(inch)	(mm)	(mm)	(W/m)	(mm)	(W/m)	(mm)	(W/m)	(mm)	(W/m)	(mm)	(W/m)
1/2	21.3	*30	31.4	*30	38.3	*30	46.0	*30	54.3	*30	63.4
1	33.7	*30	40.0	*30	48.8	*30	58.7	*30	69.3	*30	78.2
1½	48.3	*30	49.0	*30	59.9	*30	72.0	*30	82.3	*30	95.8
2	60.3	*30	55.9	*30	68.3	*30	79.6	*30	93.8	30	109.0
3	88.9	*30	70.7	*30	84.1	*30	100.6	30	118.3	40	116.2
4	114.3	*30	80.4	*30	98.1	30	117.2	30	137.6	40	135.4
6	168.3	*30	101.9	30	124.0	30	147.8	40	147.9	50	150.9
8	219.1	*40	119.3	*40	145.0	40	148.2	50	153.5	60	160.1
10	273.0	*40	135.6	40	142.4	50	150.1	60	158.8	70	168.3
12	323.9	*40	149.5	40	157.8	50	166.9	70	162.1	80	173.7
14	355.6	*40	157.5	40	166.8	60	160.1	70	172.1	90	172.2
16	406.4	*40	169.5	50	161.4	60	174.2	80	173.7	100	176.5
20	508.0	*40	169.4	50	184.7	70	184.4	90	187.6	110	193.2
24	609.6	*40	187.8	60	188.5	80	192.7	100	199.1	130	197.4

* Reduced thickness would be sufficient to fulfil the 90°C surface temperature requirement.
The heat transfer calculated is expressed in Watt per meter (W/m).

III.2.6 TANK INSULATION

Depending on the function of the tank, the requirements may be different and the calculation of thickness varies accordingly.

If a tank is used to store a liquid during a long period of time, a typical requirement will be the limitation of the heat transfer. For tanks containing liquefied gases, the requirement is often expressed as a boil off rate limitation per unit of time (usually per 24 hours). It means that only a limited percentage of the liquid stored in the tank can vaporise in a given period of time due to energy entering the tank. It is clear that the boil off is fixed by the heat transfer through all parts of the tank: base, walls and roof.

In the opposite direction, if a tank is used in a receiving terminal where liquefied gas has to be re-gasified, the limitation of boil off may not be a severe requirement and the acceptable heat transfer can be fairly high. A requirement for the base insulation would be to avoid the heat transfer becoming so high that freezing of the support structure creates mechanical problems. Finally a specialised requirement consists in protecting materials of the construction from being brought to temperatures at which they would lose their mechanical strength, should a major leakage occur.

A similar type of requirement can be found for high temperature storage tanks.

Needless to say, the requirement for the limitation of surface condensation can apply for cold temperature tanks and the limitation of surface temperature can apply for hot tank. They should be treated in the same method as suggested for pipings.

III.3

FOAMGLAS® CELLULAR GLASS FABRICATED ELEMENTS

III.3.1 NUMBER OF LAYERS

Although the question of the number of layers will be discussed later in some particular cases, it may be useful to present already here the general principles determining them.

Several reasons can be given for increasing the number of layers: a multi-layer system allows the designer to select staggered joints and consequently to have longer Z joints, reducing the danger of water vapour passing from the exterior to the structure in the case of a cold insulation. Also the inevitable small thermal short circuit due to the joints is reduced when Z joints are used. This point should not be excessively emphasised since the thermal conductivity of bitumen or adhesive or sealer when used, is low and the percentage of joints is limited.

In some unusual conditions, increasing the number of layers by one unit is useful to reduce the temperature variation in the outside layer and avoid extreme temperature in the joints of the outside layer that could cause deterioration of the sealers.

Finally increasing the number of layers helps to avoid cracks due to thermal shocks when sudden and very large temperature changes take place.

As opposed to that, several good reasons can be given for reducing the number of layers to a minimum. First of all, thick layers are mechanically stronger than thin ones. Should a slab or a pipe covering be locally unsupported and should someone walk on it or apply other loads, the mechanical resistance of the thicker element will obviously be greater. It may be worth while mentioning that thickness smaller than 40 mm should be avoided in cellular glass, except for pipe covering up to an IPS lower than 6 inches where 30 mm is the limit.

The second reason to limit the number of layers in the case of cellular glass is based on the dimensional stability of the product. There is no need to compensate for material movement causing joints to open by longer Z joints: cellular glass is dimensionally stable and its coefficient of thermal expansion is low and close to the value of steel: $9 \times 10^{-6} K^{-1}$ for FOAMGLAS® cellular glass compared to $12 \times 10^{-6} K^{-1}$ for carbon steel. Moreover the moderate length of the slabs or pipe coverings – 600 mm – limits the effect of thermal movements.

Applied cost constitutes the final and very important reason to limit the number of layers. Except for flat slabs, the material cost of the two layer system is somewhat higher than the cost of a single layer. It may also include increased consumption of accessory products. The difference increases with the complexity of the cellular glass elements and can reach 30% for small and complex shapes. Moreover, a two layer system entails more or less double the installation time. The corresponding labour cost represents often the major part of the installed cost.

When reviewing major FOAMGLAS® cellular glass applications, we will indicate the number of layers generally specified but the reasons given to increase or decrease the number of layers should be kept in mind for specific situations.

III.3.2 GEOMETRY OF THE FOAMGLAS® CELLULAR GLASS ELEMENTS TO BE SELECTED AS A FUNCTION OF THE SURFACE TO BE INSULATED

III.3.2.1 General Principles

Cellular glass being rigid, its geometry has to be adapted to the surface in order to avoid wide open joints between the FOAMGLAS® cellular glass elements or excessive voids between them and the surface to be insulated.
Too wide open joints would be difficult to seal in cold applications and would mean higher thermal transfer.
Poor fitting to the surface could result into rupture if people walk on the insulation, specially when small thickness elements are used.
To be able to meet the application requirements for the avoidance of moisture penetration with the present adhesives and sealers, experience has

shown that the theoretical distance between a circular surface and the cellular glass element should not exceed a maximum of about 2 mm and that the maximum joint opening should not exceed about 3 mm.

Figure III.2 and associated formula indicate how the values of b and x should be calculated to the given limits.

Distance to the surface

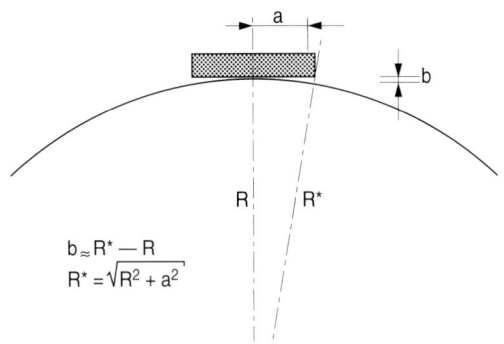

$$b \approx R^* - R$$
$$R^* = \sqrt{R^2 + a^2}$$

The following table has been elaborated in order to limit the theoretical distance to a maximun of 2 mm.

width of flat FOAMGLAS® element	minimum diameter of cylindrical vessel
150 mm	3 000 mm
225 mm	7 000 mm
300 mm	12 000 mm
450 mm	25 000 mm

Axial joint opening

It is recommended to limit the opening (x) to 3 mm at the outside. In order to achieve this, at least the longitudinal side should be bevelled in most cases.

$$\frac{x/2}{d} = \frac{a}{R}$$
$$x \approx \frac{2a \times d}{R}$$

Fig. III.2. Calculation of the maximum distance and of the joint opening.

These two values b and x should be considered for guidance but not as absolute limits since many factors can moderately influence them. For instance, the presence of an adhesive or a sealer that is soft and resilient helps to compensate for the lack of geometry of the FOAMGLAS® cellular glass elements.

PART III. HOW TO DESIGN AND INSTALL FOAMGLAS® INSULATION FOR INDUSTRIAL APPLICATIONS

For a dual temperature insulation cycling frequently from low to high temperatures the given values are an absolute maximum, since the adhesives and sealers corresponding to the imposed temperature range are difficult, if not impossible to find and generally present problems with ageing. For such an application, the geometry of the system should be good enough to avoid relying excessively on adhesives and sealers.

III.3.2.2 Fabrication Method

The shaping of the cellular glass elements can be carried out on the jobsite or in a fabrication shop. In several European countries, one or more fabrication shops having the know how to shape FOAMGLAS® cellular glass have been installed. Such fabrication shops also exist in other countries.

As a general rule, this work is preferably carried out in a fabrication shop when a large series of pieces are required and on the jobsite for special elements that have to be adapted to complicated surfaces. In the case of very large jobs, such as LNG terminals, the two methods complement each other.

In specialised fabrication shops, the FOAMGLAS® cellular glass elements are obtained by abrasion or by cutting pieces out of flat slabs or out of several flat slabs adhered together with an appropriate adhesive to form a billet. It should be remembered that the operating temperature has to be given to enable the fabrication shop to choose the adhesive accordingly. Bitumen is generally adopted for low and ambient temperatures up to 120°C (pipe operating temperature [1]). For higher temperatures, at which bitumen would lose its adhesive strength, a high temperature adhesive is used, for instance a modified gypsum powder PC® HIGH TEMPERATURE ANTI-ABRASIVE.

The type of machinery and its level of complexity is adapted to the series of pieces to be fabricated.

Fabrication on the jobsite is made with several specific tools, also following the abrasion or cutting technique. For instance special shapes can be obtained by pressing the FOAMGLAS® cellular glass elements on a rotating form on which carborundum or other abrasive has been adhered. Cutting is made with simple machines or even with saw and master plate. The advantage of site fabrication is the possibility to check the shape and di-

---

[1] The bitumen joint is not totally at this temperature. Only the part close to the pipe can reach this temperature.

mensions on the surface to be insulated and avoid the preparation of drawings for elements that have to be made only a few times. Moreover last minute modification of the installation is easily dealt with.

Regarding adaptation of the FOAMGLAS® cellular glass elements to local irregularities of the surface, such as welding borns, overlapping of steel plates, etc., the work is generally carried out on the jobsite using the abrasion technique.

III.3.2.3 Definition of some Available FOAMGLAS® Cellular Glass Elements

General

Two standardisation systems are generally adopted:

- The system for which the standardisation of the insulation thicknesses is such that the outer diameter of a FOAMGLAS® cellular glass shell always corresponds to the outer diameter of a standard pipe size. This system has the merit of allowing perfect nesting of the pipe shell, even when a further layer is added but the thicknesses vary in each case and are not rounded off values. On the basis of the size of the FOAMGLAS® cellular glass slabs, 450 × 600 mm, the standard length of pipe shell is always 450 or 600 mm; to save labour on the jobsite, the most common length is 600 mm.
- The classical one in which insulation thicknesses increase in steps of 10 mm. (for instance: 40, 50, 60, 70, 80, 90 or 100 mm).

The inside diameter of the fabricated elements corresponds to the actual pipe size, increased by its tolerance. When anti-abrasive is foreseen, the inside diameter of the pipe is increased to an amount including twice the anti-abrasive layer thickness; a value of 2 × 2 mm is adequate for PC® HIGH TEMPERATURE ANTI-ABRASIVE.

Pipe Shells (PSH)

Shells for straight pipes having an outside insulated diameter not exceeding 300 mm (in fact 298.5 mm) can be cut out of a 15 cm thick slabs. Because adhesive is not needed to fabricate them, these shells are in one piece; they can be applied to the whole service temperature range of FOAMGLAS® cellular glass (Fig. III.3).

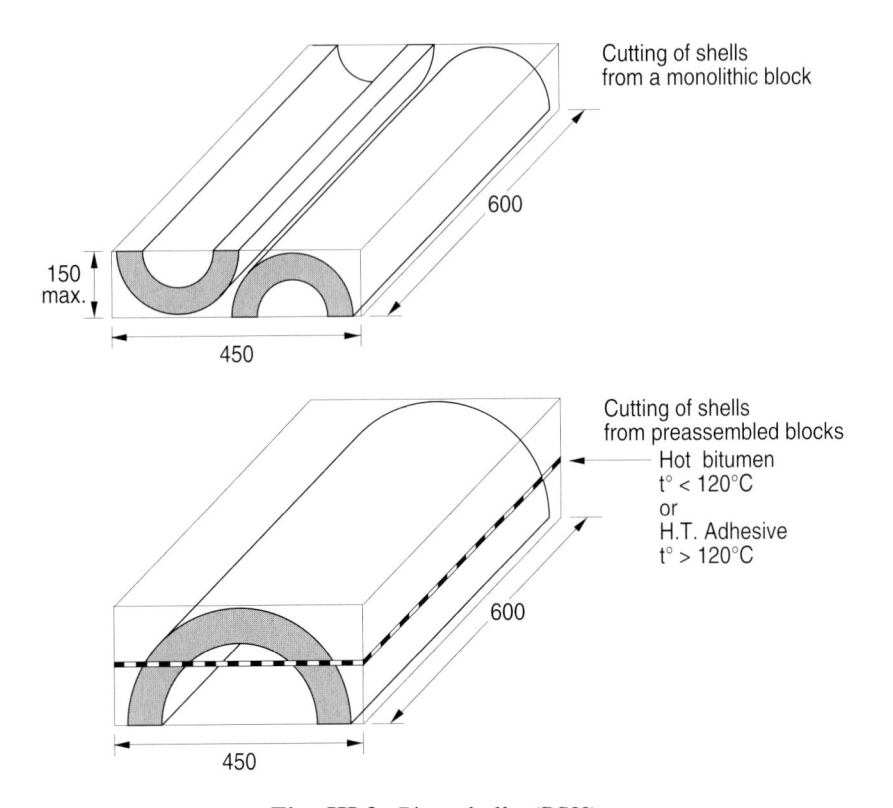

Fig. III.3. Pipe shells (PSH).

When the insulation outside diameter ranges from 300 mm to 450 mm, FOAMGLAS® cellular glass slabs have to be preassembled to form a billet; larger elements may be cut from this (Fig. III.3).

As previously indicated, hot bitumen is used as the adhesive when the pipe operating temperature does not exceed 120°C while a high temperature adhesive, for instance PC® HIGH TEMPERATURE ANTI-ABRASIVE, is used for higher temperatures.

This system that consists of only two pieces for the total circumference can only be used when the outside insulation diameter does not exceed 450 mm. The above types of pipe shell are designated PSH (pipe shell half).

Pipe Segments (PSG)

When the insulated shell exterior diameter ranges from 450 mm to 920 mm, the two piece configuration is no longer possible and is replaced by curved segments for straight pipes. These segments are manufactured from FOAMGLAS® cellular glass slabs cut in two pieces lengthways to make 225 mm wide elements. Their curvature corresponds to the pipe diameter. Dependant on the thickness, they may include a joint in the direction perpendicular to the diameter (so-called non-through joints). The edges of these elements are bevelled, to achieve perfect fitting.

Since the circumference of the insulated pipe is rarely a multiple of 225 mm, the last segment will be cut on the jobsite to fit exactly (Fig. III.4 and Fig. III.5).

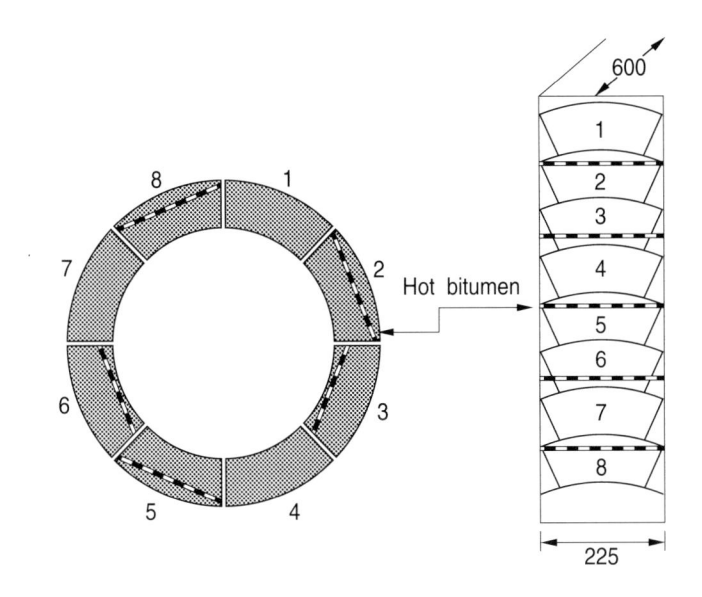

Fig. III.4. Pipe segments (PSG).

If the cut off exceeds 110 mm in width, it will generally be reused.

173

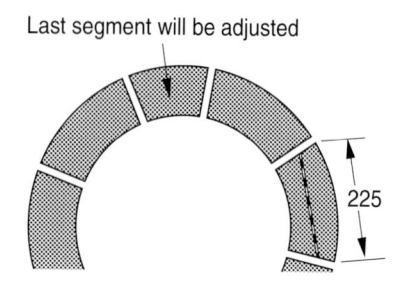

Fig. III.5. Last segment of a circumference.

These elements are called type PSG (pipe segments).

Tank Segments (TSG)

For insulated diameter exceeding 920 mm, vessel or tank segments can be used. In fabrication shops, they are produced for diameters ranging from 920 to 7000 mm and more. They are basically the same as the pipe segments having a length of 600 mm and a maximum width of 220 mm but they often incorporate a joint perpendicular to the diameter (Fig. III.6a). Depending on the operating temperature, the adhesive can be bitumen (up to 120°C) or a high temperature adhesive.

Here again, the last segment will have its width cut on the jobsite to fit the remaining space. These elements are called TSG (tank segments).

A first alternative solution, often used for high temperature applications consists of bevelled segments cut from half slabs, measuring 295 by 450 mm (Fig. III.6b).

An other alternative solution in the choice of tank segments consists of using flat slabs cut in elements small enough to comply with the rules given previously.

To take into account the maximum distance from the flat element to the surface to be insulated, the minimum diameter of the cylindrical vessel should be as follows:

Maximum width of flat FOAMGLAS® cellular glass elements (mm)	Minimum vessel diameter (mm)
150	3,000
225	7,000
300	12,000
450	25,000

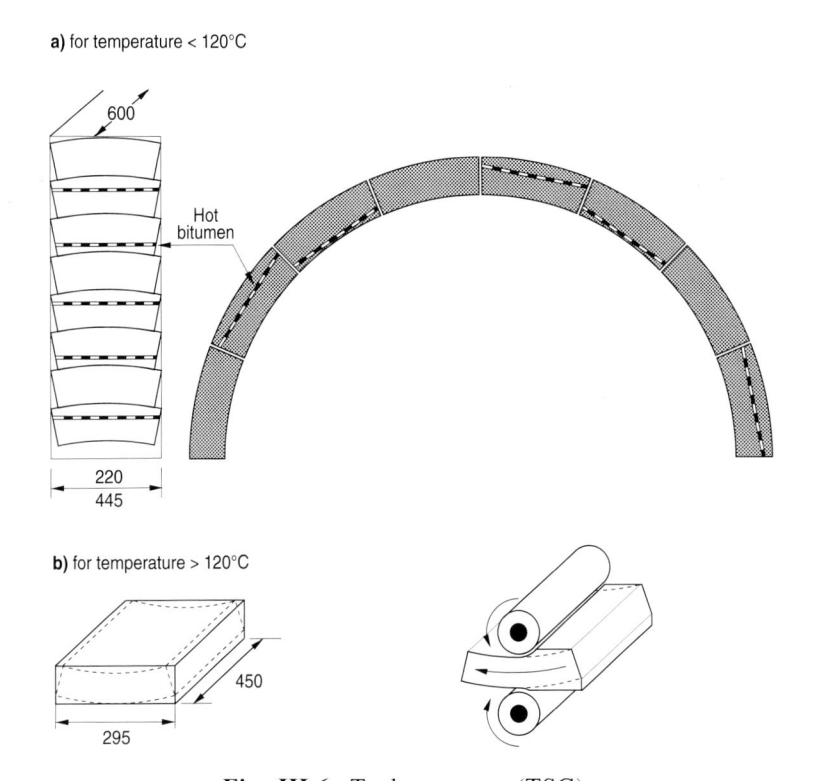

a) for temperature < 120°C

b) for temperature > 120°C

Fig. III.6. Tank segment (TSG).

As far as the axial joint is concerned, the opening depends on the tank surface diameter and on the cellular glass thickness. The Table III.10 summarises the minimum tank diameter on which a FOAMGLAS® cellular glass element of a given thickness can be applied to comply with the 3 mm maximum joint opening.

TABLE III.10

Minimum tank diameter in m

Width of flat FOAMGLAS® cellular glass elements (mm)	FOAMGLAS® cellular glass element thickness (mm)								
	40	50	60	70	80	90	100	120	130	140
150	8	10	12	14	16	18	20	24	26	28
225	12	15	18	21	24	27	30	36	39	42
300	16	20	24	28	32	36	40	48	52	56
450	24	30	36	42	48	54	60	72	78	84

The values given in this table constitute the most stringent situation since it is always possible to slightly abrade the edges of the FOAMGLAS® cellular glass elements on the jobsite.

Some contractors are using a push box tool (Fig. III.7) made of a box with the bevelled edges covered with abrasive material like carborundum to facilitate this operation.

Owing to the rather large diameter needed to meet the requirement for the limitation of the joint openings, one can understand easily that tank seg-

Fig. III.7. Push box.

ments are often used for diameters of 10 to 20 meters, dependant upon the thickness. Moreover the use of 150 mm wide flat slabs requires more elements, more installation time – and more sealer for low temperature application – than the use of tank segments. Generally the total installed cost is higher with 150 mm wide flat slabs.

As a final conclusion, prefabricated tank segments will be preferred in most cases, except for large diameters and thicknesses.

Elbows: E90 and E45

As far as FOAMGLAS® cellular glass elbows are concerned, they generally have a radius equal to 1.5 D, D being the pipe diameter and are then designated type 3D. They exist for 90° and 45° angle and are respectively called E90 and E45. When the outside insulation diameter does not exceed 298.5 mm, they are cut from two monolithic cellular glass slabs and consequently can be used in the full cellular glass temperature range.

When the outside insulation diameter exceeds 298.5 mm, the FOAMGLAS® cellular glass elbows are fabricated by assembling trapezoidal sections of straight pipe shells or straight pipe segments preassembled together. Here again the choice of adhesive is made in function of the service temperature. Usually FOAMGLAS® cellular glass elbows are made in a single insulation layer since staggering the joints between the layers is very difficult (Fig. III.8).

Flanges: F15 and F30

For the insulation of flanges, a set of shells and/or segments of the appropriate diameters and lengths are fabricated. Two types of flange insulation exist for pipes: firstly for class 150 and secondly for class 300. They are respectively called F15 and F30. Previous comments on the choice of adhesive in relation to the service temperature and the possible application of anti-abrasive with its corresponding influence on the insulation interior diameter apply (Fig. III.9).

Valves: V15 and V30

In the case of valves, a set of shells, segments and flat slabs are prepared in a fabrication shop or on the jobsite.

V15 corresponds to piping of class 150 and V30 corresponds to piping of class 300. Owing to the number of cases, only a single example of what can be made is given in Fig. III.10.

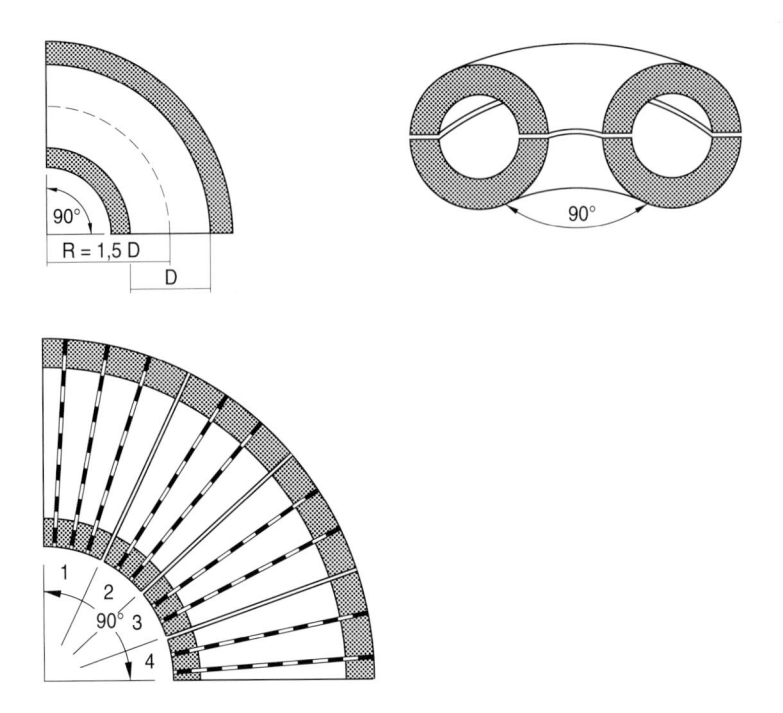

Fig. III.8. Elbows: E90.

A. 1 half shell to be cut in 4 pieces for the dutchman

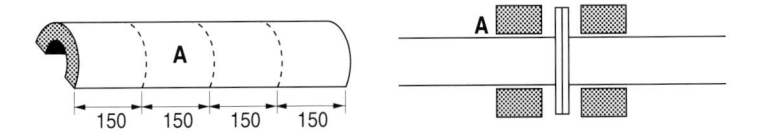

B. 2 half shells to be applied over the dutchman and to be adjusted in lenght

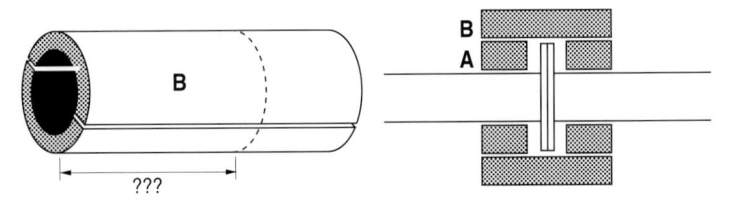

Fig. III.9. Flanges: F15.

How to apply FOAMGLAS® valve insulation

Example
Pipe diameter : 3" (88.9 mm)
Valve class 150
FOAMGLAS® Thnickness 2" (52 mm).

Material delivered
- 1 half shell of 600 mm type PSH
  dim. 3" x 2" - pieces A.
- 4 half shells of 600 mm type PSH
  dim. 7" x 2" - pieces B1, B2, B3, B4.
- 1 flat block of 300 x 450 mm
  thick 50 mm - piece C.

A 1 half shell to be cut into 4 pieces of 150 mm each for the packing piece (dutchman).

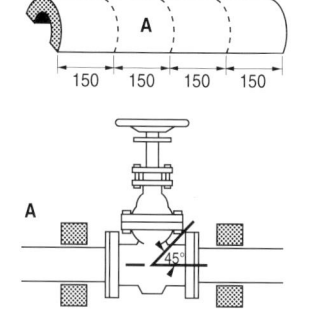

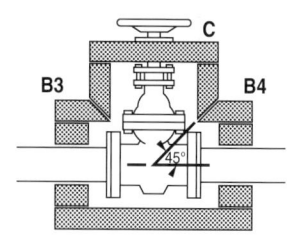

B1 1 half shell to be cut at 45° using a mitre-box.

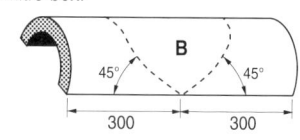

B2 1 half shell to be placed at the bottom.

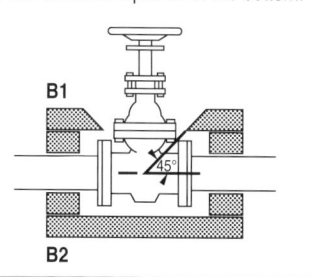

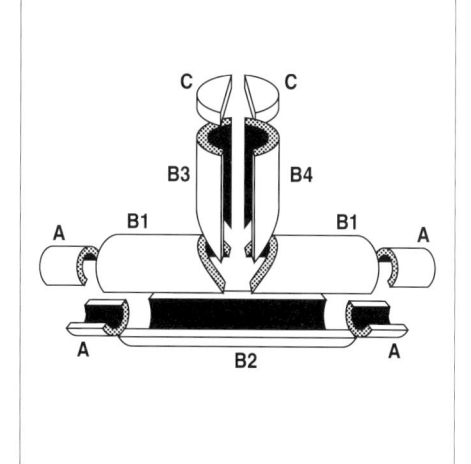

B3/B4 2 half shells to be cut at 45° using a mitre-box and to be adjusted in length.

C 1 flat block to be cut in 2 and to be adjusted around spindle.

Fig. III.10. Valve: V15.

Vessel Heads

For the insulation of vessel heads, special segments are manufactured to fit the geometry of the vessel heads.

Vessel head segments HEH correspond to $R = D$ and $r = 0.1\ D$ and HKH corresponds to $R = 0.8\ D$ and $r = 0.154\ D$.

This is illustrated by Fig. III.11.

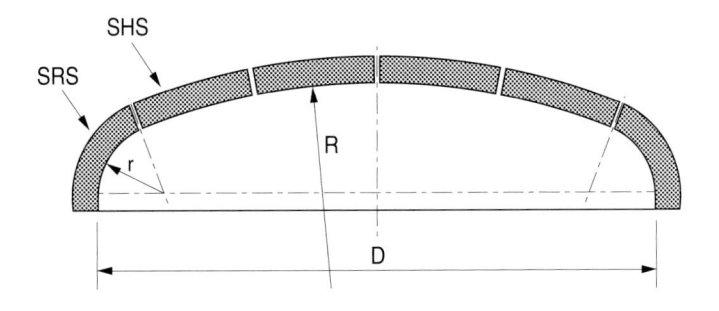

Fig. III.11. Vessel head segments (Example of HEH).

Standardised dimensions for the central part of the vessel head segments (SHS: Spherical Head Segments) are as follows:

Radius of curvature (mm)	Dimensions of vessel head segments (mm)
750 to 900	215 × 295
901 to 5000	295 × 440

For the peripheral part, the SRS (Small Radius Segments) segment width at the base does not exceed 145 mm (Fig. III.12).

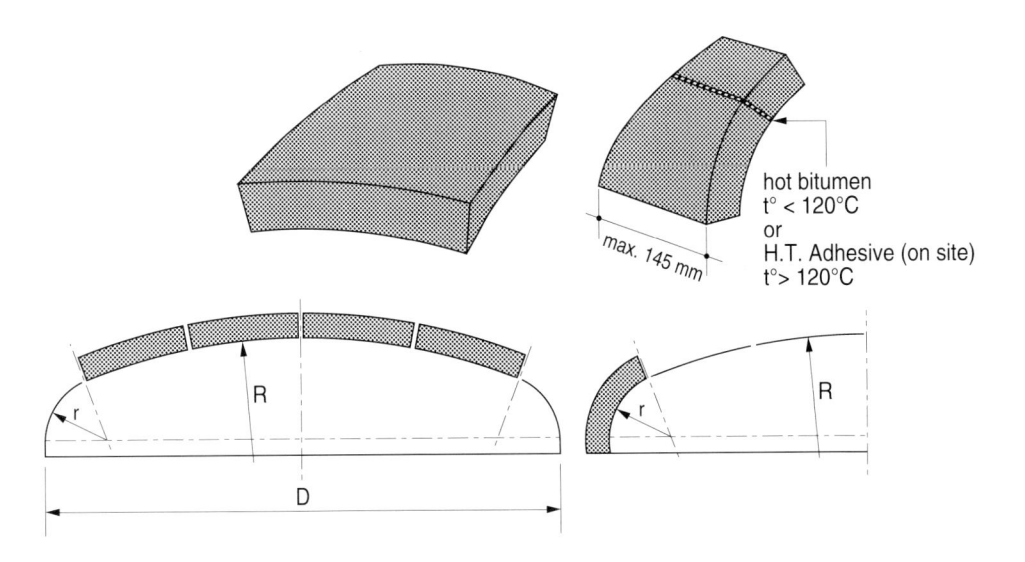

CHAPTER III.3 FOAMGLAS® CELLULAR GLASS FABRICATED ELEMENTS

Fig. III.12. SRS and SHS segments.

Special non standard vessel head segments can be fabricated for non classical vessels.

Cone Segment: CSG

Cone segments (CSG) can be fabricated to insulate conical vessel bottoms and concentric reducers when their insulated diameter exceeds 450 mm.

III.4

REVIEW OF THE MAIN FOAMGLAS® INSULATION CONCEPTS

III.4.1 PRELIMINARY CONDITIONS BEFORE INSULATION APPLICATION

Before starting to install FOAMGLAS® insulation in the industrial field, some preliminary conditions should be met so that a high quality application may be achieved.

1. FOAMGLAS® cellular glass slabs and elements should be transported and stored vertically. The packages should be protected from the weather during storage.

Time and money can certainly be saved by avoiding breakage of FOAMGLAS® cellular glass elements although in many cases, a broken element can be used provided the parts still fit together and allow a correct application making a well-butted or well-filled joint, with a sealer suitable for the required conditions.

2. As it is well known, cellular glass does not absorb moisture but a dry outer surface is necessary to allow a good adherence of the adhesives and coatings generally used in the insulation system.

3. For the same reasons, the surface to insulate and accessory materials should also be dry. Depending on the local conditions, this may be a requirement that the temperature of the surface to be insulated is not lower than a certain level to avoid condensation of water. A minimum temperature may also be needed to obtain the adhesive workability.

4. The surface should be clean and free from traces of grease, rust and dust. If the surface is corroded or otherwise unsuitable, it shall be sandblasted to grey metal and primed with a suitable primer. The sandblasted area shall not exceed the surface that can be primed the same day. The primer must be thoroughly dry and the surface shall be inspected before the application of the insulation.

The primer shall be compatible with the adhesive if an adhesive is used. As indicated in the corresponding data sheet a 10 times diluted PC® 56 ADHESIVE is used as PC® 56 ADHESIVE own primer on absorbing surface. PC® 56 ADHESIVE should not be applied with other primers. Regarding PC® 88 ADHESIVE, no primer is needed.

When needed, the primer shall be applied on attachments, supports and other parts before the application of the insulation. These requirements are more important for an adhered specification than for a dry applied system that is kept in place by mechanical fixings such as bands or straps.

5. Hydrostatic, radiographic and other tests should have been completed before the insulation is applied. It is clear that access to the insulated surface may be needed for these operations and that dismantling and repairing insulation does not improve its quality. One notable exception to this rule consists in the large tank base ring insulations which of necessity are installed before the hydrostatic tests of the tanks.

6. The temperature limits of the accessory products should be respected during storage and application. With some products like PC® 88 ADHESIVE that support negative temperatures but require positive temperatures, preferably around 20°C, to be worked with easily, a classic method consists in having a small closed storage room electrically heated in which the next day consumption is kept at the required temperature. Depending on the drum dimensions and the type of product, half a day to one day is needed to heat the cold product to about +20°C.

7. Scaffoldings, if needed, will be of such a design and such a stability that they allow the insulation work to be carried out in good conditions. The same will apply to tarpaulins or equivalent.

8. Water and moisture due to rain, snow and more insidiously surface condensation being the classic enemy of any insulation system, particularly those operating at low temperature, it has to be decided what type of protection is needed.

When good weather conditions are expected, it is generally sufficient to seal the final edges of the insulation at the end of the working day, before a shower and before any interruption of the work. But in difficult cases, tarpaulins may be necessary. A typical example could be a vessel roof to be insulated during rainy conditions.

9. Pipings and other elements bridging long spans should be adequately supported to avoid undue movements or stresses on the insulation. In the same way, expansion joints, bellows, etc. shall be positioned to avoid any undue stresses in the insulation.

The preliminary conditions apply to practically all types of FOAMGLAS® insulation systems, whatever the temperature, although the absence of water is more important for low temperature applications.

III.4.2 COLD APPLICATIONS: PIPING AND EQUIPMENT

Requirements

The main requirements are usually:

- Thermal requirements:
  - Heat transfer limitation
  - Surface condensation limitation
  - Eventually temperature process control
  - Fire protection
  - Possibly non absorption of flammable liquids

  Obviously only some of these requirements apply in each specific case
- Moisture resistance
- Corrosion resistance
- Possibly chemical resistance
- Mechanical resistance.

FOAMGLAS® Insulation System

The importance of moisture resistance should be pointed out for low temperature applications.

The insulation system usually consists of

- A possible anti-corrosion layer applied on the insulated surface
- A possible anti-abrasive layer if needed
- One or several layers of FOAMGLAS® cellular glass dry applied with staggered joints
- A sealer applied in the joints of the outside layer

- Straps to keep the system in place
- A fabric reinforced coating or metal jacket.

Fully adhered systems are not common for low temperature systems applied on piping, with the exception of chilled water lines. This is explained by the fact that adhesives which work at very low temperature are very unusual and are expensive. The only industrially produced adhesive that can be applied with FOAMGLAS® insulation down to service temperatures of –50°C is PC® 88 ADHESIVE; a few other adhesives are found acceptable for temperatures down to –30°C. Of course, this problem does not exist for chilled water lines working at about +2°C. It is easy and safe to maintain pipe coverings on a pipe line by using straps at the rate of two straps for one standard length (0.60 m) of insulation.

For equipment, such as vessels working at low temperatures, the same situation applies.

In the case of vertical axis vessels, adhered insulations are often considered for the top part of the vessel as a reinforced protection against water intrusion particularly if some mechanical abuse is foreseen. Of course this assumes that the service temperature does not preclude the use of adhesive.

Anti-corrosion Layer

Anti-corrosion layers or other anti-corrosion systems are applied at the discretion of the owner or the design engineer who are in the best position to evaluate all the factors involved in the decision.

Some general remarks can be made:

- Corrosion development is slower at low temperature
- Water or moisture should not penetrate a well designed and installed FOAMGLAS® cellular glass system provided it is not subject to damages or to conditions for which it was not designed
- In the case of an adhered system, water or moisture penetration is theoretically impossible
- The period of time during which the installation will stay unprotected before the insulation system is installed and completed is also a factor to take into consideration, in relation to the prevailing climatic conditions.

In severe service conditions applying an anti-corrosion layer may be a very reasonable choice, its cost being moderate compared to the potential problems associated with a corroded piping and equipment.

Anti-abrasive Layer

In each case, it has to be decided if an anti-abrasive layer is required. Its function is double: first to reinforce the walls of the surface open cells of FOAMGLAS® insulation and secondly to protect the possible anti-corrosion layer applied on the surface from being damaged by the cell walls.

The real need of these two functions remains questionable in the majority of the cases: when the relative movements of the insulation and the support are moderate and only take place occasionally, the anti-abrasive layer can be spared. But the anti-abrasive layer should be applied in two cases:

1. When the piping and the insulation are subjected to permanent or frequent mechanical vibrations, for instance on a piping connecting a piston compressor.
2. When frequent important relative movements between the support and the insulation are expected, for instance in the case of insulation systems cycling frequently with large temperature variations: a process line cycling from ambient temperature to low temperature would be a typical example.

The need of an anti-abrasive layer should certainly be analysed when the temperature is lower than −100°C.

PC® ANTI-ABRASIVE COMPOUND 2A is generally specified except in the cases, such as liquid oxygen line, where incombustible products are imposed. In this case, PC® HIGH TEMPERATURE ANTI-ABRASIVE should be used.

Thickness, Number of Layers and Straps

The thickness will be determined in relation with those thermal requirements previously outlined applied to the specific case. If no specific criterion is available, one can consult the general chapter on thickness determination.

As explained in III.3.1, no absolute rule can be given to specify the number of layers, but general guidance can be presented as follows:

Down to −40°C, a single layer insulation system is generally sufficient; for temperature ranging from −40°C to −170°C, a two layer insulation system is generally advisable, provided sealer in the joints of the outside layer is not exposed to excessively low temperatures.

For lower temperature, a three layer insulation system has generally to be considered. It is exceptional that a system of more than three layers has to be adopted.

It should always be verified that the relative thickness of the outside layer leads to a temperature profile in winter conditions which does not exceed the technical limitations of the sealer placed in this outside layer.

The joints between the different layers should be staggered.

If staggering the joints in the same layer presents some advantages like lengthening the path of a possible water entry, it should not be used when rigidity of the insulation system presents drawbacks such as near fixing points, tees and other particular points.

Obviously, each layer has to be separately fixed with straps; they should be in stainless steel (ANSI type 304 or BSI 304 S16) at least for the outside layer. Their section should not be less than 12.7 mm × 0.4 mm.

If fire requirements are not imposed self-adhesive fibre reinforced tapes are generally acceptable for inner layers. Vessels and equipment should preferably have stainless steel straps.

Sealer

The resistance to moisture penetration certainly represents one of the most important requirements for cold applications. Since cellular glass itself does not absorb either water or water vapour, the only possibility for water or vapour penetration concerns the joints. Due to the dimensional stability of FOAMGLAS® insulation and its low coefficient of thermal expansion, the joints are fairly stable and do not experience substantial movement. For this reason, it is generally sufficient to seal the joints of the outside layer (or the only layer if there is only one). PITTSEAL® 444 is usually specified. It should be remembered here that the resistance of a well-filled joint with a good sealer substantially exceeds the resistance of a classical water vapour barrier or water vapour retarder since the path to cross corresponds to the full thickness of the exterior insulation layer – i.e. generally 40 mm or more – compared to a few millimetres for a vapour barrier. Moreover, the sealer in the joint is better protected from mechanical damages than a thin layer applied on the exterior face of the insulation. Finally, the old practice of applying sealer between the different layers of insulation should not be followed in the case of FOAMGLAS® insulation. It was advisable when using permeable insulation materials that had to be protected from moisture as much as possible but it is not needed when using FOAMGLAS® cellular glass. Moreover, especially for very low temperature applications, the sealers could be brought to very low temperatures at which they may become brittle and fragile and could create problems. They give an artificial impression of safety.

Sealing the joints of an internal layer should only be considered in exceptional cases when damage is expected to the outside layer. Moreover, it is generally very difficult to find sealers that will keep their flexibility at the temperature of an internal layer.

For the same reason, no sealer should be applied between layers.

Reinforced Coating or Metal Jacket

Since a properly installed exterior layer of FOAMGLAS® cellular glass with the joints filled with sealer constitutes an excellent vapour and weather barrier, it is in principle sufficient to protect a low temperature pipeline with a weather barrier. A reinforced mastic coating or a metal jacket correspond to the requirements of the system. The reasons to choose a reinforced coating or a metal jacket has been outlined in II.2.3. It should nevertheless be mentioned that some designers prefer to install an actual vapour barrier such as a low permeability reinforced mastic since they estimate that, at least for some jobsites, the filling of the joints of the exterior layer with sealer will not be executed with enough care. Such a position should be taken only in extreme cases and not as a rule.

Some other designers also specify a cell filler (generally PITTCOTE® 300) on the FOAMGLAS® cellular glass surface before the metal jacket is applied. This solution is advisable in high wind areas or when many significant temperature variations are expected.

III.4.3 HIGH TEMPERATURE APPLICATIONS: PIPING AND EQUIPMENT

Requirements

The main requirements are usually

- Thermal requirements:
    - Heat transfer limitation
    - Protection from burns
    - Temperature process control
    - Resistance to rapid temperature variations
    - Protection of structural material against excessive temperature
    - Fire protection
    - Non absorption of flammable liquids

Obviously only some of these requirements apply in each specific case
- Moisture resistance if the temperature is lower than 100°C
- Corrosion resistance
- Chemical resistance
- Mechanical resistance.

It should be noted that the moisture requirement is not mentioned for temperatures exceeding 100°C. Even if moisture should penetrate an idling high temperature FOAMGLAS® cellular glass application, it would be able to escape when the installation goes on stream provided there is no exterior perfect vapour barrier and no joint sealer.

FOAMGLAS® Insulation System

The insulation system usually consists of

- A possible anti-corrosion layer applied on the insulated surface
- A possible anti-abrasive layer if required
- One or several layers of FOAMGLAS® cellular glass dry applied with staggered joints
- Straps to keep the system in place
- A fabric reinforced coating or metal jacket.

In the case of piping and equipment, fully adhered systems are practically never used for high temperature applications. The fixing method based on the use of straps at the rate of two per standard length (0.60 m) is easier and faster to apply.

In the case of vertical vessels, the adhered specifications may be used particularly for the roof when the temperature does not exceed the range of the adhesive and when special conditions, like possible mechanical abuse or many inlets or outlets, nozzles, etc. are foreseen. PC® 88 ADHESIVE can be used as a bedding compound between FOAMGLAS® cellular glass and the vessel head, to obtain a compact system. In this case, the tensile stresses are low and PC® 88 ADHESIVE temperature limit can be increased from 80°C (as an adhesive) to higher values. Successful applications operating at 150°C have been reported.

Anti-corrosion Layer

The decision to apply an anti-corrosion layer is made by the owner or the design engineer who can appreciate in each particular case, whether it is necessary. But consideration should be given to the fact that corrosion is accelerated by higher temperatures – particularly in the range from 60 to 100°C – and therefore anti-corrosion layers should often be used for hot applications.

Anti-abrasive Layer

The reasons to use anti-abrasive layers are basically the same as for low temperature applications: when mechanical vibrations or frequent important relative movements between the surface and the insulation are expected. PC® HIGH TEMPERATURE ANTI-ABRASIVE is used to fulfil this function when necessary.

Thickness, Number of Layers and Straps

The thickness determined according to the quoted requirements influences the number of layers. The choice of the number of layers should not be governed by the idea of avoiding moisture penetration by long Z joints. On the contrary fewer numbers of layers will allow the moisture to escape more easily, should it enter the system. The number of layers is also influenced by the consideration that an excessive temperature difference between the two main faces of a FOAMGLAS® cellular glass slab or pipe covering will result in a thermal stress, particularly if the slab or fabricated element is restrained on the edges. This may cause cracking.

When FOAMGLAS® cellular glass is not restrained on the edges or by tight strappings, a sudden temperature difference of about 120°C is acceptable between the two faces of the element. Up to about 150°C pipe operating temperature, a single layer is generally recommended. Pipe insulation may be secured by fibre-reinforced tape if a metal jacket is used or by stainless steel bands if a reinforced weather barrier coating is used. Two straps are applied per standard length, their section should be at least 12.7 mm × 0.4 mm.

If several layers are applied, the joints are staggered between them to reduce the potential small thermal short circuit.

As already explained, sealer should not be applied in the joints.

Reinforced Coating or Metal Jacket

As already explained, the moisture resistance requirement is not necessary for FOAMGLAS® cellular glass installations operating at temperatures higher than 100°C and consequently a weather barrier is sufficient. Thus a low permeability mastic is not needed and an ordinary fabric reinforced coating or a metal jacket can be used.

III.4.4　DUAL TEMPERATURE APPLICATIONS: PIPING AND EQUIPMENT

Requirements

In these installations, the temperature frequently fluctuates from high to low temperature and back.

These insulation systems should thus be designed to meet the requirements of both low and high temperature applications and are among the most delicate to design and to install.

FOAMGLAS® Insulation System

The insulation system usually consists of

- A possible anti-corrosion layer applied on the surface to be insulated
- A possible anti-abrasive layer
- One or several layers of FOAMGLAS® cellular glass dry applied with staggered joints
- A sealer applied in the joints of the outside layer
- A fabric reinforced coating or a metal jacket.

Anti-corrosion Layer

These installations operate in medium or high temperatures that are prone to corrosion and also experience low temperatures that favour moisture penetration. Consequently corrosion resistance constitutes a stringent criterion and it would generally be a false saving to eliminate a good anti-corrosion layer on the steel surface.

Anti-abrasive Layer

Provided the operating temperature range is wide enough – for instance more than 150°C – an anti-abrasive layer should be recommended since thermal movements will be frequent.

Moreover, damage of the anti-corrosion layer by the surface open cells of FOAMGLAS® cellular glass could lead to corrosion problems and should be avoided.

Thickness, Number of Layers and Straps

To choose the insulation thickness, the designer should consider all the relevant criteria and determine the value that meets all of them. Depending on the low and high operating temperatures, the dominating requirements will be those associated with one case or the other. Possibly, conflict between the values provided by the different requirements may arise: this should be solved by engineering evaluation.

Since the joints of the exterior layer have to be filled by sealer, the temperature limitation of the sealer when operating at low and at high temperature will have to be verified and may impose a modification of the number of layers or their respective thickness to adapt the temperature profile to the sealers possibilities.

When more than a single layer is installed, the joints between the different layers should be staggered. Within one layer, the joints are usually not staggered to avoid too much rigidity in a system aimed at withstanding frequent temperature variations. For the straps, stainless steel will be used for the outside layer and for the others if fire requirements are severe.

If there is no fire requirement, self adhesive fibre reinforced straps are acceptable for the inside layers.

Sealer

The requirement of moisture resistance for the low part of the temperature cycle is considered as predominant and consequently the joints of the outside layer are filled with a sealer. Usually PITTSEAL® 444 is applied, preferably with a gun. Special attention will be required to make sure that the joints are filled.

Reinforced Coating or Metal Jacket

Due to the stringent requirements on moisture resistance, there is a trend to prefer a fabric reinforced coating to a metal jacket.

III.4.5 CHILLED WATER PIPELINES

Short Description and Requirements

These lines carry water for air-conditioning cooling and usually operate at temperatures around +2°C. These are mainly installed within buildings such as office buildings, administrative centres, hospitals, etc. For this very specific application, five requirements generally dominate:

- Heat transfer limitation
- Surface condensation limitation
- Resistance to moisture penetration
- Resistance to corrosion of the pipe
- Good fire resistance, since these pipelines are installed within buildings and could act as a wick to propagate fire through walls and floors.

Other requirements may be added but they are normally less critical.

FOAMGLAS® Insulation

Due to the importance of the moisture resistance, a compact system is often foreseen. With this concept, an adhesive is used to adhere the pipe coverings to the pipeline and to fill the joints between the pipe coverings. When using an adhesive with a high resistance against water vapour transmission, a system presenting a high resistance against moisture penetration is obtained.

It consists of:

- An anti-corrosion layer applied on the insulated surface
- A single FOAMGLAS® cellular glass layer adhered to the pipeline with filled joints
- Straps to keep the system in place, as a supplementary precaution
- A fabric reinforced coating or a metal jacket.

Anti-corrosion Layer

Due to the corrosion possibility, the anti-corrosion layer is often used. National standards even recommend it in some countries in spite of the fact that a well installed, fully adhered system does not allow moisture to penetrate it.

Anti-abrasive Layer

Since this system is adhered, no anti-abrasive layer is required.

Thickness, Number of Layers and Straps

Owing to the very moderate operating temperature, the thickness usually amounts to 30 to 50 mm, 30 mm corresponding to very small diameters for which the difference between actual thickness and equivalent thickness is substantial.

When the specific conditions apply, Tables III.5a and III.5b, column 0°C of III.2.1 can be used as a guideline.
Greater thicknesses are only needed for exceptionally high relative humidity.

In these conditions, the system consists in only one layer.

When using an adhered system, one could question the need of straps. Actually they should be considered as complementary to safety or a help during installation. Except if fire requirements are extreme, fibre-reinforced tapes are acceptable.

Reinforced Coating or Metal Jacket

The majority of chilled water pipelines are installed within buildings; a special coating PC® 74A has been developed which offers the advantage of outstanding fire behaviour: it is classified A in Germany according to DIN 4102. It is applied with a glass fabric reinforcement. A metal jacket can be applied as an alternative protection.

Important notes:

1. For about two years, an alternative cellular glass based system has been developed: FOAMGLAS® Alu pipe. This product consists of FOAMGLAS®

cellular glass pipe covering pre-applied on an 0.2 mm thick aluminium foil that is used as a jacket. An overlap is provided on the longitudinal joint and aluminium straps are used to hold the system.

FOAMGLAS® Alu pipes are installed with strips of PC® 18 applied with a gun to fill the longitudinal and the circumferential joints.

During this work a moderate excess of PC® 18 is applied near the part of the joints in contact with the pipes with a view of adhering locally to the pipe and creating partitions of the reduced space between the pipe and the pipe covering. This helps to limit the migration of water and water vapour along the pipe to a large extent and to reduce the corrosion danger accordingly, should a water vapour penetration occur.

Straps – preferably in aluminium to avoid compatibility problems with the aluminium jacket – are then placed to secure the system in place at the rate of two straps per standard length. For aesthetical reasons, self adhering aluminium tapes are applied on the circumferential joints.

Obviously when choosing FOAMGLAS® Alu pipe, no further coating or metal jacket is used.

This system enables fast application, but does not present the same redundant safety (anti-corrosion layer, adhered system, jacket) as the compact system (Fig. III.13).

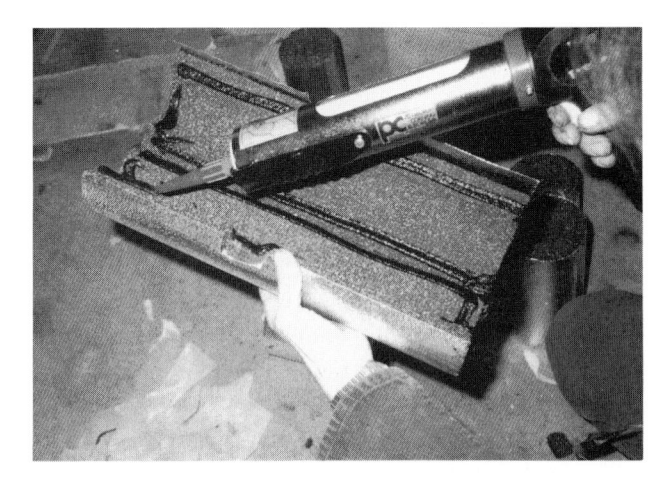

Fig. III.13. FOAMGLAS® Alu pipe.

2. To be complete, we should mention that FOAMGLAS® cellular glass pipe coverings on chilled lines are also installed with the classic system: not an adhered system but sealer in the joints. Provided the sealer is well chosen, for instance PITTSEAL 444, and correctly applied on the total depth of the joints, this system works very satisfactorily.

III.4.6 OVERFIT/RETROFIT

Description and Requirements

This system consists in applying an outside layer of FOAMGLAS® cellular glass on a high temperature insulation made of permeable insulation. It allows a reduction of heat transfer, a diminution of surface temperature and, if needed, a drying out of the moisture that possibly has penetrated into the permeable insulant.

Three requirements generally dominate the others: to reduce the heat transfer to the desired level, to limit the outside surface temperature to an acceptable level and to have an interface FOAMGLAS® cellular glass/permeable insulating material temperature of at least 100°C to make sure that the existing layer lowest temperature exceeds 100°C.

The reason of the third requirement is explained in some details in V.4.4 and V.4.5.

Other requirements may apply.

FOAMGLAS® Insulation System

The insulation system is somewhat different from the previous ones since the purpose is to retrofit an existing insulation system and achieve an open system enabling moisture to be driven out of it without allowing water to penetrate it. After having taken away the existing metal jacket or the fibre reinforced coating, the following layers are applied:

- A layer of FOAMGLAS® cellular glass dry applied on the existing material
- Dry joints, no sealer in them
- Straps to keep the new layer in place
- Metal jacket installed in such a way (for instance regarding sheet overlapping) that water will not penetrate the system but that the water vapour can escape.

Anti-corrosion Layer

The principle of overfit/retrofit being to install a complementary layer of insulation on an already insulated line, no access is given to apply an anti-corrosion layer on the steel line. This implies that the system should not be installed on already dangerously corroded lines. It can be assumed that the drying out process will eliminate enough water to avoid further extensive corrosion but the system cannot repair already corroded lines.

Anti-abrasive Layer

No anti-abrasive layer is needed.

Thickness, Number of Layers and Straps

As explained in III.2.4, a thickness of 50 mm is generally required to obtain the drying out phenomena. Somewhat higher values could be needed if the heat transfer requirements or the surface temperature limitation requirements are particularly stringent but it is very unusual to exceed 120 mm.

Consequently a single layer application is required. Due to the desire to favour moisture elimination it would be wrong to install two layers, lengthening the route of escape.

Since we are considering an outside layer and since the pressure built up during the drying out process can be considerable, straps should be only in stainless steel. The dimensions of these straps and the type of stainless steel have been previously defined.

Sealer

From the principle of the system, it becomes evident that no sealer should be applied. It could lead to serious troubles, if the permeable insulation contains large amounts of moisture or water, particularly when it operates at high temperature.

Reinforced Coating or Metal Jacket

A tight fabric reinforced coating should be avoided to facilitate vapour water elimination.

A metal jacket should be installed making sure that it is water tight but not vapour tight. To obtain this, the joints on the higher part of it or where the water will normally enter, should be tightly fixed. The joints on the lower part or where the water – even taking into consideration the driving force of the wind or other similar factors – will not enter, should be kept sufficiently open to enable water vapour to escape.

Drain holes at the lower part of the metal jacket to facilitate the evacuation of water are advisable.

III.4.7 UNDERGROUND LINES

Description and Requirements

The underground insulation system using cellular glass insulation, consists in applying the FOAMGLAS® cellular glass pipe covering on the pipeline and after having applied a coating, in burying the whole system directly in the ground. This method, based on cellular glass mechanical strength, eliminates the need for concrete tunnels and provides the corresponding safety.

This application has the following requirements:

- Limitation of the coating temperature to an acceptable value
- Limitation of the heat transfer
- Limitation of the temperature variations of the fluid between the beginning and the end of the line for process reasons
- Resistance to moisture penetration in the system
- Resistance to corrosion of the pipe
- High mechanical strength able to carry the expected load.

FOAMGLAS® Insulation System

It consists of a drainage if the water table level requires it and

- An anti-corrosion layer
- An anti-abrasive layer
- One or several layers of FOAMGLAS® cellular glass dry applied with staggered joints
- Exceptionally, a sealer
- Straps to keep the system in place
- The coating, usually PITTWRAP®
- The backfill.

Drainage

If the ground water table is higher than the lowest part of the insulated pipe, a drainage should be provided.

It can be made with a thick layer of gravel on which a thick layer of sand is placed. The purpose is to reduce the amount of water in the ground to limit moisture penetration danger, should a weak point exist in the coating. It should be remembered (see II.1.5) that the resistance of cellular glass to water decreases when temperature increases. It is consequently not recommended to subject cellular glass to prolonged contact with hot water or even more with steam. Adequate drainage, is a useful complement to a well-applied high quality coating.

The purpose of the sand layer, under and around the insulated pipe consists of protecting its coating from gravel or stones that could damage it and create weak points for water penetration. It also creates a slip layer allowing small movements when the system is put into service.

Anti-corrosion Layer

Although the decision to specify an anti-corrosion layer and/or a galvanic corrosion protection has to be made by the design engineer, it should be mentioned that applying an anti-corrosion layer is generally a wise decision since these systems work at temperatures where corrosion presents a real danger. Moreover, should water penetrate them, it would only be seen too late when a problem arises.

Although water entry should not occur when a properly designed system is installed, it can always be subjected to penetration due to mechanical abuse often caused by civil works in the neighbourhood.

Finally, access to underground lines to repair them is obviously more difficult than for above ground lines.

Anti-abrasive Layer

In the FOAMGLAS® cellular glass underground piping system, the cellular glass and its coating are in close contact with the ground at least when the ground has had the time to settle. Consequently when the pipe experiences temperature variations, it expands and contracts, moving within the FOAMGLAS® cellular glass pipe covering.

For this reason, it is recommended to apply a layer of anti-abrasive on the inner face of the FOAMGLAS® cellular glass pipe covering unless the temperature variations are very seldom. It is important to avoid any deterioration of the anti-corrosion layer.

Thickness, Number of Layers and Straps

The thickness will be determined according to the method presented in III.2.5. Special attention should be paid to the choice of the ground thermal conductivity. In the case of long underground lines, several different thermal conductivities on various locations resulting in several different insulation thicknesses may have to be considered. Very often a single layer system is sufficient but applications in two layers are not unusual when the operating temperature associated to the requirements justifies it (Fig. III.14).

Straps are usually made of glass fabric reinforced self adhesive tape except in the case of the inner layer of a double layer system, when temperature requires stainless steel straps.

Fig. III.14. *Norenco Corporation.* Insulation of one of the longest high temperature underground steam lines serving one user.

Sealers

Normally no sealer is applied, the high temperature preventing it. If the temperature remains moderate, lower than about 80°C, sealers may be applied in the circumferential and longitudinal joints, as an extra safety. It is not often necessary. PITTSEAL® 444 would be chosen and be applied with a gun.

Coating

PITTWRAP® is applied in a cigarette type wrap and is heat-sealed at the overlap of the longitudinal direction. Butt joints are covered with a 10 cm strip that is also heat-sealed with a propane torch at the edges around the circumference.

For particular points such as elbows, tees, etc. it is not easy to apply a wrap. Consequently, the PITTWRAP® is replaced by three layers of PITTCOTE® 300 coating reinforced with two layers of glass fabric. It is important to achieve tightness. This implies that seals and overlap areas are properly installed and controlled with care. It is also essential to design them in a way that minimises the danger of water entry, should they fail. Alternatively the PITTCOTE® 300 solution can be applied for the total line.

Important notes:

1. Since access to underground lines is difficult as soon as they are on stream, a special inspection of the work should be carried out before the backfill is poured in place.

2. For the same reasons, special attention should be paid to particular points such as expansion loops, hot boxes, tees, etc.

III.4.8 LOW TEMPERATURE TANK BOTTOM

Description

In this section, large vertical axis tanks used to store liquefied gases such as LPG, butane or propane, ammonia, ethylene, methane (LNG), oxygen or nitrogen are considered.

These tanks usually have a base insulation made of several layers of FOAMGLAS® cellular glass HLB type. Since the insulation system developed for these applications differs from those previously described, the presentation of this section will not be directly comparable to the others.

Requirements

The main requirements can be listed as follows:

- Heat transfer limitation
- Mechanical strength, mainly compressive strength requirement due to the load that the tank base insulation has to carry
- Structural material protection against extreme temperatures. This requirement takes a decisive importance for the liquefied gas, such as LNG, oxygen or nitrogen which are stored at such temperature that normal containment products in direct contact would become brittle
- No ground frost heaving as would occur if a tank was built directly on the ground
- Fire protection
- No absorption of flammable liquid
- In the case of liquefied oxygen or gas stored at a lower temperature, compatibility with liquid oxygen
- Moisture resistance
- Chemical resistance against the stored liquid, should a leak occur.

The number of requirements and the critical aspect of several of them indicate the stringent character of this application.

FOAMGLAS® Insulation System

As explained in II.1.15, the compressive strength of cellular glass is related to the type of capping material selected to fill the surface open cells and provide the full strength of the product.

Different potential capping materials have enabled the development of several tank base insulation systems.

The insulation is applied between the concrete foundation slab and the tank steel base plate; the typical system consists of:

- Concrete foundation slab
- Approved interleaving layer and FOAMGLAS® cellular glass HLB type
- Approved interleaving layer and FOAMGLAS® cellular glass HLB type (As often as required)

- ...
- Approved interleaving layer
- Slip layer
- Distribution slab
- Inner tank steel base plate.

Depending of the type of tank, three interleaving layers have been tested and confirmed by numerous existing constructions:

- Hot bitumen applied in sufficient quantities to fill the FOAMGLAS® cellular glass surface open cells is the more frequent system
- A well defined roofing felt, Pluvex no: 1, Damp Proof Course Class A manufactured to BS 6398 Class A, with a hessian base
- A system composed of PC® 85 POWDER, Marglass Loomstate 327 fabric and PC® 85 POWDER for the case in which strictly inorganic products are required such as liquid oxygen or nitrogen tank bases.

As already mentioned FOAMGLAS® cellular glass HLB type has been specifically developed for tank base insulation. This HLB type (High Load Bearing) is characterised by a very stringent quality control commensurate with the technical risk associated with high tank bottom insulation. Moreover a special purchase agreement has been prepared that details how the acceptance test program should be carried out in the presence of the customer or his representative.

The type of FOAMGLAS® cellular glass HLB to select for a given tank base is based on the maximum working stress that it may experience. It is usual to design for one of the following situations:

- During normal operations (when the stored liquid has a high density)
- During the water test (when the stored liquid has a low density)
- During an earthquake in areas where this possibility has to be taken into account.

Historically, many load bearing applications of FOAMGLAS® cellular glass, and specially tank bases, have been designed with a safety factor of three, based on the maximum working stress and the average published compressive strength.

For very thick insulation requiring several layers, it is advisable to take into consideration a reduction factor corresponding to the "column" effect. This means that several slabs applied as many layers on top of each other, have a lower compressive strength than the average compressive strength of the same slabs tested independently. This is an illustration of the principle that a chain breaks at its weakest point.

Obviously the design engineer should make the final calculations and decisions.

The properties of the different types of FOAMGLAS® cellular glass HLB have been given in II.1.13.

Thickness, Number of Layers and Application Techniques

The thickness will be determined according to the previously expressed requirements. Although they vary substantially in function of the local conditions, the following range of values may be used as guidelines:

Products	Temperature (°C)	Thickness (cm)
Butane	−1	5-10
Ammonia	−33	12-15
Propane	−44	15-20
Propylene	−47	15-25
Ethane	−88	30-50
Ethylene	−104	30-50
Methane	−161	50-80
Oxygen	−183	70-120
Nitrogen	−196	70-120

In the case of some cryogenic liquids, a difference is generally made between two types of gas terminals. Liquefaction gas terminals have storage tanks for which one has to use a liquefaction process and is interested in the reduction of heat transfer to reduce the amount of liquefaction needed. Regasification gas terminal storage tanks for which the reduction of heat transfer is less important since the liquefied gas will have to be regasified, anyhow. In this latter case, avoiding low temperature in the concrete foundation slab or possible freezing and heaving of the ground normally constitute the most important criteria. Needless to say, the failure to meet this requirement could lead to a dramatic accident of the tank foundation. The number of layers can be chosen after considering the total thickness required and the maximum possible thickness of one layer. For FOAMGLAS® cellular glass HLB 115 and 135 types, the maximum thickness is 120 mm or 5 inches, and for FOAMGLAS® cellular glass HLB 155 and 175, the maximum is 100 mm or 4 inches.

When applying FOAMGLAS® cellular glass HLB with hot bitumen, the best technique often called the "swimming in technique" (Fig. III.15) consists in

Fig. III.15. "Swimming in technique".

pouring bitumen on an area slightly larger than one slab of the concrete foundation or on the previous FOAMGLAS® cellular glass layer at the recommended temperature (160°C to 180°C for bitumen 85/25) and in pushing gently the next slab in the bitumen in a way that allows the bitumen to fill up the two joints between the slab and those already installed. Then the slab will be kept in place for about one minute to achieve a bitumen temperature low enough to provide good adhesion and good filling of the surface open cells. The next layer is then applied with staggered joints. This method is aimed at obtaining the FOAMGLAS® cellular glass full compressive strength and well-filled joints between the slabs as narrow and as tight as possible.

When applying FOAMGLAS® cellular glass HLB with Pluvex N°: 1 Damp Proof Course Class A, this roofing felt is dry applied with butt joints, and no overlaps. Following this method, FOAMGLAS® cellular glass slabs are not partially unsupported when bridging areas at different levels. Due to the heavy loads, this could lead to rupture because of excessive flexural stresses. The FOAMGLAS® cellular glass slabs are dry applied on the Pluvex layer, a new Pluvex layer is then applied and the next FOAMGLAS® cellular glass layer is dry applied with staggered joints. The final layer is a Pluvex layer.

When comparing the two methods, it can be said that choosing the bitumen as the interleaving layer provides better tightness, should a leak occur.

The Pluvex method is somewhat faster to apply and does not require hot bitumen being brought in a tank, avoiding also the resultant smoke and the danger of burns.

Recently the hot bitumen method seems to have been favoured in several cases due to the increased importance given to the tightness in case of possible leaks.

Finally it should be mentioned that the settlement of the base is more important when using the Pluvex system than when using hot bitumen.

This is explained by the fact that the Pluvex has to fill the surface open cells (generally during the water test of the tank) before becoming suitable for use compared to the bitumen system that immediately fills the surface open cells.

This settlement is not detrimental provided that the design is prepared accordingly. On the other hand, it would generally be complicated to mix the two types of application technique due to this difference of settlement.

The third method is more specifically aimed at liquefied oxygen or nitrogen tank base insulation for which it is necessary to use only inorganic products (compatibility with liquid oxygen). The reason is obvious in the case of liquid oxygen but it also applies for the other liquefied gases having a liquefaction temperature lower than oxygen. This is caused by the fact that condensation of the liquid oxygen of the air on the support is always possible, if the insulation system is not absolutely tight. Consequently many designers and some national codes impose the oxygen compatibility requirements to all applications operating at temperatures lower than –183°C.

The PC® 85 POWDER is a dry one-component mix of modified calcium sulphate with some fillers. It is actually a special version of the PC® HIGH TEMPERATURE ANTI-ABRASIVE, for which the organic content has been reduced to practically zero.

The Marglass 327 Loomstate fabric, finish interleaving layer is an inorganic glass cloth felt woven without an organic binder.

Without going into details, the application technique consists in applying PC® 85 POWDER on the two faces of the FOAMGLAS® cellular glass slab, levelling them with a rubber scraper to achieve the filling of all surface open cells. The Marglass 327 Loomstate fabric is then dry applied with butt joints before the first layer of PC® 85 POWDER prepared FOAMGLAS® cellular glass, between each layer and after the last layer.

FOAMGLAS® cellular glass slabs are applied in layers with staggered joints.

PART III. HOW TO DESIGN AND INSTALL FOAMGLAS® INSULATION FOR INDUSTRIAL APPLICATIONS

A distribution layer should be used on the last interleaving layer applied on the last insulation layer. It can consist in a layer of fine, dry sand. It is important to choose a sand having a granularity such that no grain could result in a local point load on the distribution layer or even more on the FOAMGLAS® cellular glass layer. For obvious reasons, it should be dry. The distribution layer can also be a concrete layer. In this case a slip layer between the interleaving layer and the concrete layer is needed. This can be achieved by two thin plastic foils or in the case of liquefied oxygen or nitrogen two sheets of copper or aluminium having a thickness of 0.2 mm each.

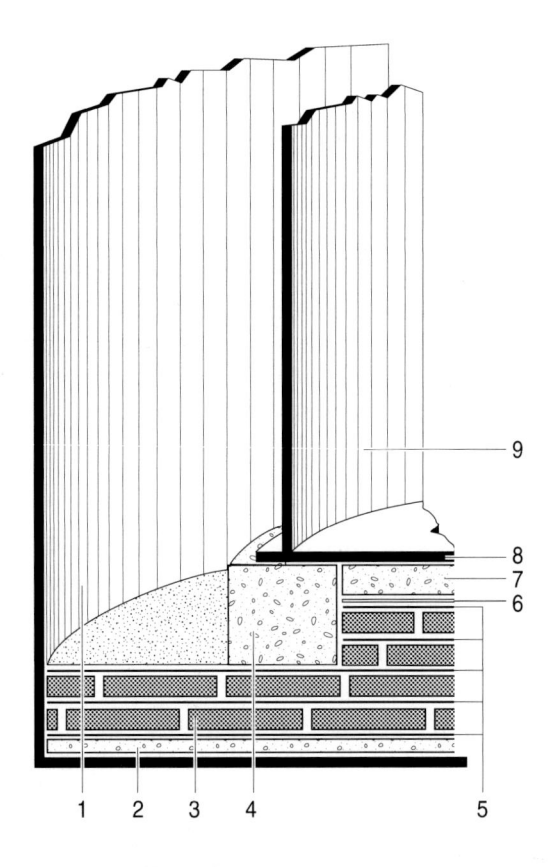

Fig. III.16. Low tanks.

1. Outer tank wall.
2. Levelling screed.
3. FOAMGLAS® HLB insulation.
4. Concrete ring beam.
5. Approved interleaving layer.
6. Slip layer.
7. Repartition slab.
8. Inner tank bottom.
9. Inner tank wall.

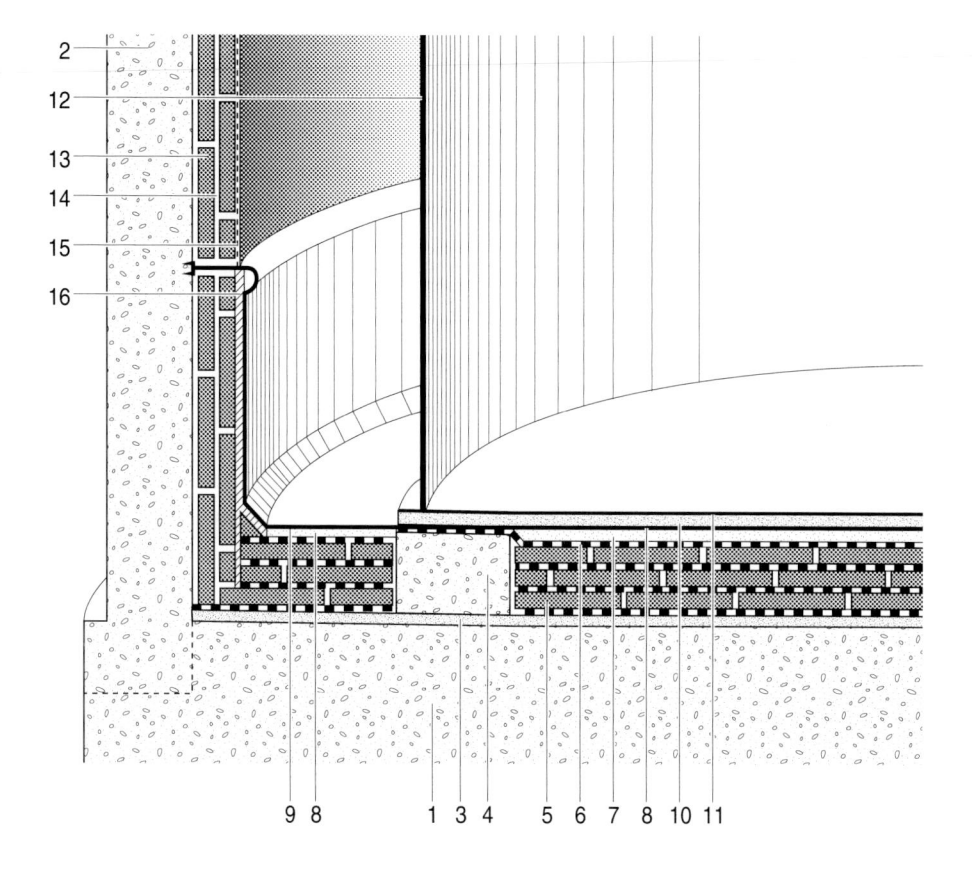

Fig. III.17. High tanks.

1. Concrete or metal base.
2. Concrete or metal outside tank wall (optional).
3. Levelling screed.
4. Lightweight concrete ring beam.
5. FOAMGLAS® HLB tank base insulation.
6. Approved interleaving material.
7. Slip layer.
8. Fine dry sand (optional).
9. Emergency complementary protection (optional).
10. Concrete slab.
11. Tank bottom.
12. Tank wall.
13. FOAMGLAS® tank wall insulation system.
14. Approved flexible adhesives to adhere and seal FOAMGLAS® slabs together and to the concrete outside wall or eventual liner.
15. Optional finish.
16. Slip layer.

Note: The sketch only outlines the general principle of the connection between bottom and wall insulation. The design engineer should prepare in each case the appropriate drawing.

A very important point related to tank base insulation concerns the insulation of the peripheral ring. Under the shell of tank, a stress concentration takes place implying that, except for small tanks (Fig. III.16), the FOAMGLAS® cellular glass slabs should be locally replaced by a special type of FOAMGLAS® cellular glass when possible or an aerated concrete ring beam. For medium size tanks, the ring beam acts as a distribution element.

For high tanks (Fig. III.17) the stress level is generally such that the safe solution consists in letting the concrete (often aerated concrete) go through the total insulation thickness, accepting a local thermal short circuit for mechanical reasons.

A study should be carried out for each tank to choose the most suitable solution. In this, not only the vertical loads and forces should be taken into consideration but also the horizontal forces that can be generated by movements of the tank due to temperature changes and other possible forces.

Often a FOAMGLAS® cellular glass HLB type with higher compressive strength is chosen for the ring than for the central part.

III.4.9 HOT TANK BASE

Description

In this section, the bottom insulation of vertical axis tanks used to store high temperature liquids is considered. Generally speaking, the size of these types of tanks is substantially smaller than the large low temperature storage tanks, but in some cases, hot tanks can be fairly large.

Requirements

The main requirements for this application can be listed as follows:

- Heat transfer limitation
- Protection of structural material against high temperatures
- Fire safety
- No absorption of flammable liquid
- Resistance to corrosion of the tank base
- Chemical resistance against the stored liquid, should a leak occur
- Mechanical strength, mainly compressive strength, since the tank base insulation has to carry the weight of the tank and its contents.

FOAMGLAS® Insulation System

Three FOAMGLAS® insulation systems have been developed for hot tank bottom insulation. They apply respectively for tanks having operating temperatures up to 80°C, 150°C and 250°C. Since they only differ by the number of layers and the type of interleaving material, they can be presented together. Thus starting from the concrete foundation slab, the FOAMGLAS® insulating system is composed of:

- Interleaving material aimed at obtaining a correct FOAMGLAS® cellular glass compressive strength and withstanding the expected temperature
- FOAMGLAS® cellular glass slab type HLB or S3
- Interleaving material
- Possibly another FOAMGLAS® cellular glass layer
- Interleaving layer
- Slip layer
- Distribution slab
- Inner tank bottom.

Depending on its service temperature, the interleaving layer can be

- Bitumen 85/25 or 110/30
- PC® 86T bitumen based product
- PC® 85 POWDER/Marglass 327 Loomstate fabric/PC® 85 POWDER

The distribution layer can be a specific oil-sand mix or fine dry sand.

The choice of the type of FOAMGLAS® cellular glass depends on the level of stress that it will have to support.

For tanks having a height not exceeding about 12 meters and containing a product with a density not exceeding about 1.0, FOAMGLAS® cellular glass S3 can generally be accepted. For higher tanks or tanks containing a product of higher density, the choice of a FOAMGLAS® cellular glass HLB type commensurate with the expected stress level is recommended. It should be noted that in deciding the type of FOAMGLAS® cellular glass, one should take into consideration not only the stress situation in the central part of the tank base but also under the tank wall if FOAMGLAS® cellular glass is applied there.

Thickness, Number of Layers and Application Techniques

The thickness is normally determined on the basis of the heat transfer. Although sophisticated methods exist taking into account, for instance lateral heat transfer, it is generally sufficient to limit the heat transfer by a calculation concerning the central part. One assumes a permanent constant temperature, say 5 or 10°C, in the ground at a depth of for instance 1 meter. In choosing the temperature of the ground, its thermal conductivity and the depth at which temperature is assumed to be constant, the water content of the ground and the level of the water table should be taken into consideration. This being said, the usual FOAMGLAS® cellular glass thicknesses for the different types of hot tank bases are generally as follows:

- Tank operating at temperature up to   80°C: 5 to 10 cm
- Tank operating at temperature up to 150°C: 10 to 20 cm
- Tank operating at temperature up to 250°C: 20 to 30 cm.

In choosing the total thickness it may be economical to directly consider the number of layers since it has a strong influence on the installed cost of the tank base insulation.

For tank operating at temperatures up to 80°C (Fig. III.18), one layer of FOAMGLAS® insulation is generally sufficient; it can be applied with hot bitumen on the concrete foundation slab with the typical "swimming in" method already described for low temperature tanks. With this method, the joints between the FOAMGLAS® cellular glass slab should be thin, tight and well-filled. An alternative solution for the interleaving material between the concrete foundation slab and the FOAMGLAS® cellular glass layer consists in using PC® 86T, a bitumen based product known for withstanding temperatures up to 90°C. On top of the FOAMGLAS® cellular glass layer the distribution layer will have a minimum thickness of 4 cm and will consist of fine oil-sand mix. It is a product made of washed and dried M34 salt free sand mixed on site with approximately 75 litre/m^3 mineral lubrication oil without additives, such as for instance Shell Carnea 100. It is important to select an oil that has a flash point substantially higher than the tank operating temperature. Moreover, no naked flame or welding should be present in the vicinity when applying the oil-sand mix and after.

Since the granularity of the sand has an influence on FOAMGLAS® cellular glass compressive strength, it should be the M34 sand that has following characteristics:

Mesh size (μm)	Sieving retaining (%)
100	99.5
160	60.5
200	11.5
315	0.5

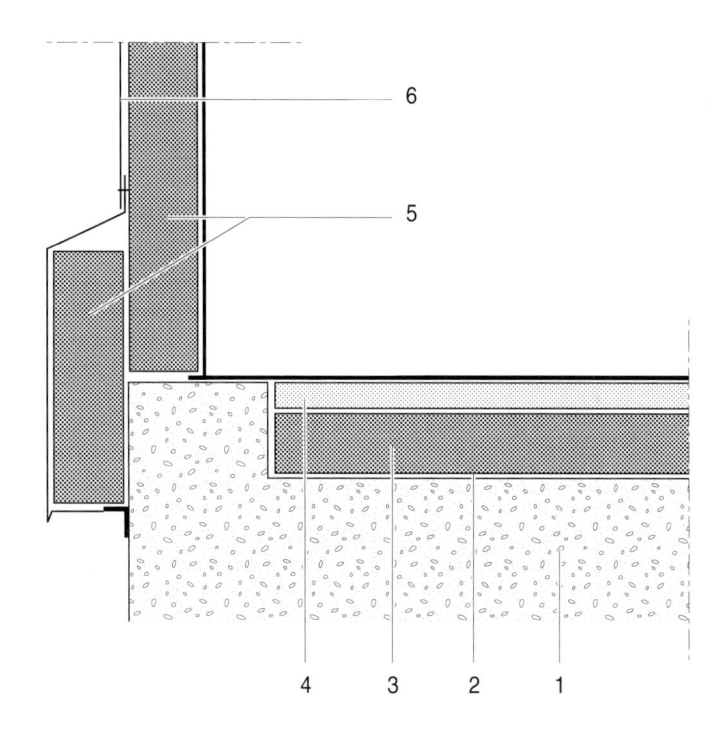

Fig. III.18. Tank base: up to 80°C.

1. Concrete.
2. Hot bitumen 85/25 or 110/30 or PC® 86T bitumen based product.
3. FOAMGLAS® bottom insulation.
4. Oil-sand mix (dried M34 sand + lubrication oil – mineral without additifs).
5. FOAMGLAS® wall insulation.
6. Cladding.

Obviously direct contact of oil sand mix with FOAMGLAS® cellular glass surface open cells cannot provide a compressive strength comparable to hot bitumen or Pluvex.

Tests have shown that with the prescribed mixture, the compressive strength is reduced by a factor of two; this point being considered, the allowable working stress when using FOAMGLAS® cellular glass type S3 with a traditional safety factor of three, should be $0.9/(3 \times 2) = 0.15$ N/mm^2.

Finally it is important to install a ring beam if the stress concentration at the shell / bottom connection justifies it or if not, to provide a device to prevent the load distribution layer from creeping away at the edges.

For tank operating at temperatures up to 150°C (Fig. III.19), one or two FOAMGLAS® cellular glass layers will be specified according to the selected total thickness.

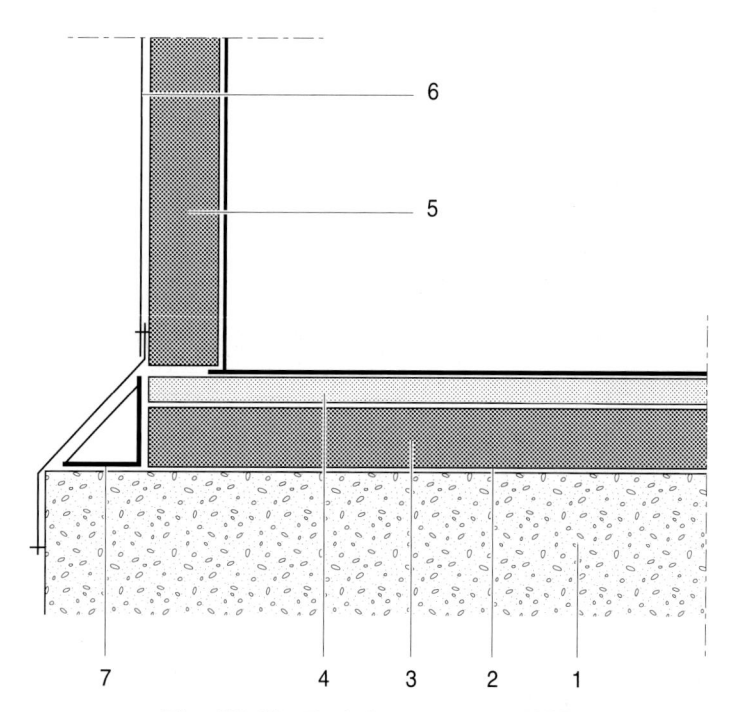

Fig. III.19. Tank base: up to 150°C.

1. Concrete.
2. PC® 86T bitumen based product.
3. FOAMGLAS® bottom insulation.
4. Oil-sand mix (dried M34 sand + lubrication oil – mineral without additifs).
5. FOAMGLAS® wall insulation.
6. Cladding.
7. Annular L-steel.

The first interleaving layer between the concrete foundation slab and the FOAMGLAS® cellular glass layer will be PC® 86T bitumen based product. The same will be used for the interleaving layer between the first FOAMGLAS® cellular glass layer and the second one, if there is a second one. The distribution slab will be an oil-sand mix. Basically the system is similar to the one for tanks operating at a temperature lower than 80°C, except that the use of blown oxidised bitumen is not recommended for the first interleaving layer.

For tank operating at temperatures up to 250°C (Fig. III.20), two or three FOAMGLAS® insulating layers are generally needed to meet the total insula-

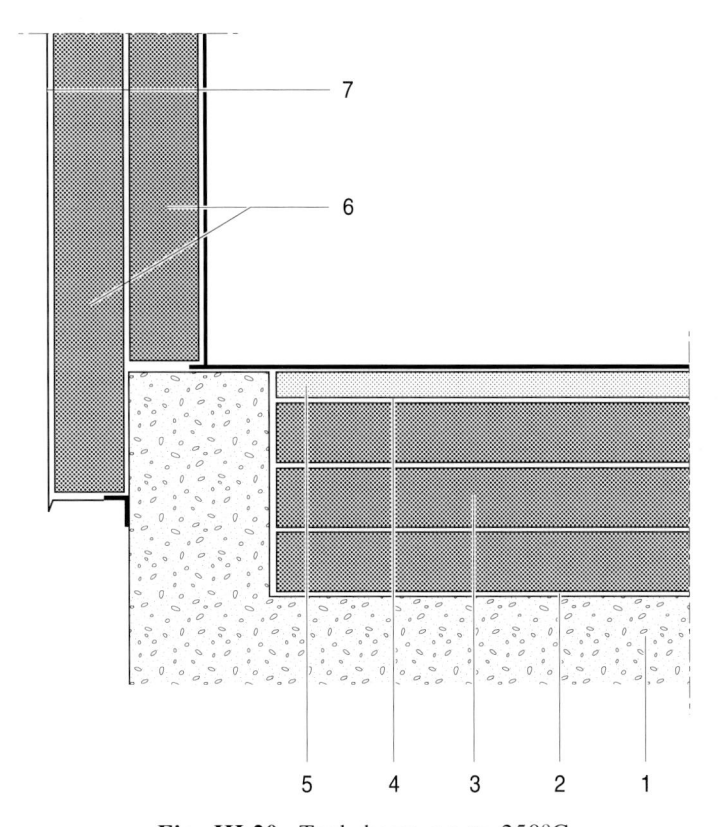

Fig. III.20. Tank base: up to 250°C.

1. Concrete.
2. PC® 86T bitumen based product or Marglass 327 Loomstate fabric + PC® 85 POWDER.
3. FOAMGLAS® bottom insulation.
4. Marglass 327 Loomstate fabric + PC® 85 POWDER.
5. Fine dry sand.
6. FOAMGLAS® wall insulation.
7. Cladding.

tion thickness requirement. The system of interleaving layers to use is different. Before the first interleaving layer is applied between the concrete foundation slab and the first layer of FOAMGLAS® cellular glass, a calculation of the temperature profile has to be made to see if PC® 86T temperature remains, in the worst conditions, lower than 90°C. If the temperature is higher, the PC® 86T cannot be used and has to be replaced by a PC® 85 POWDER/Marglass 327 Loomstate fabric/PC® 85 POWDER system, similar to the one used for liquid oxygen tank base insulation. The same PC® 85 POWDER/Marglass 327 Loomstate fabric/PC® 85 POWDER system is used as interleaving layer between the FOAMGLAS® cellular glass layers and on top of the final layer. It is complemented by a distribution layer of dry fine sand. It may appear surprising that the solution selected for tank bases operating at temperatures approaching 250°C is similar to the solution used for liquid oxygen and liquid nitrogen. The reason is that organic materials are forbidden in this case due to the danger of explosion if contact would occur with liquid oxygen and in the other case because the temperature is simply too high for them.

What has been said about compressive behaviour of FOAMGLAS® cellular glass in contact with oil-sand mix applies for tanks at 150°C or at 250°C as it does for tanks operating at temperatures not exceeding 80°C.

In the case of capping with PC® 85 POWDER/Marglass 327 Loomstate fabric/PC® 85 POWDER, the compressive strength of FOAMGLAS® cellular glass is only reduced by about 10 to 20% compared to the reference method based on bitumen.

III.4.10 LOW TEMPERATURE AND CRYOGENIC TANK WALL AND ROOF

This section is aimed at presenting the FOAMGLAS® insulation systems for the wall and the roof of large vertical storage tanks.

Short Review of the Different Types of Low Temperature and Cryogenic Storage Tanks

The type of tanks has a decisive influence on the method of insulating them; it is therefore necessary to quickly review the most common tank concepts before presenting the FOAMGLAS® insulation systems.

A first distinction concerns tanks with operating temperature not lower than about −50°C often called low temperature storage tanks (butane, propane, ammonia, propylene, etc.) and those with operating temperatures lower than −50°C (ethane, ethylene, LNG, oxygen, nitrogen, etc.) often called cryogenic storage tanks at least when the operating temperature is lower than −100°C.

Regarding the walls, the tanks operating at temperatures not lower than −50°C are generally insulated with a single or multi-layer adhered system.

For tanks operating at temperature lower than −50°C, it would be very difficult and very expensive to find an adhesive performing satisfactorily at these temperatures and the total system would be very expensive. The generally adopted solution consists in building a tank with two walls between which expanded perlite is poured. The insulation can easily be one meter thick and even more.

For the low temperature storage tanks (Fig. III.21), the usual designs are:

- The single wall tank without further protection. The wall insulation is directly installed on the outside face of the tank. This design is phasing out
- The single wall tank with a bund wall built at a distance of 1 to 2 meters from the tank. The bund wall has generally a height that will enable it to store the liquid contained in the tank, should a tank failure occur. It has also the merit to protect the tank against outside aggression (fire, flying objects, etc.). The wall insulation is installed on the outer face of the tank
- The double containment (also called double integrity tank) in which the inner steel tank is completely surrounded – base, wall and roof – by an exterior concrete tank having several functions among which the two main are to contain the liquid in case of a failure of the inside tank and to protect it from outside aggression.

Since a concrete tank is not necessarily absolutely tight, a liner is applied on the inner face of the outside concrete tank. This liner is usually made of steel sheets welded together. Sometimes it is made of an organic coating chosen to be gas tight.

In this case the insulation system can be adhered on the outside face of the inside steel tank or on the inside face of the outside tank (the liner).

In the case of tanks operating at temperatures lower than −50°C (Fig. III.22), one finds more or less the same concepts but adapted to poured in situ insulation:

Single wall tank

Single wall tank with bund wall

CONCRETE
FOAMGLAS® INSULATION
LINER

Double containment tank

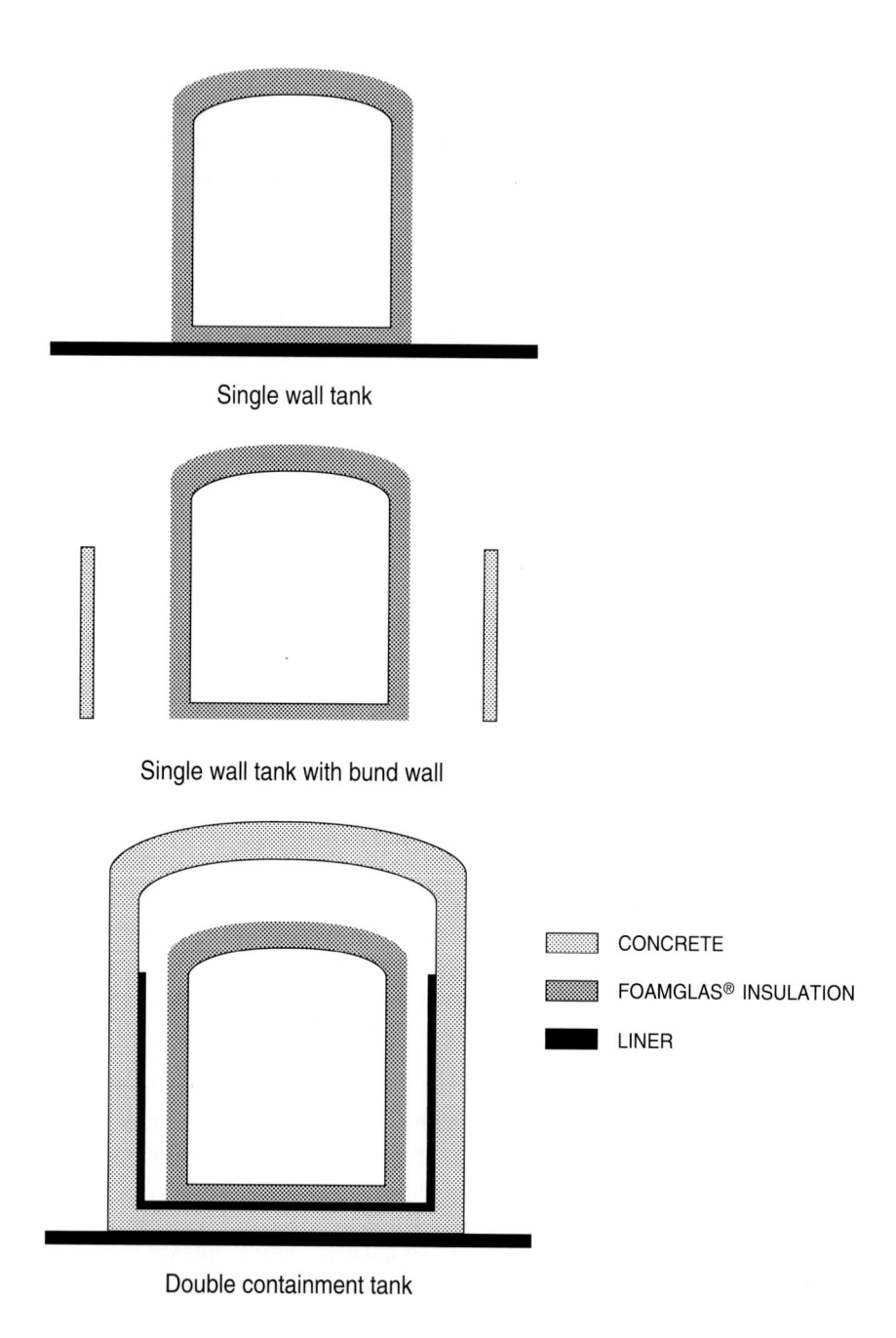

Fig. III.21. Type of tanks operating at temperatures not lower than –50°C.

CHAPTER III.4 REVIEW OF THE MAIN FOAMGLAS® INSULATION CONCEPTS

Double steel tank

Double steel tank with bund wall

INSULATION
CONCRETE
PERLITE
FOAMGLAS® INSULATION
LINER

Double containment tank

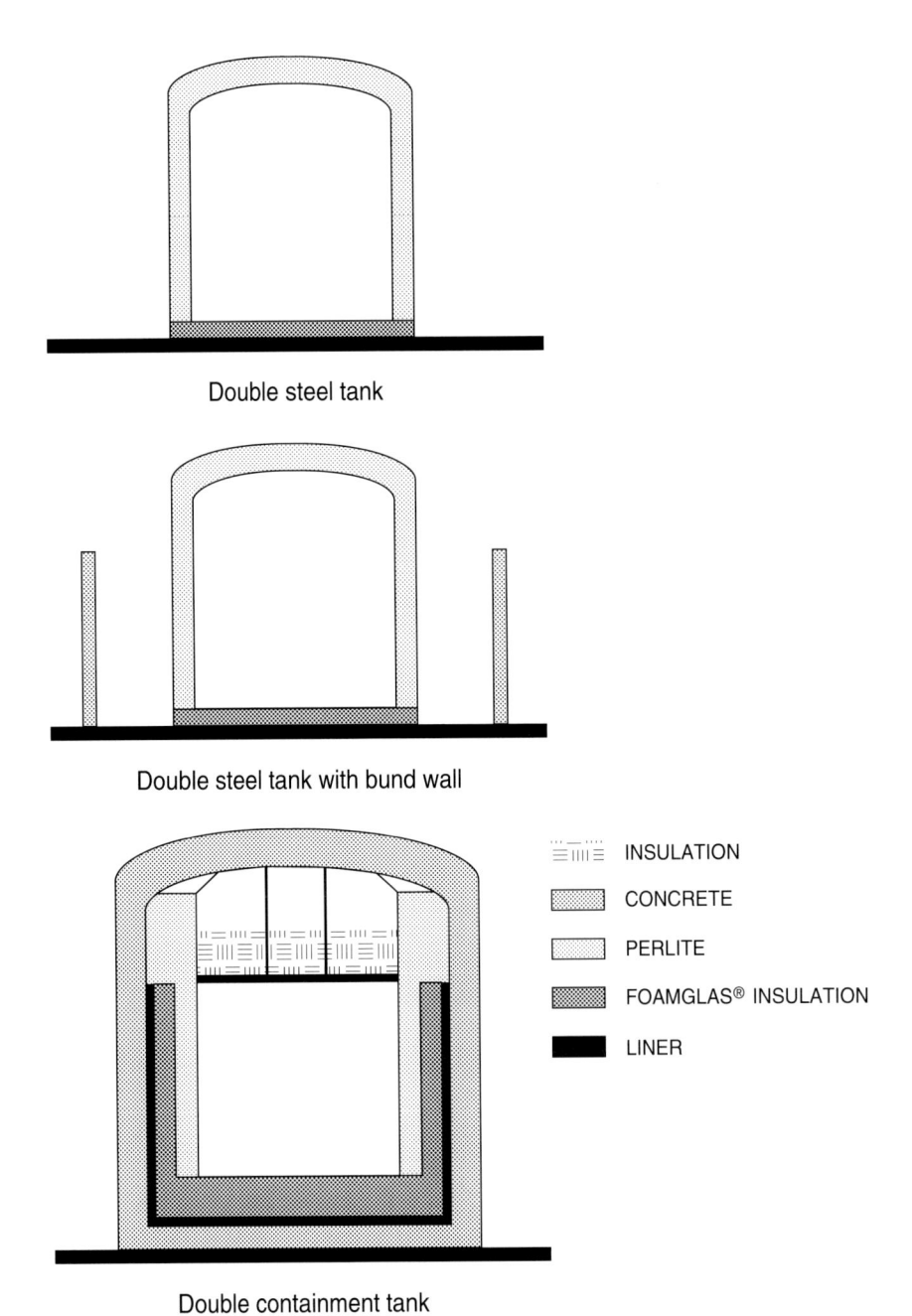

Fig. III.22. Type of tanks operating at temperatures lower than −50°C.

- The double steel tank – the inside tank in special steel to withstand cryogenic temperatures and an exterior tank in normal carbon steel, with poured in situ perlite and resilient blanket between the two tanks
- The same concept with a bund wall around it
- The double containment tank, in which the inside tank is built in cryogenic steel and the exterior tank in concrete with a normal carbon steel or coating liner on its inside face. In this case the insulation is again perlite poured in situ between the two tanks, sometimes completed by a solid insulation/protection system applied on the liner of the outside concrete tank. This secondary insulation is aimed at protecting the liner and the concrete against the cryogenic thermal shock that it will have to withstand if the inside tank should have a major failure.

Regarding the tank roof (Fig. III.23), two concepts can be used: the fixed roof built on the walls or the suspended roof that can slightly move in height within the walls and which is supported by the outer tank.

▓▓▓ FOAMGLAS® INSULATION

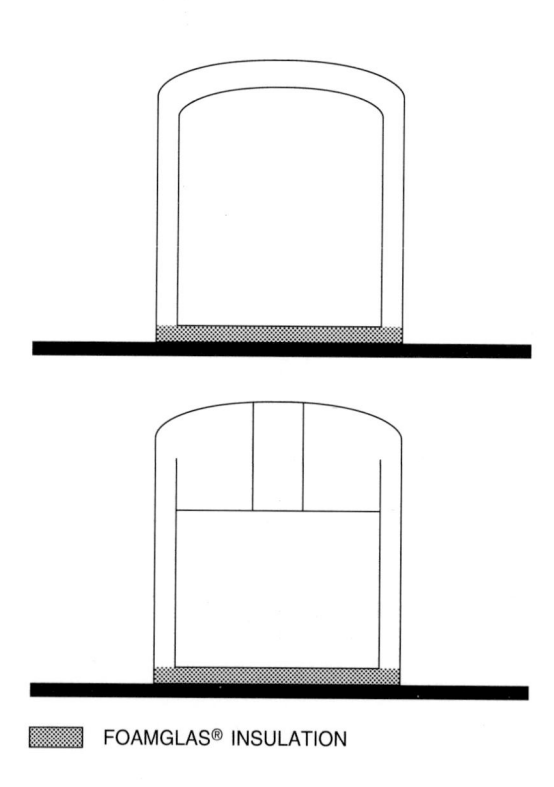

Fig. III.23. Fixed roof, suspended roof.

Depending on the design, the space between the two tanks is filled with air, with inert gas or with the vapour of the stored liquefied gas.

The most common types of tank concepts having been presented, the FOAMGLAS® insulation system corresponding to these tank designs can now be described.

III.4.11 LOW TEMPERATURE TANK WALL AND ROOF

Firstly a low temperature tank on which FOAMGLAS® insulation system is applied on the single tank wall or on the outside face of the inside tank will be considered.

Requirements

The main requirements are usually the following ones:

- Limitation of heat transfer or boil off rate
- Limitation of surface condensation
- Fire protection
- Non-absorption of flammable liquids
- Moisture resistance
- Corrosion resistance
- Chemical resistance
- Mechanical resistance.

It may be worthwhile to explain the mechanical resistance requirement in the case of tank wall insulation. There is obviously very little compressive load on a tank wall insulation. Even for the highest tanks, the weight of the FOAMGLAS® insulation and accessory products amounts to a negligible load compared to cellular glass strength. As an example the pressure exerted by a 40 m column of FOAMGLAS® cellular glass T4 having a density of 125 kg/m^3 will be

$$\begin{aligned} 40 \times 125 &= 5000 \text{ kg/m}^2 \\ &= 0.5 \text{ kg/cm}^2 \\ &= 0.05 \text{ N/mm}^2 \end{aligned}$$

compared to compressive strength of 0.7 N/mm^2 for FOAMGLAS® cellular glass T4.

But it should be kept in mind that steel tanks are "living structures" that deform, under various influences, the main ones being:

- The pressure of the liquid when filling and emptying the tank, or when the liquid level in the tank is moving up and down
- The contraction and expansion of the tank when it is cooled or warmed; particularly on the top part of the tank, the temperature can vary with the liquid level in the tank
- For single tanks, the influence of the sun radiation changing with the position of the sun.

Consequently the mechanical resistance requires that the insulation system is able to follow the surface movement without significant deterioration.

Since both steel and FOAMGLAS® cellular glass slabs are rigid, the requirement can be met by the choice of an adhesive that remains flexible at the operating temperatures. PC® 88 ADHESIVE meets this requirement.

FOAMGLAS® Insulation System

The FOAMGLAS® insulation system consists of:

- A possible anti-corrosion layer on the surface to be insulated
- One or several layers of FOAMGLAS® cellular glass adhered with PC® 88 ADHESIVE, the joints being filled with the same product
- Depending on the situation, a fabric reinforced coating or a metal jacket with the needed straps to keep it in place.

In the past, before the development of PC® 88 ADHESIVE, tank wall insulation has been carried out with dry FOAMGLAS® cellular glass slabs having only the available sealers at this time in the joints.

The system was kept in place by horizontal straps. This system was less safe than the present adhered solution that offers excellent resistance to moisture and water penetration, when properly applied.

Anti-corrosion Layer

Steel plates usually come with a factory applied anti-corrosion layer that is generally somewhat degraded during the construction, due to the manipulation and the welding of the plates together to build the tank. In spite of these deteriorations, it is generally not considered necessary to apply an anti-corrosion layer on the jobsite since the "compact" adhered system should prevent water or moisture ingress to the steel plate. Moreover the operating temperature does not favour corrosion as water turns into ice when the temperature is negative.

Anti-abrasive Layer

With an adhered system, no anti-abrasive layer is needed.

Thickness, Number of Layers and Straps

The thickness is determined according to the thermal requirements previously outlined which apply to the specific case.

The following values can be taken as guidance only, in general conditions:

Products	Temperature (°C)	Thickness (cm)
Butane	−1	5-10
Ammonia	−33	10-15
Propane	−44	15-20
Propylene	−47	15-25

As it can be seen, the thickness is often less than or about 15 cm. It allows the designer to choose a one-layer system since FOAMGLAS® cellular glass T4 is produced in a maximum thickness of 15 cm. Even if the economic thickness is slightly higher than 15 cm, a single layer system in 15 cm should be considered before making the final decision. The cost saving in replacing a two-layer by a single layer system is very considerable: about half of the installation labour and a substantial part of the adhesive (slightly less than half of it) can be saved.

The first layer is adhered with PC® 88 ADHESIVE by the full bed method to the steel tank with the joints filled with PC® 88 ADHESIVE. To obtain good adherence, the steel must be dry and clean and the anti-corrosion layer – generally a factory applied one – should be compatible with PC® 88 ADHESIVE. This implies among others that it should not be based on bitumen, tar, polyurethane or silicone. Zinc chromate for instance is acceptable with PC® 88 ADHESIVE. Since the FOAMGLAS® cellular glass layer has to follow the tank movements and deformations that have been described previously, the thickness of the joints should not be reduced to less than about 2 mm to provide the necessary flexibility and enable them to act as small expansion/contraction joints. Within the layer, the joints should not be staggered. Even at −50°C, the modulus of elasticity of PC® 88 ADHESIVE remains low enough under normal conditions to avoid generation of excessive stresses in the system.

Mathematical models have been developed to allow the designer to verify the stress situation in specific cases (classical method or finite element method).

If a second layer is needed, it should be applied in the same manner, all joints being staggered to the first layer.

Normally expansion joints are not made in the surface, the PC® 88 ADHESIVE joints being sufficient. The question of an expansion/movement joint between the wall insulation and a fixed roof insulation connected to it should be examined in each specific case.

Inlet, outlet and fixing straps should be treated in a way that allows them to move as necessary without causing deterioration of the insulation. Soft sealer, such as PITTSEAL® 444, generously applied, can easily solve this problem.

No general vapour barrier is needed but it is advisable to apply a vapour retarder at particular points like expansion joints, inlet and outlet connections, protrusions and to cover fixing strap boxes.

Steel plates overlapping at welding lines should be treated by locally abrading the FOAMGLAS® cellular glass slabs to follow the surface variations and steps.

Sealer

Usually no sealer is needed except for particular locations such as those previously mentioned.

Reinforced Coating or Metal Jacket

On single tanks, a reinforced coating or a metal jacket is needed as weather protection. The coating can be PITTCOTE® 404 reinforced by polyester fabric. It is important to give the coating enough drying time and to avoid excessive thickness in one layer. After the required drying time, a layer of Chemglaze can be applied on PITTCOTE® 404 to reinforce its surface.

The metal jacket is applied in long lengths placed vertically with Z metallic supports and metal straps to hold them in place, suitable for the worst conditions. In areas where high winds are expected, a thin layer of PITTCOTE® 300 can be applied as a cell filler at the rate of about 2 kg/m² before the installation of the metal jacket to reinforce the FOAMGLAS® cellular glass surface in case of vibration of the jacket. This cell filler must be dry before the metal jacket is installed.

In the case of a simple tank protected by a bund wall, the wall insulation is practically the same as for a single tank except that the cell filler used in the case of metal jacket can generally be omitted, the tank wall being already protected from high winds.

In the case of double containment tank, the wall insulation system is already protected from the weather and consequently the coating can generally be avoided, particularly if the cavity between the two tanks is purged by an inert gas.

Low Temperature Tank Roof Insulation

As already mentioned, two designs exist for roofs: the fixed roofs and the suspended roofs.

The insulation of fixed roofs can be made with FOAMGLAS® cellular glass following closely the method used for walls with a fully adhered insulation protected by a reinforced coating.

In the case of suspended roofs, FOAMGLAS® cellular glass could also be used but it is not often selected for two reasons: mineral wool can be specified in lower density and thus reduce the weight to be supported by the roof and is more flexible to be applied around the suspension elements.

III.4.12 CRYOGENIC TANK WALL

Description

Due to the operating temperatures, there is practically no adhesive that would be suitable to adhere FOAMGLAS® cellular glass slabs to the inside tank. Moreover the required thickness to meet thermal requirements generally ranges from 80 cm to 120 cm, sometimes more. In this thickness range, the price difference between FOAMGLAS® insulation and expanded perlite poured in situ is wide and makes a solution based on FOAMGLAS® cellular glass difficult to contemplate. But FOAMGLAS® cellular glass can be used as a secondary barrier, complementary to perlite insulation. The perlite poured in situ brings the main part of the thermal resistance and the FOAMGLAS® insulation system

225

provides a modest part of the thermal resistance but FOAMGLAS® cellular glass also gives a protection of the liner and the exterior concrete tank against thermal shock, should the interior tank experience a major failure.

Requirements

Thus, one requirement dominates the other: to protect the liner and the exterior tank against a failure of the inside tank, particularly the so-called "brittle failure", during a given period of time.

This is an example of the requirement given in III.1.1.7.

For some designers, the protection should only concern the first few metres of the outside tank wall. For others, the protection should be applied to such a height that the total volume of liquefied gas stored in the inside tank could be contained in the outside tank during a certain period of time, sufficient to take action. For instance the liquid could be pumped into another tank, a ship or whatever container that could be made available quickly. This explains why the protection requirement is expressed for a certain period of time decided in function of the specific jobsite, the proximity of another tank, ship, etc. It usually ranges from 24 hours to a few weeks.

It should also be mentioned that the FOAMGLAS® insulation/protection system is aimed at double containment tanks that do not have an earth berm directly connected to the outside tank. With an earth berm, the temperature at the liner interface may become too low for the adhesive after the accident and consequently this system is not recommended.

FOAMGLAS® Insulation System

The FOAMGLAS® insulation/protection system has to be adapted to the relevant requirement including the type of stored liquefied gas, the imposed protection period, etc. Typically, in the case of an LNG tank to be protected for a few days, the insulation system described from the exterior concrete tank liner towards the inside tank consists of

- The liner, usually in steel, possibly made of a gas tight coating
- A first layer of FOAMGLAS® cellular glass adhered with PC® 88 ADHESIVE, the joints being tight and filled with PC® 88 ADHESIVE
- A second layer of FOAMGLAS® cellular glass adhered with PC® 80M MORTAR, the joints being tight and filled with PC® 80M MORTAR and a glass fabric being unbedded in the interface PC® 80M MORTAR layer (the so-called laminated system)

- A third layer of FOAMGLAS® cellular glass adhered in a PC® 80M MORTAR laminated exactly as the second layer
- No coating on the third layer.

The principle of this insulation system is based on the fact that if a single slab of FOAMGLAS® cellular glass does not provided a high resistance against a thermal shock, the system called "laminate" consisting of FOAMGLAS® cellular glass slab, an adequate adhesive in which a glass fabric is incorporated and another FOAMGLAS® cellular glass slab presents a much better resistance against thermal shock.

The resistance is again increased when two laminates are used, three FOAMGLAS® cellular glass layers and two adhesive layers reinforced with glass fabrics (Fig. III.24).

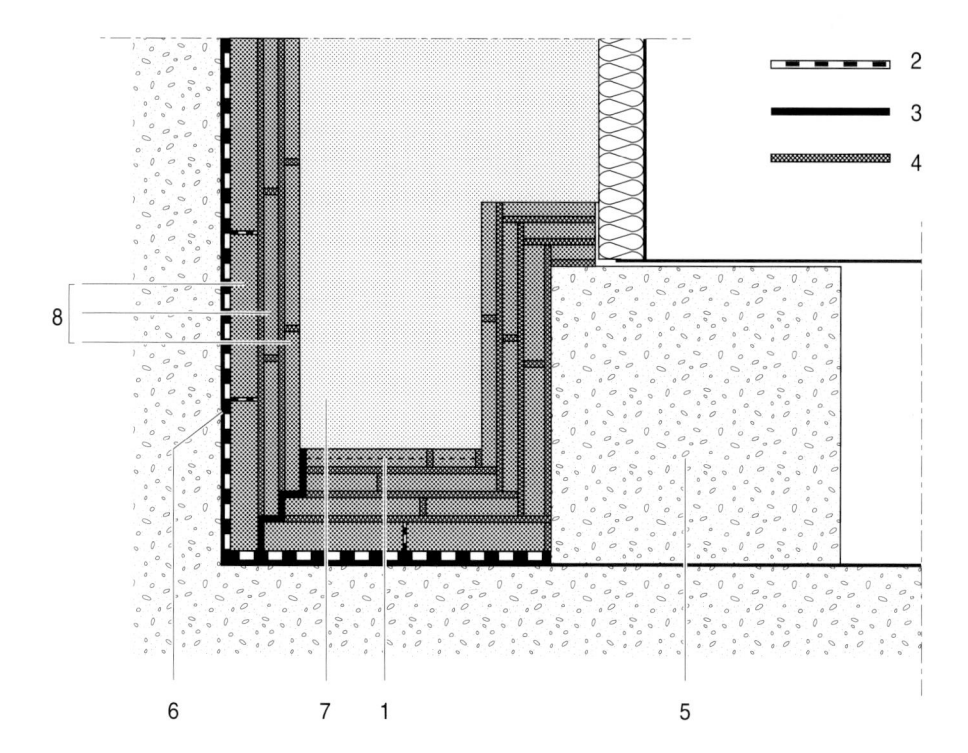

Fig. III.24. Laminated system in three layers.
1. Complementary local protection.
2. PC® 88 ADHESIVE.
3. PITTSEAL® 444.
4. PC® 80M MORTAR + glass fabric.
5. Perlite concrete.
6. Steel liner.
7. Loose perlite.
8. FOAMGLAS® insulation.

PC® 88 ADHESIVE is used to adhere the first layer to the tank liner since at this interface, the temperature should not be reduced to a too low value for it and since its adherence to steel is outstanding.

PC® 80M MORTAR is used for the other layers since it can support cryogenic temperature and has a coefficient of thermal contraction almost identical to FOAMGLAS® cellular glass over a large temperature range.

Thickness, Number of Layers and Installation Technique

FOAMGLAS® cellular glass secondary insulation/protection only plays a minor role in the heat transfer limitation compared to the expanded perlite poured in situ insulation that has to be at least 90 cm thick. This minimum thickness is imposed for practical reasons: the workers need a free space of about 90 cm to be able to install a movable scaffolding and adhere the three layer system. But the FOAMGLAS® insulation system plays by far the major role in the protection of the liner and the outside concrete tank in case of a major failure of the inside tank. Consequently, its thickness and particularly its number of layers – and thus of laminates – has to be commensurate with the possible thermal shock and the period of time during which protection is required.

As an example a three-layer system having a total thickness of 6 + 6 + 10 = 22 cm has been found sufficient to offer a several day protection in the case of a brittle failure occurring to an LNG tank. This was shown in a sophisticated test carried out with liquid nitrogen. Other cases can be handled by interpolation or by other tests.

Due to the very specific character of this application, installation details are not given here. They are indicated in the relevant specification.

Sealer

No sealer is used except in possible expansion/movement joints.

Finish

Since the FOAMGLAS® cellular glass secondary insulation system is applied inside the cavity, no finish is necessary. Moreover, except PC® 80M MORTAR, no coating is known which would survive a cryogenic thermal shock without shearing from the FOAMGLAS® cellular glass.

III.4.13 LOW TEMPERATURE SPHERE

Introduction

Spheres are used to store low temperature liquid and present the advantage to offer a reduced exterior surface for a given storage volume compared to all other possible shapes but they require spherical shapes for the steel elements used to build them. Since it is not easy to form these elements in cryogenic steel, the large majority of spheres operating at low temperature are not cooled down to cryogenic temperature. An operating temperature of –50°C represents by far the lowest usual value and often gases are stored in vapour phase, under pressure at a temperature not as low as the liquefaction temperature.

Since it is difficult to form metal jacket to the spherical shape, fabric reinforced coatings are often the preferred solution.

Generally spheres are limited to a single surface, the double envelope concept being practically not often used for spheres. Finally, due to the sphere total symmetry, the method of applying cellular glass is the same for all the spherical surface with the only exception of straps.

Requirements

The main requirements for low temperature sphere insulation are usually the following:

- Heat transfer limitation
- Surface condensation limitation
- Fire protection
- Non absorption of flammable liquid
- Moisture resistance
- Corrosion resistance
- Possibly, chemical resistance
- Mechanical strength.

It should be pointed out that some designers insist very much on fire resistance since the double containment concept is not often used for spheres.

In V.1, a number of fire resistance tests of FOAMGLAS® insulation systems are given. Special attention should be paid to the reference [1] here below which corresponds to a fire test run to simulate a fire coming from the outside of a tank or a sphere.

FOAMGLAS® Insulation System

Assuming that the operating temperature is not below –50°C, the insulation system for a sphere consists of:

- A possible anti-corrosion layer applied on the support
- One or several layers of FOAMGLAS® cellular glass adhered with PC® 88 ADHESIVE
- If required, straps as an extra-safety
- A fabric reinforced coating or a metal jacket.

Anti-corrosion Layer

Anti-corrosion layers or anti-corrosion systems are applied at the discretion of the owner or the design engineer who are in the best position to evaluate all the factors involved in the decision.

But what has been said for low temperature tank wall insulation applies to a certain extent to spheres, provided special care is given to make the insulation system water and vapour tight, particularly the top of the sphere. Specially at inlets, outlets, or protrusions that are connected to the top of the sphere, the insulation around them has to be carefully installed and controlled to avoid any water penetration.

Anti-abrasive Layer

With an adhered system, no anti-abrasive layer is needed.

Thickness, Number of Layers, Straps and Installation Technique

The thickness will be determined on the base of the previously outlined requirements: heat transfer limitation, surface condensation limitation and

---

[1] "Ad Hoc Fire Resistance Test", Yarsley Technical Center, C 72756/2, 13 December (1982).

possibly fire protection requirements. As general guidance the following thicknesses can be considered in normal conditions:

Temperature (°C)	Thickness (cm)
−10	6 to 10
−10 to −20	8 to 15
−20 to −30	10 to 20
−30 to −40	15 to 20

As recommended for low temperature tank wall insulation, the possibility of working in one layer, whenever acceptable, should be considered.

Even if the economic thickness is slightly higher than the maximum one-layer thickness, the possibility of a single layer insulation system should be kept in mind before making the final decision since the cost saving in replacing a two-layer system by a single layer is considerable: about half of the installation labour and a substantial part of the adhesive (slightly less than half of it). Moreover, it should be recognised that, in the case of a sphere, the costs associated with the scaffolding and the labour are usually fairly high.

Due to the shape of a sphere, geometrical adaptation of the flat slabs is generally needed to obtain a good fitting to the sphere, narrow joints and a regular outside surface, good support of the exterior coating.

In the case of spheres having a diameter smaller than 20 m, prefabricated spherical segments should be used. They are manufactured from half slab in a plant or fabrication shop and their dimensions 295 × 440 mm are slightly lower than half slab. The two main faces are curved and the four edges are bevelled.

In the case of spheres having a diameter greater than 20 m, on site preparation of the elements is sufficient. The half slabs are adjusted by abrasion to fit the curvature of the sphere and the four edges are bevelled. For large spheres, the outside face of the segments is generally not treated.

The application of one or two FOAMGLAS® cellular glass layers with PC® 88 ADHESIVE is carried out following the same general method as for low temperature tank wall. The main difference is that for the lower part of the sphere, the adhesive has to support the weight of the slab and is not helped by the adjacent slabs. This can be achieved by two methods:

- Either by using temporary banding that should remain in place until the adhesive is set. The temporary banding is generally in rubber or similar elastic material allowing to slide in the proposed segments one after the other.
- Or by using the hold catalyst that accelerates the setting of the PC® 88 ADHESIVE to the point that if one can keep the segments in place for about one minute at ambient temperature, they are then adhered. It should be mentioned that some health and hygiene questions have arisen about the hold catalyst. It can only be used by trained workers, aware of the possible dangers, knowing the required precautions and after having verified the current health regulations in the jobsite country.

From a general standpoint, the solution based on temporary banding should be favoured. Incidentally the use of hold catalyst should be avoided in the cavity of double containment walls.

Since the curved spherical segments are rectangular, it is necessary to cut some of them to the shape of the sphere. Several methods exist to limit the cutting as much as possible and make the work as easy as possible. According to one of them, generally considered as the most efficient, the work should be carried out as follows.

One should mark the sphere's equator, divide the lower and upper hemispheres into six identical sections by striking six equidistant meridians from the equator to the bottom and the top poles and apply the spherical segments as indicated in Fig. III.25.

Regarding straps, they should be considered as an optional additional safety measure since the mechanical strength of PC® 88 ADHESIVE is such that FOAMGLAS® cellular glass slabs properly adhered should not fall and have not fallen in practice. When using straps they should be in stainless steel AISI type 304 (BSI 304 S16) and their section should measure 19 mm × 0.8 mm. The main advantages and disadvantages of straps are as follows:

The straps provide some increased safety in case of fire or at least during the initial period of a fire since their resistance to high temperature is better than PC® 88 ADHESIVE or even cellular glass. But this statement has to be balanced by the fact that the coefficient of thermal expansion of stainless steel is higher than that of cellular glass and the temperature of the strap will increase much more and much faster than that of cellular glass. Consequently straps will expand and tend to become loose in case of a fire coming from the outside of the sphere.

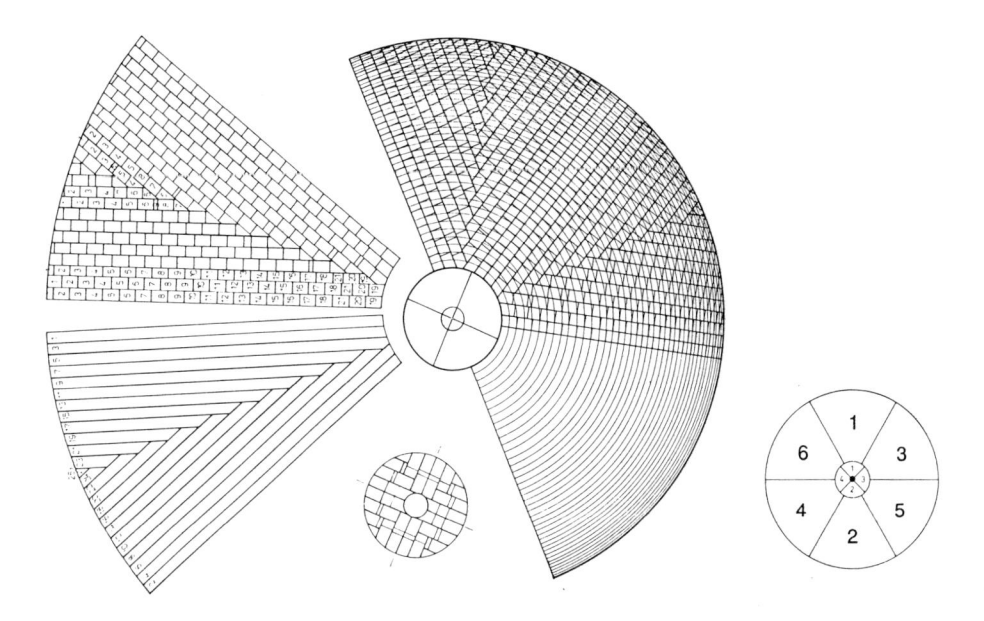

Fig. III.25. Application order for the upper hemisphere starting at the equator.

The main disadvantage of straps is related to the fact that in normal service they expand and contract with the temperature variations that are considerable on the surface of an insulating layer and their movements tend to cut and deteriorate the fabric reinforced coating.

Except for very small spheres, straps are only applied – when specified – on the lower hemisphere. They are placed between a bottom ring made of a 10 mm diameter steel tube having a diameter of about two metres and an equatorial ring fixed to the sphere.

As it can be seen, their fixing imposes some thermal short circuits, although moderate to fix the equatorial ring to the sphere.

Sealer

After the application of the FOAMGLAS® insulation system, it is generally necessary to repair some defects like gaps between slabs, unfilled joints etc. and to abrade too sharp angles in the outside face to avoid aesthetic defects. Filling gaps or local defects is carried out, using PITTSEAL® 444.

Coating

The fabric reinforced coating is applied in the same way as for low temperature single wall tank wall and roof.

III.4.14 DIGESTER

Digesters are usually medium or large size, vertical, cylindrical tanks made in heavy water tight concrete, sometimes in steel. Often they have a conical bottom and a conical, spherical or flat roof. They contain liquid at a moderate temperature in which biological and chemical reactions take place. The operating temperature is often around 30 to 40°C. The conical bottom is normally below the ground level. (See colour photo 15 between pages 64 and 65).

Their diameters range from 5 to 30 m, even more. Their heights range from a few meters to 30 m, sometimes more.

Requirements

The main requirements are usually:

- Heat transfer limitation
- Constancy of operating temperature within its strict limits
- Moisture resistance
- Chemical resistance. FOAMGLAS® insulation is normally not in contact with the liquid processed in the digester but it can be in case of an overfill or an accident
- Particularly for the base insulation: mechanical strength
- Resistance against vermin and rodents
- Non combustibility, especially when the process implies the production of methane.

Regarding mechanical resistance, two remarks should be made: firstly, the stress level on the bottom insulation can be fairly high. It should be remembered that the digesters are often built in heavy concrete and contain a liquid with a density generally slightly greater than one. At equal size, the pressure on the bottom can be higher than for a liquefied petrochemical gas storage tank.

Secondly, when they have a conical bottom, the load is not applied perpendicular to the main face of the slabs but at an angle. It can be decomposed in a load perpendicular to the main faces and in a load parallel to the main faces.

When making the choice of the type of FOAMGLAS® cellular glass for the bottom insulation, these points should be kept in mind.

FOAMGLAS® Insulation System

Due to the moderate operating temperatures, the chosen insulation system is similar to those used in the building field. It consists of:

- For the bottom:
  - An adhesive which can be hot bitumen or PC® 56 ADHESIVE also used to fill the joints
  - FOAMGLAS® cellular glass slabs of the appropriate type
  - A bitumen or PC® 56 ADHESIVE facing
- For the wall and the roof:
  - A primer
  - An adhesive which can be hot bitumen or PC® 56 ADHESIVE also used to fill the joints
  - FOAMGLAS® cellular glass T4 or T2
  - A coating or a jacket for the wall or a coating or roofing membranes for the roof depending on its shape.

Based on what has been said about the mechanical requirements, FOAMGLAS® cellular glass type S3 or HLB or even type F [2] will be selected for the bottom insulation. For the wall and the roof insulation, FOAMGLAS® cellular glass T4 or T2 has a sufficient compressive strength. The general principle of the system consists in obtaining a compact layer.

Anti-corrosion Layer

No anti-corrosion layer is necessary on a concrete layer.

---

[2] See specific data sheet for this type of FOAMGLAS® cellular glass which is aimed at the insulation of building foundations more than at industrial applications. Its main characteristic is an average ultimate compression strength of 1.6 N/mm^2.

Anti-abrasive Layer

No anti-abrasive layer is necessary in an adhered specification.

Thickness, Number of Layers, Installation Technique

Due to the moderate operating temperature of the digesters, the thicknesses for this application usually range from 5 to 12 cm with a tendency to be slightly larger for the roofs than for the walls and for the walls than for the bottom.

Due to the thickness range, one layer of FOAMGLAS® cellular glass is generally sufficient. For the bottom insulation, generally of conical shape, the support is usually a non reinforced, levelling concrete layer or a structural concrete layer that should have no irregularity exceeding 2 mm under a 60 cm rule.

When PC® 56 ADHESIVE is specified, the primer consists of the same product, diluted in the ratio of 10 to 1 by volume.

When hot bitumen is used, the primer is normally PC® EM (bitumen emulsion).

The geometry of FOAMGLAS® cellular glass elements is adapted to the diameter of the conical base or to the roof.

For diameters not exceeding 7 metres, conical segments are used.

When the diameters range from 7 to 20 metres, half slabs of 300 × 450 mm are used.

When the diameters exceed 20 metres, normal slabs of 600 × 450 mm are used.

It is thus normal to mix segments, half slabs and eventually full slabs in the same base insulation.

For the base, the application starts with the lowest part having the smallest diameter and progressively climbs to the greatest diameter. Conical segments are applied in circles. Half slabs and slabs are applied in lines. The joints between the FOAMGLAS® cellular glass elements are narrow and well filled. As soon as the FOAMGLAS® cellular glass layer is applied, a PC® 56 ADHESIVE layer or a bitumen layer is placed on it to fill the surface open cells and ensure that the product has its full compressive strength (Fig. III.26).

For the wall insulation, the geometry of the elements is also adapted to the cylindrical shape of the wall: FOAMGLAS® cellular glass curved wall seg-

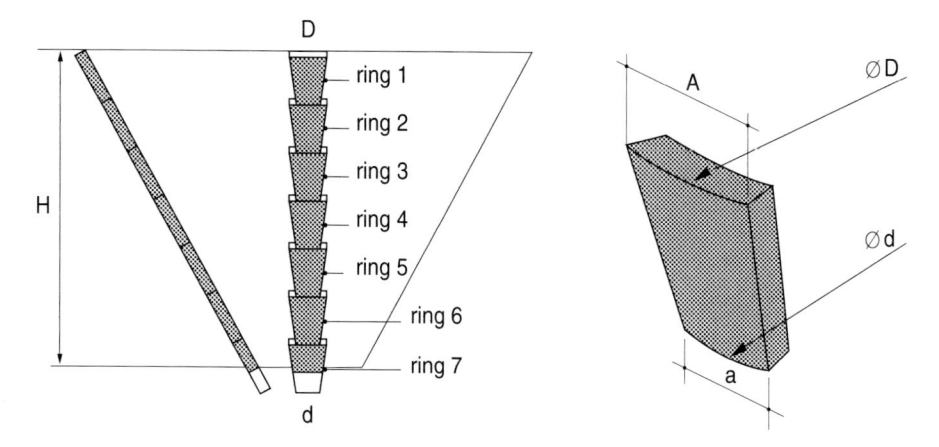

Cone sizes: D = 5.000 m d = 1.000 m H = 2.000 m
angle = 45° FOAMGLAS® thickness = 0.100 m

	D (m)	d (m)	A (mm)	a (mm)
Ring 1 = 56 segments	5.1414	4.5050	288	252
Ring 2 = 49 segments	4.5050	3.8686	288	248
Ring 3 = 42 segments	3.8686	3.2322	289	241
Ring 4 = 36 segments	3.2322	2.5958	282	226
Ring 5 = 29 segments	2.5958	1.9594	281	212
Ring 6 = 22 segments	1.9594	1.3230	279	188
Ring 7 = 15 segments	1.3230	0.6866	277	143

Number of rings = 7   Total number of segments = 249.
Theoretical outside surface insulation = 27.91 m².

Fig. III.26. Example of cone insulation.

ments are used for diameters not exceeding 7 meters, half slabs are used for diameters ranging from 7 to 20 meters and full size slabs are used for diameters exceeding 20 meters, their length being parallel to the cylindrical axis.

The FOAMGLAS® cellular glass element installation is similar to the method used for the bottom. Complementary to it, mechanical fixings in Z shape are shot in the concrete wall and embedded in the cellular glass, at the rate of one per element. Alternatively, stainless steel straps can be used at the rate of two per standard length of 60 cm.

The wall insulation can be finished with a coating or a jacket or when aesthetic and environmental consideration imposes it, a brick wall installed in keeping an air gap to create an artificial cavity wall. The coating can be two layers of PITTCOTE® 300 reinforced with a glass or polyester fabric or two layers of PITTCOTE® 404 reinforced with a polyester fabric. The jacket can be in aluminium, galvanised or stainless steel installed on supports fixed through the insulation in the concrete wall. Another possibility, providing a more aesthetical aspect, consists in using mineral fibre reinforced cement plates instead of a metal jacket.

For the roof insulation, the installation technique for the cellular glass layer is similar to the bottom, except that one starts with the lowest part of the roof that has the largest diameter. If possible, flat slabs are used before coming to the reduced diameters for which cellular glass segments are needed. Especially when the slope is considerable, it is important to fix a wooden sleeper or a steel corner at the lowest part of the roof to avoid the sliding of the insulation layer. When selecting a steel profile, it should not exceed half of the insulation thickness to avoid a local thermal short circuit. To finish the roof insulation, one can specify a reinforced coating such as for the wall, based on PITTCOTE® 300 or PITTCOTE® 404 or roof membranes, as generally adopted for flat roofed buildings.

Finally good flashing is needed to avoid water penetration into the insulation system.

III.4.15 INDUSTRIAL CHIMNEY

Description

This section presents the method for insulating large size industrial chimneys having a concrete stack. They can have one or several flue pipes. Their exterior diameters range from 2 to 12 meters, sometimes more. Chimneys that are incorporated within a building such as office buildings, hospitals and more general institutional buildings are not considered here.

Industrial chimney construction being a very specialised field, this section only outlines general principles and, even more than other sections, does not enter into details.

The general design consists in building a concrete tube with a concentric smaller diameter refractory brick tube insulated with FOAMGLAS® cellular glass leaving a cavity for inspection purpose. The FOAMGLAS® insulation is built in sections of about 12 metres in height as separate self supporting walls, able to expand and contract freely (Fig. III.27).

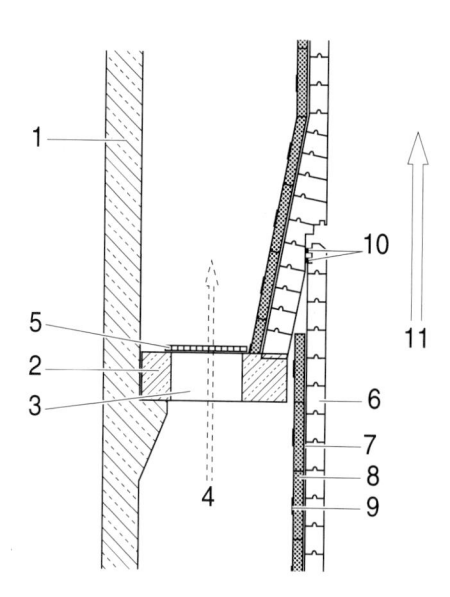

Fig. III.27. Principle of the insulation of an industrial chimney.
1. Outside wall of stack.
2. Concrete ring.
3. Air gap.
4. Aeration stream.
5. Access walk.
6. Acid resistant brick lining.
7. Distance layer, about 10 mm (cardboard).
8. FOAMGLAS® insulation.
9. Glass fabric strips.
10. Packing cord.
11. Flue gases.

Requirements

The main requirements are usually:

- Limitation of the temperature drop of flue gases to avoid or strongly reduce the condensation of acid gases near the top of the chimney
- Limitation of the temperature difference between the two faces of the concrete stack to avoid excessive thermal stresses in the concrete
- Limitation of the temperature difference between the two faces of the refractory brick masonry of the inner lining to reduce the thermal stresses to an acceptable level

- Limitation of the heat loss to keep the temperature in the cavity at an acceptable level
- Chemical resistance, particularly to the acids and reagents contained in the flue gases. In principle the cellular glass layer should not be subjected to chemical attack but it can be if cracks occur in the refractory bricks. FOAMGLAS® insulation is not supposed to be an anti-acid material but it offers a certain protection when the normal refractory brick layer has aged and failed
- Resistance to high temperature. Due to the thermal resistance of the refractory brick layer and its inertia, the flue gases temperature can exceed the limit of 430°C generally given for cellular glass. Some FOAMGLAS® cellular glass insulated chimneys have been operating at temperatures around 500°C with peaks at even higher temperature
- Mechanical strength to enable the construction of self supporting walls
- Coefficient of thermal expansion not too different from the values for concrete and refractory bricks
- Non combustibility.

FOAMGLAS® Insulation System

The FOAMGLAS® insulation system consists of a self supporting wall built near the refractory brick stack but at such a distance from the concrete tube that a space for inspection is available (about 80 cm wide). At regular intervals of about 12 meters, the insulation system and the refractory brick walls are supported by corbels fixed in the concrete structure. An expansion joint is placed at the end of each section.

Thickness, Number of Layers, Straps

Depending on the requirements previously outlined, particularly on heat transfer and temperature limitations, the FOAMGLAS® cellular glass thickness is calculated with an appropriate computer programme. It usually varies from 5 to about 12 cm. The 5 cm thickness is considered as a strict minimum to be able to build a self supporting wall. This is erected near the exterior face of the inner flue tube constructed in acid resistant bricks. The distance between the exterior face of the inner tube and the inside face of the insulation is approximately 10 mm, the corresponding air gap being obtained by sandwiching a corrugated cardboard or equivalent combustible material of the equivalent thickness during the erection of the FOAMGLAS® cellular glass wall. It helps to maintain the distance when erecting the free standing cellular glass wall.

The mortar can be Hoechst acid-proof cement HB or Hoechst acid-proof cement HFR mixed with the necessary quantity of potassium silicate and possibly some water (for HFR type). Due to the geometry of the FOAMGLAS® cellular glass wall, radiused and bevelled segments are chosen for small diameters, flat slabs being only suitable for very large chimneys.

In many cases, the slabs are cut longitudinally in elements narrow enough to approach the needed geometry (Fig. III.28).

The width of the elements is given in relation to minimum diameter of the FOAMGLAS® cellular glass stack by the Table III.11:

TABLE III.11

Width of flat FOAMGLAS® cellular glass element (mm)	Minimum diameter of stack (m)
150	1.50
225	3.00
300	5.00
450	12.50

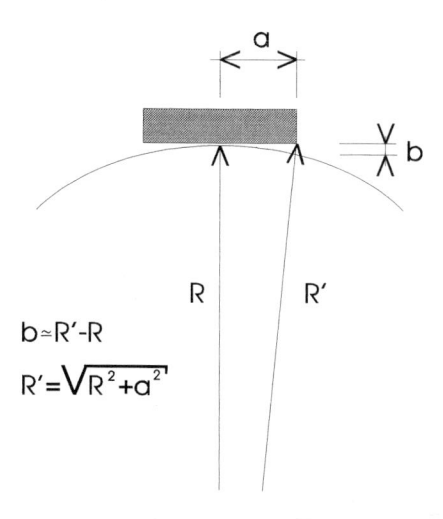

$b \approx R' - R$

$R' = \sqrt{R^2 + a^2}$

Fig. III.28. Calculation of the maximum distance.

To improve the stability of the FOAMGLAS® cellular glass self supporting wall, alkali resistant glass fabric strips of medium weight (150 g/m²), 100 mm wide are applied as horizontal straps at the rate of one per each row of insulation slabs. They are held in place by an adhesive or with metal straps. The adhesive is usually PC® 74A, a two-component product of hydraulic setting cement, containing copolymers in emulsion. The metal straps are in annealed stainless steel (AISI type 304) and usually have a 13 mm width and a 0.4 mm thickness.

Sealer

The joints between the FOAMGLAS® cellular glass slabs are filled with mortar, no sealer is required.

Reinforced Coating or Metal Jacket

The FOAMGLAS® cellular glass self supporting wall being actually inside the chimney, no reinforced coating or metal jacket is necessary.

Note:

An alternative construction technique consists in applying the FOAMGLAS® cellular glass directly against the internal face of the concrete wall of the chimney and keeping only a small air gap between the insulation and the refractory brick wall. The adhesive is then generally a modified bitumen emulsion, provided the temperature is suitable.

This technique is somewhat cheaper since it enables the designer to reduce the exterior concrete tube diameter for a given inside flue tube section but it presents two drawbacks: it is not possible to visit the chimney and repair it when needed. Moreover, should the refractory bricks leak, the danger of condensation of the gas is much greater due to the absence of the large ventilation cavity.

III.4.16 OFFSHORE

Description

There is a very special case for offshore applications; the particular conditions prevailing justify some special precautions and make the choice of FOAMGLAS® cellular glass even more necessary than for onshore applications.

The offshore applications are usually classified in
- Cold insulations
- Personnel protection
- Hot insulations.

Requirements

Depending on the type of application, the previously presented requirements apply.

Special emphasis should be given to the following points that practically concern all offshore applications:

- The relative humidity is always very high
- The corrosion risk is increased by the local conditions such as the saline atmosphere, frequent deluge tests, etc.
- Fire risks are considerable and in spite of all the precautions taken, the danger of having some gases or flammable liquids in contact with the insulation systems or even with high temperature pipes and equipment is dramatically present. For offshore installations, the fire requirements are very often expressed in terms of non combustibility of materials, fire resistance rating of systems often determined according to the hydrocarbon fire test method and last but not least no absorption of combustible liquids
- High mechanical stresses on systems, for instance jacket, due to frequent high winds, storms, vibrations, etc.

Although they are not strictly application requirements but very important considerations for the platform owner or operator, two other points should be mentioned:

- The cost of the shut down of a platform is particularly high
- When an insulation system is installed on a platform, the delivery time for all the components has to be short and perfectly respected. This remark is of a special importance in the case of repair or maintenance work.

Some FOAMGLAS® Insulation System Particular Points in the Case of Offshore Applications

The general concept of specifications previously outlined remains valid for offshore applications but they have to be reviewed taking into consideration the particularly stringent requirements of offshore applications.

Offshore conditions impose or strongly suggest a particular choice in the following cases.

At the design level, the remark previously made (See III.2.1), that it is sometimes inefficient to choose thicknesses that would almost always prevent condensation applies to offshore conditions. Often, the choice of thickness preventing surface condensation would result in impractical values.

Regarding the corrosion danger, the systems should be designed as water and water vapour tight as possible to reduce corrosion danger to the minimum. Due to the ambient conditions, this necessity is even greater than for onshore systems. Moreover, the anti-corrosion layer and similar devices that may be considered as optional for onshore applications in normal conditions become necessary on offshore jobs.

Regarding the anti-abrasive layer, it should be always applied on offshore insulation except for adhered systems. The specifications should take into account the possible vibrations of platforms due to the process or to high winds and also the need to protect the anti-corrosion layer.

In liaison with the high fire resistance rating requirements, reinforced organic coatings should be replaced by metal jackets, preferably stainless steel jackets, which offer better behaviour in case of fire.
As another example of the influence of the working conditions, metal jackets should have circumferential joints ball-swaged and sealed with mastic. On vertical piping, a minimum of 50 mm vertical laps should also be sealed with mastic.

Moreover all accessory materials associated with the stainless steel jackets such as bands, seals, screws, clips, etc. should also be in stainless steel and should be dimensioned to survive without excessive stresses the problems caused by high winds.

III.4.17 STAINLESS STEEL SUPPORT

Insulation is applied on stainless steel surface in many cases, typically in food industry, breweries and often in nuclear industry. The insulation systems used on stainless steel are similar to those on carbon steel operating in the same conditions except that they have to avoid the danger of stainless steel stress corrosion. If corrosion is a general threat for all steel surfaces, stainless steel, needed for other reasons, is prone to stress corrosion which means that when under mechanical stress, it can be very damaged and even cracked in a relatively short period of time when particular conditions are met. Structural problems resulting from stainless steel cracking are possible and can lead to dangerous situations.

Requirements

In addition to the other requirements imposed by the application, insulation systems applied on stainless steel have to eliminate or at least reduce substantially the risk of stress corrosion.

To elaborate somewhat this very important requirement, it is useful to outline the conditions conducive to stainless steel stress corrosion (for more details see V.6.2) as follows:

- Mechanical stress in the stainless steel must be present
- Water must be present
- Oxygen must be present. There is almost always some oxygen in water
- Free (soluble) chlorides, fluorides and halides increase the danger of stainless steel stress corrosion. In many countries, including the U.S.A., the acceptable level of free chlorides and fluorides is determined in relation to the amount of leachable (soluble) sodium and silicate ions, which act as inhibitors
- pH outside the 7 to 11.7 range is generally considered more dangerous
- Certain temperature ranges favour stress corrosion. Although it may occur from cryogenic to high temperatures, the most dangerous zone corresponds to the presence of liquid water, thus 0 to 100°C and in this zone from about 60 to 100°C since the speed of chemical reactions is accelerated at higher temperatures.

Unfortunately, many applications in the food and brewery industry are within or close to this temperature range.

FOAMGLAS® Insulation Systems

The insulation systems should be designed and installed in a way which limits as much as possible the danger of stress corrosion and its consequences.

A first method consists in reducing the level of mechanical stresses in the stainless steel surface. If this should be adequately taken into consideration by the mechanical design engineer, it has to be acknowledged that it often requires thicker sections of stainless steel and the corresponding higher cost. Moreover at the insulation system design level, little can be done regarding mechanical stresses. Attention should be given to the temperature profiles, elimination of thermal short circuits to reduce the thermal and the corresponding mechanical stresses. If not decisive, this point should still be kept in mind to avoid local points at which poor insulation would lead to high thermal stress and consequently high mechanical stresses.

The next consideration concerns the choice of insulation and accessory materials. It is well known that FOAMGLAS® insulation has passed several tests showing that it does not contribute to stress corrosion. As explained in paragraph V.6.4.3. FOAMGLAS® insulation is recognised as acceptable for stainless steel surfaces by the German norm AGI Q137 and has been successfully tested according to ASTM and other test methods. This should be complemented by an adequate choice of accessory materials that should also have been investigated for compatibility with stainless steel: this is the case for PC® 88 ADHESIVE used to adhere FOAMGLAS® cellular glass to stainless steel when "compact" systems are preferred and PITTSEAL® 444 applied as a sealer in the joints. The importance of these two accessory products should be fully acknowledged since they are or can be in contact with the stainless steel.

Clearly accessory products are an important part of the insulation system and only products having been successfully tested for compatibility with stainless steel should be accepted at the design stage and should not be replaced during the installation.

The next critical step to avoid stress corrosion consists in choosing insulation systems that virtually eliminate the danger of water or water vapour penetration.

Except overfit insulation systems that are designed to address a special situation (see III.4.6) all FOAMGLAS® insulation systems have been designed to avoid water and water vapour intrusion in normal circumstances but the level of safety differs between them.

It is clear that a one-layer system with joints sealed does offer a good resistance to water penetration but it is also obvious that a two-layer adhered compact system offers the same result with an even higher degree of safety.

If, due to poor installation, misuse of the system, mechanical abuse or for any other reason, a joint fails, there are still other joints and layers in the two-layer system to avoid direct water contact with the stainless steel surface. It is not true to the same extent in a one-layer non adhered system.

Consequently an engineering decision has to be made in choosing an insulation system that may appear somewhat "over engineered" for a normal situation but well adapted to face the danger of stress corrosion on stainless steel support. Adhered compact systems – provided temperature allows the use of a stainless steel compatible adhesive – should be preferred, whenever possible.

Last but not least, special attention should be paid to all singular points such as inlet, outlet, protrusions, valves, etc. to avoid water penetration.

As a summary, there is no special system for stainless steel surface but the choice should be made in the existing systems to meet the particular requirements of this surface.

III.4.18 NUCLEAR

Description

FOAMGLAS® insulation has been used in the nuclear field mainly in the following applications:

- Cold pipes, chilled water
- Protection against frost
- Insulation inside the containment
- Underground insulation
- Building insulation
- Chimneys in power plants (see III.4.15)
- Exposed steam lines with the overfit system.

So far, FOAMGLAS® insulation does not seem to have been used for the insulation of the primary circuit.

Requirements

Depending on the applications, the classic previously outlined requirements apply. But some specific requirements for the nuclear applications are to be added. As explained in III.1.1.11, they should be found in national and/or international regulations.

Although a review of these regulations is a weighty subject, it is possible to provide general indications about FOAMGLAS® cellular glass, its accessory products and their behaviour and compliance with regulations.

FOAMGLAS® insulation complies with NRC guide 1.36 *(Nuclear Regulatory Commission)*.

Moreover FOAMGLAS® cellular glass, PC® 88 ADHESIVE, PITTSEAL® 444 and PC® FABRIC 79P coated with PITTCOTE® 404 have been exposed to various levels of gamma radiation following a procedure developed by the *American Institute of Electrical and Electronics Engineers* (Standard 323-1974) and tested afterwards.

The products were subjected to doses of 10 megarads and 200 megarads gamma radiation. IEEE Standard 323 relates such exposure to the following conditions:

- Test conditions for pressurised water reactors, typical in-containment design basis event test conditions:
  4 megarads after 1 hour
  20 megarads after 12 hours
  110 megarads after 6 months
  150 megarads after 1 year
- Test conditions for boiling water reactors, typical in-containment design basis event test conditions:
  26 megarads integrated over the accident

Test Programme

The FOAMGLAS® insulation was tested for possible effects on thermal conductivity, water vapour permeability, and compressive strength. The sealers and adhesives were tested for possible effects on water vapour permeability, tensile strength, elongation under tension, and brittle point (flexibility as a function of temperature).

Test Results

FOAMGLAS® Insulation: Exposure of 10 megarads had no noticeable effect on the properties. Exposure of 200 megarads caused a slight apparent increase in thermal conductivity but no effect on the other properties.

PC® 88 ADHESIVE: The radiation exposure increased the tensile strength with a corresponding decrease in elongation as the material became stiffer and more brittle.

PITTSEAL® 444: The sample exposed to 10 megarads appeared slightly softer in consistency than the controls; however, there was no noticeable change due to exposure to 200 megarads.

PC® FABRIC 79 Coated with PITTCOTE® 404 Coating: Exposure to 10 megarads had no noticeable effect. Exposure to 200 megarads resulted in increased brittleness.

Regarding nuclear applications, it should also be remembered that pipings, vessels, etc. in stainless steel are very often used in this field. The corresponding requirements (see III.4.17) should be considered.

FOAMGLAS® Insulation Systems

FOAMGLAS® insulation systems used for applications in the nuclear field do not differ substantially from those generally specified but some of the options are more directed towards the specific needs of the applications.

For instance metal jackets, often in stainless steel, have preference. The situation is somewhat similar to offshore or applications on stainless steel surface.

III.4.19 PARTICULAR POINTS

In this section, a number of particular points like expansion joint, flange insulation, pipe hanger, cradle, support, leg, underground line expansion chamber, line anchor, wall entrance, etc. will be reviewed, principle figures being proposed and important points commented.

Obviously it is not possible to analyse all the cases and to give details for each of them but the given examples and the corresponding comments should be useful for the design of specific cases.

III.4.19.1 Expansion/Contraction Joints on Insulated Pipelines

Although the coefficient of thermal expansion of cellular glass slabs ($9 \times 10^{-6} K^{-1}$) is fairly close to those of carbon steel ($12 \times 10^{-6} K^{-1}$), austenitic stainless steel ($16 \times 10^{-6} K^{-1}$), copper ($17 \times 10^{-6} K^{-1}$) and aluminum ($22 \times 10^{-6} K^{-1}$) at ambient temperature, they are not identical and moreover the temperature variations of the surface to be insulated and of insulation are different.

Since the insulation pipe coverings or slabs are partly restrained, stresses may be generated and could eventually be excessive, resulting in cracks. This is particularly true for cold lines for which the surface contracts more than the insulation and consequently introduce compressive stresses in the insulation. To avoid this situation, contraction joints should be placed at suitable distances on a straight line and near singular points like elbows, when needed on a curved line.

A rule of thumb, which has been confirmed by many existing applications, consists in installing a contraction joint on straight line when the differential movement between the surface and the insulation exceeds 15 mm.

An example will show how the distance between two contraction joints can be calculated: an ethylene line in austenitic stainless steel operating at –104°C will be installed at a temperature of +15°C in a location where the maximum ambient temperature in summer will be +30°C and the minimum ambient temperature in winter –10°C. A simplified calculation can be made, in the case of a one layer insulation (for the sake of simplicity) as follows:

- Contraction of the pipeline per linear meter:
$$\Delta l_{steel} = \alpha \times 1 \times \Delta t$$
$$= 16 \times 10^{-6} \times 10^3 \times [15 - (-104)]$$
$$= 16 \times 10^{-6} \times 10^3 \times 119$$
$$= 1.904 \text{ mm/m}$$

- Average operating temperature of the insulation in summer conditions (from +30°C to –104°C): –37°C.

 Contraction of the insulation in summer conditions:
$$\Delta l_{FG,S} = \alpha \times 1 \times \Delta t$$
$$= 9 \times 10^{-6} \times 10^3 \times [15 - (-37)]$$
$$= 0.468 \text{ mm/m}$$

- Average operating temperature of the insulation in winter conditions (from –10°C to –104°C): –57°C.

 Contraction of the insulation in winter conditions:
$$\Delta l_{FG,W} = \alpha \times 1 \times \Delta t$$
$$= 9 \times 10^{-6} \times 10^3 \times [15 - (-57)]$$
$$= 0.648 \text{ mm/m}$$

- Differential contractions per linear meter:
  In summer conditions....... 1.904 − 0.468 = 1.436 mm/m
  In winter conditions.......... 1.904 − 0.648 = 1.256 mm/m

The summer conditions correspond to the greater differential movement.
Recommended distance between two successive contraction joints on a straight line:

$$L = \frac{15}{\Delta (\Delta l_{steel} - \Delta l_{FG,S})}$$
$$= \frac{15}{1.436}$$
$$= 10.44 \text{ m}$$

i.e. about 10.5 metres.

If high accuracy is needed, it is advisable to introduce in the calculation different values of the coefficient of thermal contraction in different temperature ranges (Table III.12).

TABLE III.12
Thermal expansion/contraction coefficients

Temperature	Aluminium	Copper	Stainless steel	Carbon steel	FOAMGLAS®
(°C)	(10^{-6}K^{-1})	(10^{-6}K^{-1})	(10^{-6}K^{-1})	(10^{-6}K^{-1})	(10^{-6}K^{-1})
200			17.28	13.44	9.28
150			17.10	13.05	9.28
100	23.46	17.30	16.78	12.76	9.28
50	22.97	17.00	16.30	12.44	9.28
20	22.50	16.74	16.20	12.00	9.00
0	22.30	16.60	15.90	12.00	9.00
−50	21.90	16.20	15.30	11.80	8.70
−100	21.20	15.80	14.70	11.50	8.30
−150	20.00	15.10	14.10	11.10	8.00
−200	18.70	14.40	13.40	10.60	7.40

To facilitate the calculations, diagrams have been precalculated for low and high temperature applications.

It is assumed that the pipe insulation is installed at an ambient temperature of 20°C. The differential expansion is calculated when the outside temperature

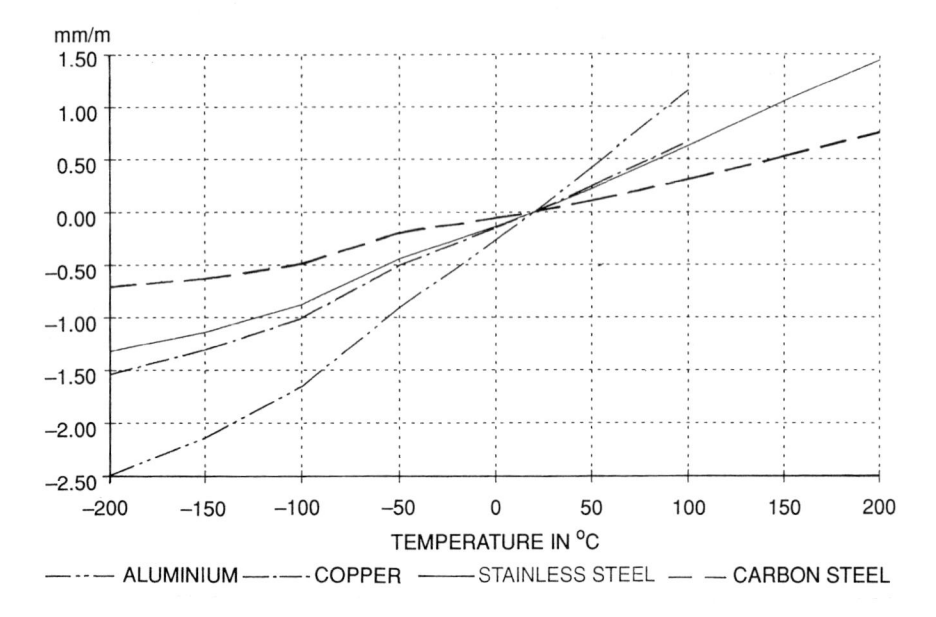

Fig. III.29. Differential expansion/contraction between FOAMGLAS® cellular glass and metal piping in mm/m of pipe run.

is also 20°C. Temperature variations resulting from sunload are not taken into consideration (Fig. III.29).

In the case of dual temperature applications, the distance between the expansion/contraction joints in straight line is determined by the more severe conditions, either by direct calculation, or by consulting the previous diagrams.

It may be worth pointing out that the previous calculations and the corresponding diagrams are based on the assumption of a solid single unbroken insulated surface and insulation layer. This assumption applies to the insulated surface but not to the insulation that is made of FOAMGLAS® cellular glass slabs or pipe coverings having a length of 60 cm. Consequently a considerable number of normal joints exist between two expansion/contraction joints. They are either dry (high temperature or internal layers) or filled with a soft sealer for the exterior layer of a low temperature application. In these cases, the normal joints present a certain possibility of absorbing relative movements without high stresses and consequently the 15 mm differential displacement used as assumption for the calculations does not occur in reality, the movement being limited to a lower value. The 15 mm differential displacement is a safe rule of thumb to calculate the distance between two consecutive expansion/contraction joints but the actual relative movement does not reach this value.

In the case of horizontal lines, the expansion/contraction joints can be made as in Figs. III.30 and III.31.

As it can be seen, the principle consists in connecting the two sides of the expansion/contraction joint only by a soft product such as PITTSEAL® 444 sealer to allow relative movement and avoid stress build up. Particularly for low temperature applications, it is important to avoid any water or water vapour penetration. This is achieved with Z joints having a long circumferential sliding compound layer, even a double Z joint in the case of three-layer insulation systems.

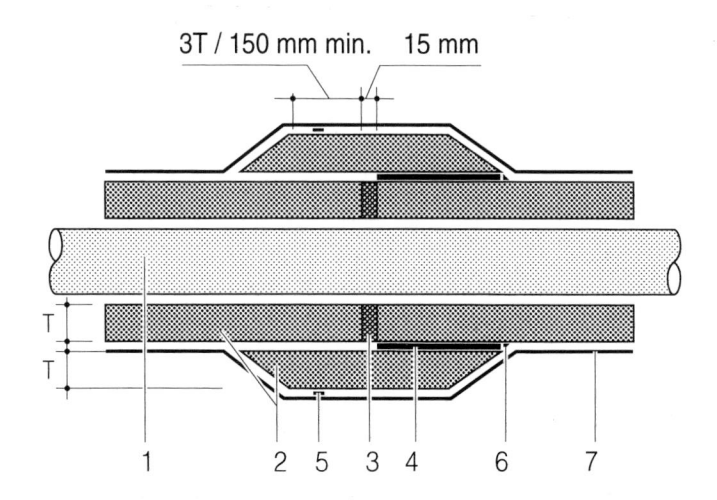

Fig. III.30. Contraction joint for a high temperature horizontal pipeline insulated with one layer.
1. Pipe.
2. FOAMGLAS® insulation.
3. Joint filled with resilient insulation material.
4. Sliding compound.
5. Fixing straps.
6. PITTSEAL® 444.
7. Reinforced mastic finish.

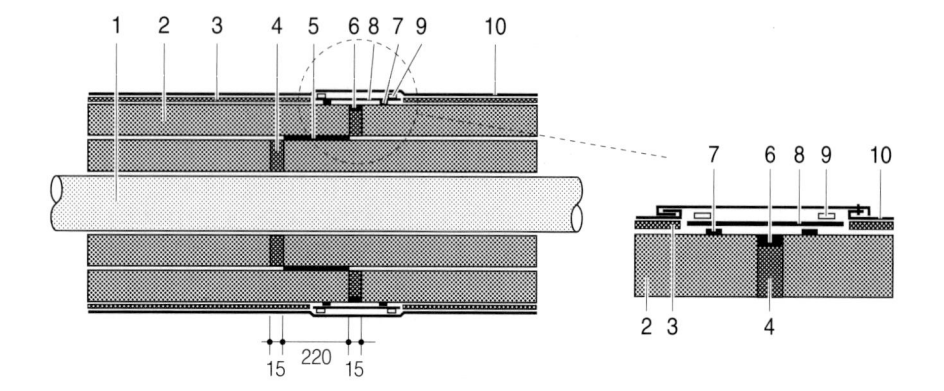

Fig. III.31. Contraction joint for a low temperature horizontal pipeline insulated with three layers.

1. Pipe.
2. FOAMGLAS® insulation.
3. Cell filler.
4. Joint filled with resilient insulation material.
5. Sliding compound compatible with temperature, for instance PITTSEAL® 444.
6. PITTSEAL® 444.
7. PITTSEAL® 444.
8. Polyisobutylene local vapour barrier.
9. Fixing straps.
10. Metal jacket allowing contraction.

In the case of a one-layer insulation system, a complementary local FOAMGLAS® cellular glass element is needed to obtain the Z joint configuration. Metal straps are applied in such a way that they do not prevent the relative movement. As a complementary protection against water and water vapour penetration, a local vapour brake is often applied in front of the expansion joint on the exterior layer. When a complementary local FOAMGLAS® cellular glass element is not installed it is made of a product – such as polyisobutylene that withstands fairly high extension before breaking. A resilient or a cushioning material, such as low density fibreglass insulation for high temperature or polyethylene foam for low temperature is used whenever possible, to avoid thermal short circuits. In the case of low temperature applications, avoiding water or water vapour entry being a predominant requirement, the cushioning material is often complemented by a partial filling with a sealer such as PITTSEAL® 444.

In the case of vertical lines, the expansion/contraction joints can be made as indicated in Figs. III.32 and III.33.

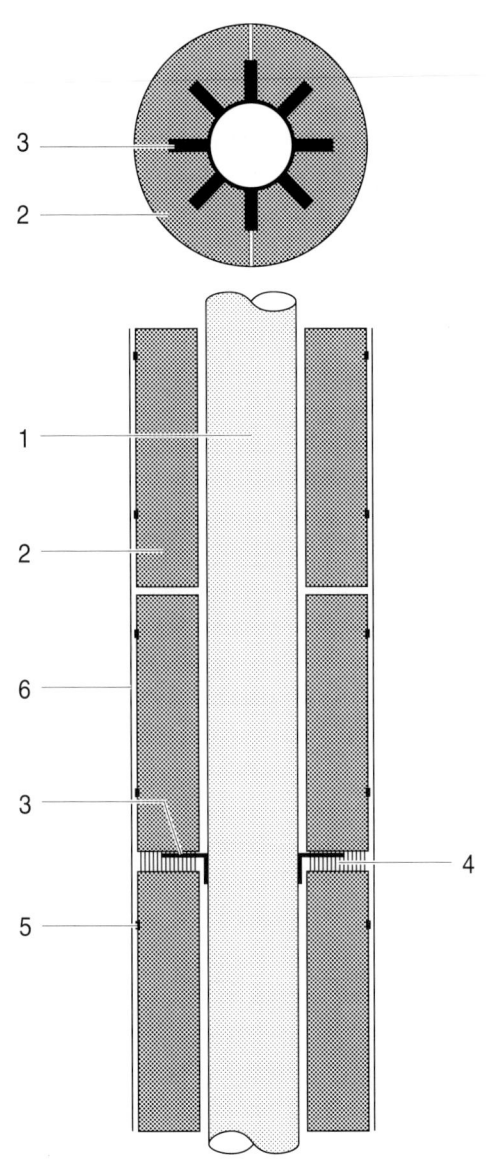

Fig. III.32. Contraction joint for a high temperature vertical pipeline insulated with a one-layer.

1. Piping.
2. FOAMGLAS® insulation.
3. Insulation support.
4. Resilient insulation material.
5. Fixing straps.
6. Metal jacket.

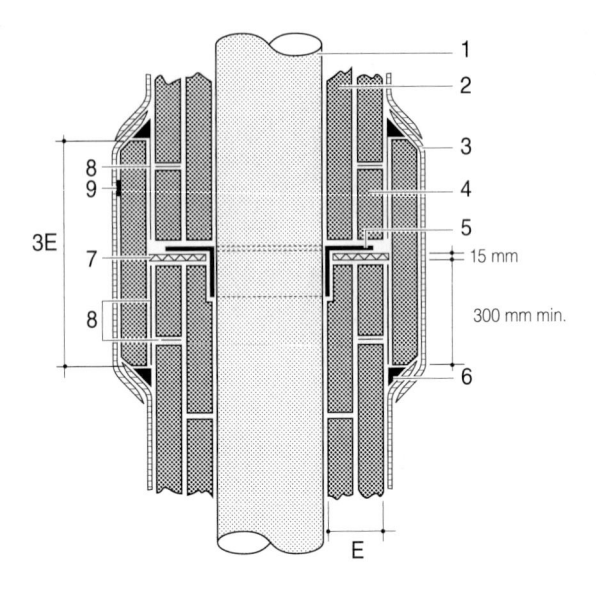

Fig. III.33. Contraction joint for a low temperature vertical pipeline insulated with two-layers.

1. Pipe.
2. First layer of FOAMGLAS® insulation.
3. Reinforced mastic finish.
4. Second layer of FOAMGLAS® insulation with sealed joints.
5. Support angle.
6. PITTSEAL® 444 sealant.
7. Joint filled with resilient insulating material.
8. PITTSEAL® 444.
9. Fixing strap (not over sliding part).

The principles are the same as for a horizontal line but an insulation support has to be fitted to restart with the insulation after the interruption caused by the expansion joint. The complementary local FOAMGLAS® cellular glass element is aimed at providing a protection against the thermal short circuit that is possible with the steel support. To reduce this problem as much as possible but still carry the weight of the insulation, the support is generally limited to half of the insulation thickness of the outer layer. The contraction joint is also provided with a sliding compound layer and eventually a local vapour brake.

III.4.19.2 Expansion/Contraction Joints of Vessels and Tank Walls

The case of vessel and tank walls is very similar to those of vertical pipelines. Provided the vessel of the tank has a certain size, an expansion/contraction joint is often placed near the wall/roof juncture where movements can be expected (Figs. III.34 and III.35).

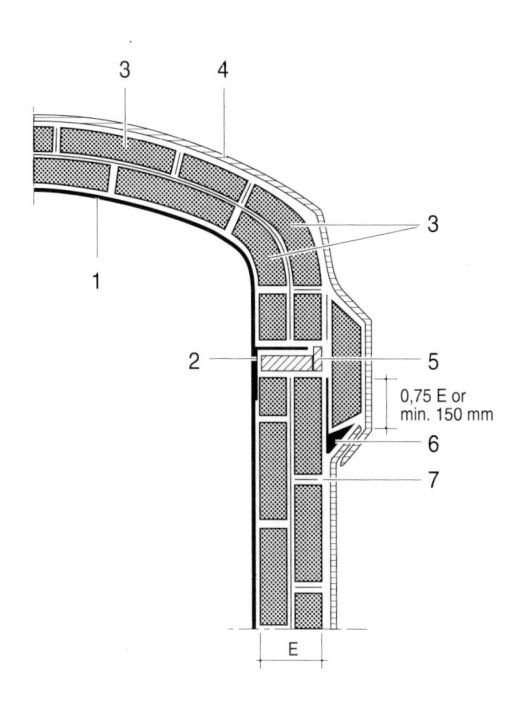

Fig. III.34. Contraction joint near the wall/roof juncture of a vertical vessel.

1. Vessel wall.
2. Insulation support angle.
3. FOAMGLAS® insulation.
4. Reinforced mastic finish.
5. Joint filled with resilient insulating material.
6. PITTSEAL® 444.
7. PITTSEAL® 444 in outer FOAMGLAS® insulation layer only.

In the case of PC® 88 ADHESIVE adhered system, the flexibility of the insulation system is such that some designers do not use any contraction joints even in the case of large size vertical tanks. Before making such a decision, a stress analysis is advisable.

III.4.19.3 Insulation of Pipeline Flanges

The principle of the insulation of pipeline flanges consists in making an oversized cylindrical box with whenever possible the same insulation thicknesses as on the pipe line. As for an expansion joint, a cushioning material is

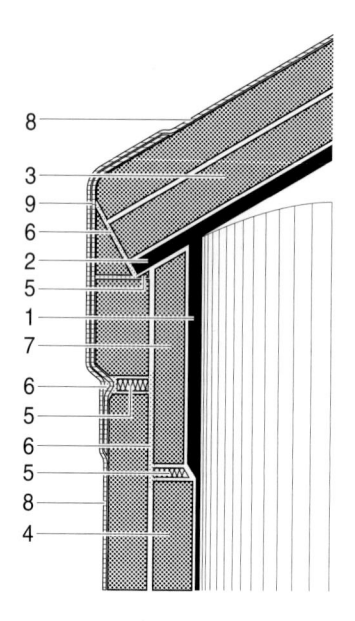

Fig. III.35. Contraction/movement joint near the wall/roof juncture of a vertical vessel.

1. Tank wall.
2. Tank roof.
3. FOAMGLAS® roof insulation adhered and joints sealed with PC® 88 ADHESIVE.
4. FOAMGLAS® wall insulation adhered and joints sealed with PC® 88 ADHESIVE.
5. Low density mineral wool.
6. PITTSEAL® 444.
7. FOAMGLAS® insulation specially cut to required shape and thickness, adhered with PC® 88 ADHESIVE.
8. PITTCOTE® 404 flexible finish reinforced with PC® FABRIC 79P.
9. Double layer of PITTCOTE® 404 flexible finish over junction and movement joint.

applied between the flange in steel and the FOAMGLAS® insulation to allow relative movement.

Depending on the thickness, it can be made with or without an intermediate element (Dutchman) between the main pipeline insulation and the oversized pipe insulation (Fig. III.36).

It is very important to obtain water and water vapour tightness between the normal and the oversized insulation. This tightness must be kept even after removing and reinstalling the oversized element for a control of the flange. This objective can be achieved by generous use of PITTSEAL® 444.

Figure III.37 shows a flange installed on an inlet/outlet near a vessel wall.

It should be noticed that the resilient or cushioning material is extended to the tank wall.

Flange is insulated with
a second shell layer

Flange insulation supported
by thicker covering

Flange insulation supported
by packing piece ("Dutchman")

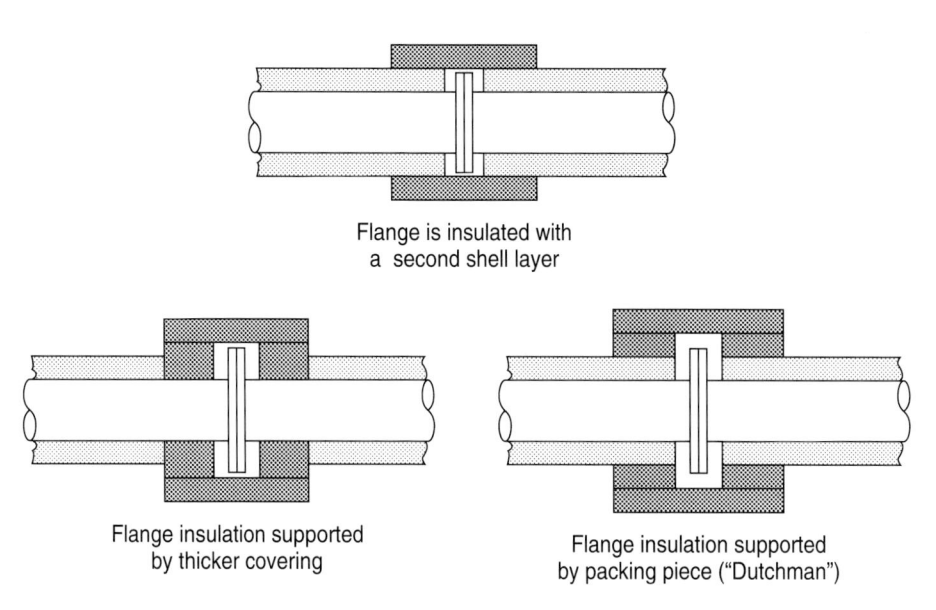

Fig. III.36. Principle of a pipeline flange insulation.

III.4.19.4 Insulation of Pipe Hangers and Support Assembly

On Fig. III.38, it can be seen that the hanger rod is insulated on a length equal to at least four times the thickness of the normal pipeline insulation. The purpose is to avoid a thermal short circuit that would result in substantial heat transfer and condensation for a low temperature application. This condensation would be detrimental since the water will tend by gravity to penetrate the hanger box.

Once again repeating ourselves, the objective of water and water vapour tightness should be pointed out. It can be achieved by adequate use of PITTSEAL® 444.

On Fig. III.38, the FOAMGLAS® insulation does not carry the weight of the line. This is directly supported by the hanger rod and a steel collar.

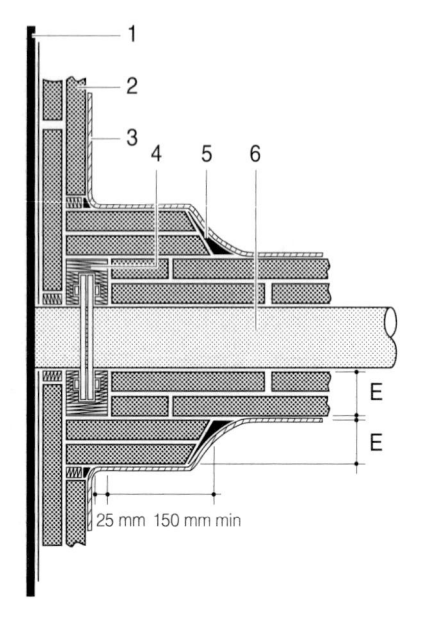

Fig. III.37. Insertion of an inlet/outlet flange near a vessel wall.

1. Vessel wall.
2. FOAMGLAS® insulation.
3. Weather coating.
4. Expansion-contraction joint filled with a resilient insulating material.
5. PITTSEAL® 444.
6. Pipe.

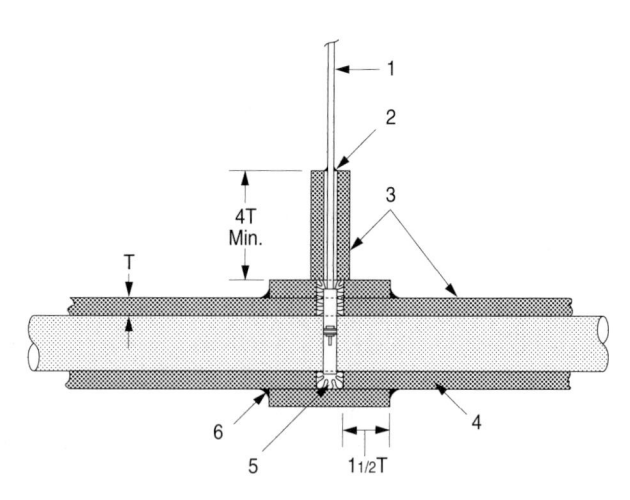

Fig. III.38. Pipe hanger assembly insulation.

1. Hanger rod.
2. Flash finish around hanger rod.
3. Reinforced mastic finish.
4. FOAMGLAS® insulation.
5. Cushioning material.
6. Heavy fillet of PITTSEAL® 444.

Due to cellular glass high compressive strength, it is possible in the majority of the cases to design a saddle type pipe support that eliminates the thermal short circuit (Fig. III.39).

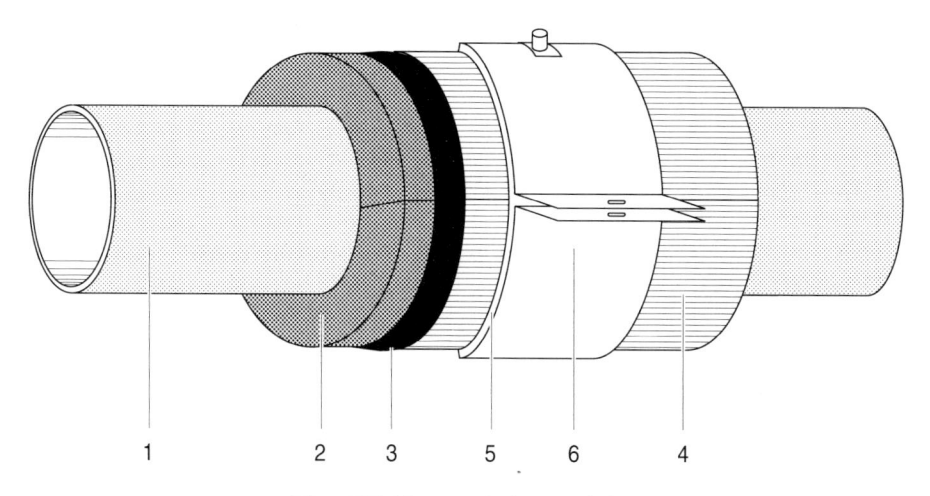

Fig. III.39. Insulating saddle.
1. Pipe.
2. FOAMGLAS® insulation.
3. PC® 88 ADHESIVE or PC® 56 ADHESIVE in fonction of the service temperature).
4. Metal cladding.
5. PITTSEAL® 444.
6. Fixing strap.

In designing the supports, the weight to consider includes the pipeline itself, its insulation and finish, the liquid inside the pipeline, the possible snow and ice load and finally the weight of the man who theoretically should not walk on the insulated line. For this last load it is careful to increase the weight by about 50% to take into account the dynamic aspect.

In the case of fixed points, forces resulting from restrained movements have also to be taken into account.

The total weight is transferred from the pipeline through the FOAMGLAS® insulation and the saddle (and possibly the steel roller) to the support itself.

The normal method of calculation consists in choosing a conventional supporting area corresponding to the contact area between the pipe and the insulation based on a 120° angle (Fig. III.40).

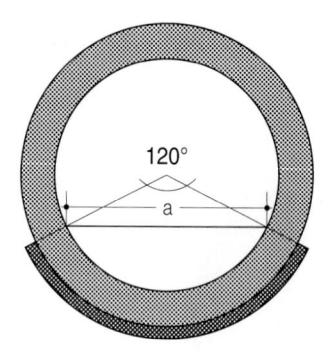

Fig. III.40. Calculation of the length of the saddle.

It is given by:

$$S = a \times L$$

where:

- a is the projection of the supporting area: $a = D \sin 60°$
- L is the length
- D is the diameter of the pipe.

The calculation to determine the length of the saddle is then made on the assumption of pure compressive strength. A safety factor of five is usually chosen. This design and this method of calculation apply for pipeline support in which only axial movements are possible. It should not be adopted for fixed points and situations in which other forces have also to be taken into account.

III.4.19.5 Vessel Supports

In this case, the weight to carry is often such that it would be difficult, if not practically impossible to support through a limited FOAMGLAS® cellular glass surface. Direct contact between the steel support and the vessel is needed. Figs. III.41 and III.42 show how to solve typical cases:

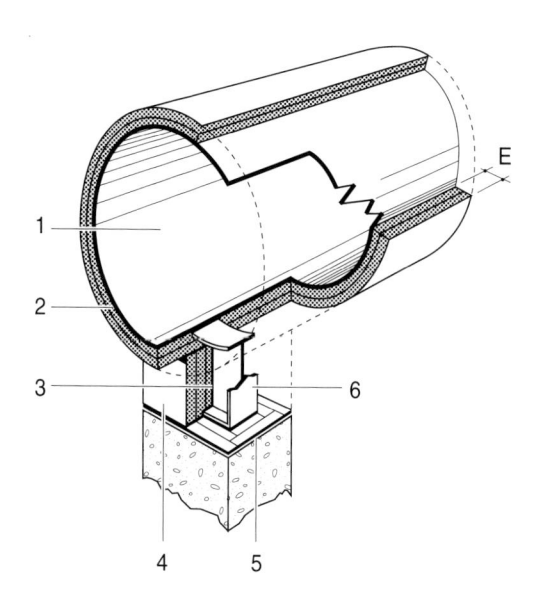

Fig. III.41. Typical example of support for low temperature insulated vessel.

1. Vessel.
2. FOAMGLAS® insulation.
3. PITTSEAL® 444.
4. Reinforced mastic finish.
5. Concrete surface coated with mastic before FOAMGLAS® insulation is applied.
6. Steel support saddle.

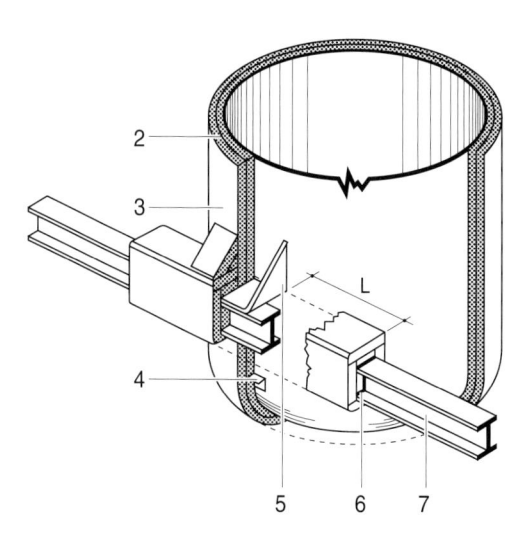

Fig. III.42. Typical example of support for low temperature insulated vessel.
L > 4 × Thickness

1. Vessel.
2. FOAMGLAS® insulation.
3. Reinforced mastic finish.
4. Insulation support angle.
5. Support lug.
6. PITTSEAL® 444.
7. Structural member supporting the vessel.

263

Here again the principle consists in insulating the support sufficiently to reduce the possible thermal short circuit and limit the condensation danger in the case of low temperature applications.

A length of four times the insulation thickness is typical. It should be pointed out that these types of supports are not designed to allow relative movement.

III.4.19.6 Sphere Legs

Figure III.43 presents the classical system to insulate a sphere leg.

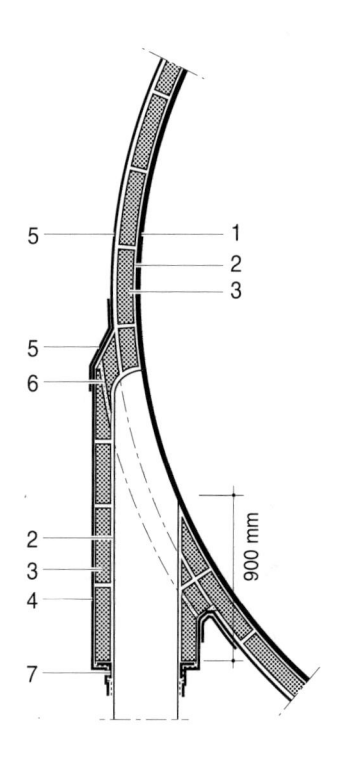

Fig. III.43. Detail of a sphere leg.

1. Sphere wall.
2. PC® 88 ADHESIVE.
3. FOAMGLAS® insulation.
4. PITTCOTE® 404 coating + PC® FABRIC 79P.
5. Supplementary layer of PITTCOTE® 404 + PC® FABRIC 79P.
6. PITTSEAL® 444 sealant.
7. Insulation support angle.

The importance of the supplementary layer of reinforced coating at the leg connection, used as local flashing should be stressed (point 5 on Fig. III.43). The distance of 900 mm has been chosen to provide an insulation of the leg on a length of about four times the normal insulation thickness.

III.4.19.7 Heat Traced Pipelines

The purpose of heat traced lines consists in achieving accurate temperature control of the line or equipment in spite of the heat loss by introducing a heat source in the line. It can be necessary for process control, to avoid freezing of the liquid or more generally solidification of material that could result in major problems. Practically the heat can be brought by electric tracing or by a tube carrying a high temperature fluid. Electric tracing is spirally wound or linear, i.e. parallel to the line. Tube tracings are usually linear. The heat traced lines have to comply with the normal requirements corresponding to the applications and to the specific requirements of accurate temperature control. When using blanket insulation or leaving a void between the pipeline and the FOAMGLAS® insulation, it is important that no water or water vapour gets into the system.

When the insulation is grooved or when a void is created between the pipeline and the insulation, it should not be broken or cracked by mechanical abuse such as heavy weight moving on the insulated pipeline. When blanket insulation is used, it should be high density mineral wool presenting good compressive strength and selected in function of the operating temperature of the tracing system (Figs. III.44 and III.45).

Grooving the FOAMGLAS® cellular glass pipe coverings or pipe segments should only be carried out when the remaining part of the FOAMGLAS® cellular glass element has a sufficient thickness to avoid breakage or cracks. Typically the remaining thickness should be equal to or exceed 4 cm for small segments and 5 cm for normal size elements (Figs. III.46 and III.47).

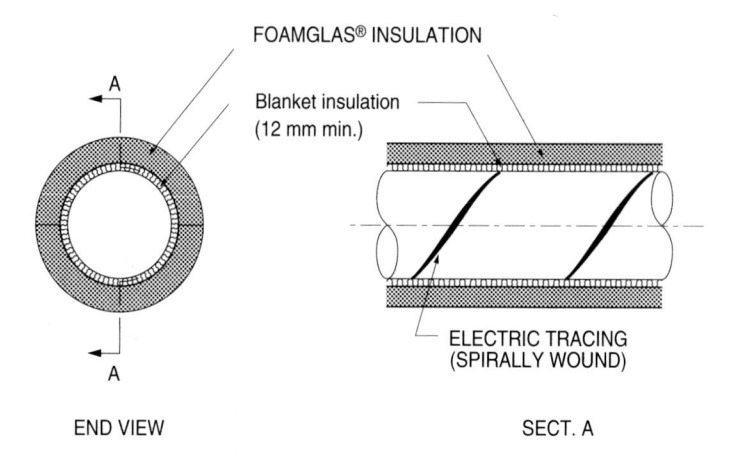

Fig. III.44. Electric tracing spirally wound.

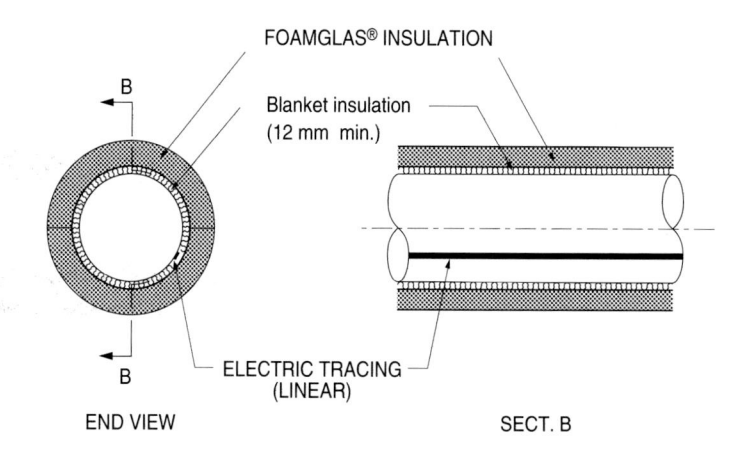

Fig. III.45. Linear electric tracing.

Although thermally not very efficient, a solution consists of placing the tracers in the joints between the insulating elements in contact with the pipeline to heat (Figs. III.48 and III.49).

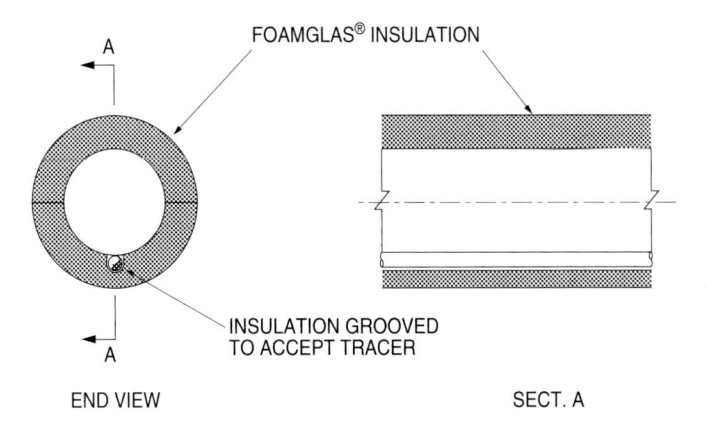

Fig. III.46. Insulation grooved to accept a tracer.

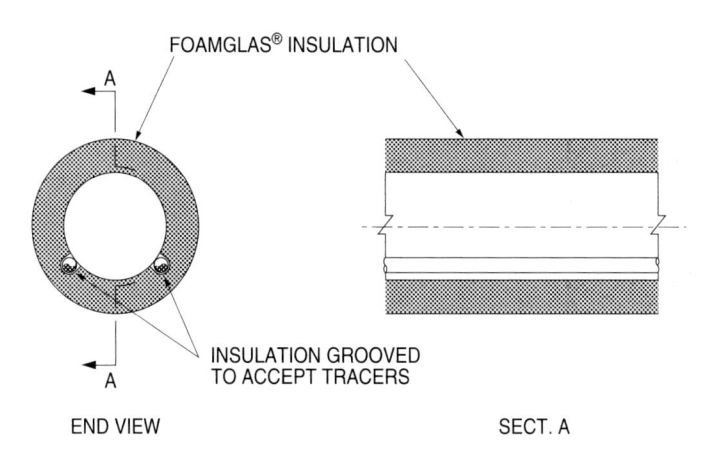

Fig. III.47. Insulation grooved to accept tracers.

When the diameter of the pipeline and its insulation thickness allow it, one oversized pipe shell may be used to enclose the tracers and the normal pipe insulation applied on the other part of the pipeline (Fig. III.50).

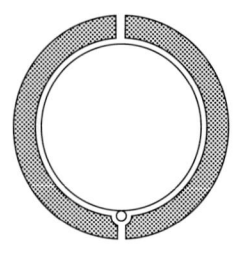

Fig. III.48. One tracer applied in the joints between pipe shells [max OD: 450 mm and FOAMGLAS® insulation thickness remaining at least 50 mm].

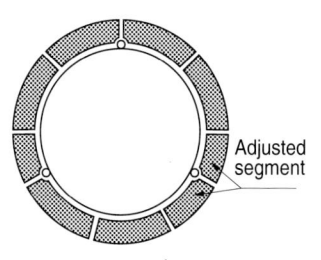

Fig. III.49. Three tracers applied in the joints between adjusted pipe segments.

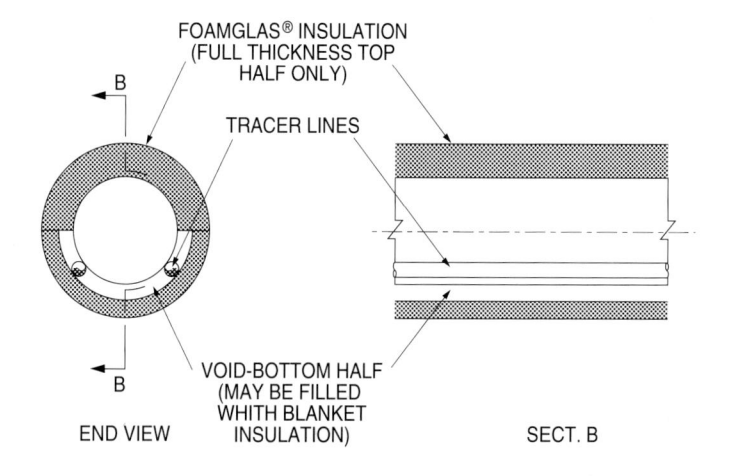

Fig. III.50. Combination of normal and oversized pipe shell.

III.4.19.8 Particular Points of Underground Insulated Pipelines

Due to their specific application conditions, such as for instance the ground reaction and the need to control their movements, underground pipelines require special attention to their particular points. As already explained, it is considered for straight line that the insulation and its jacket (normally PITTWRAP®) are more or less fixed in the ground by the backfill pressure (see III.4.7) and that the pipeline can move within the insulation, following its dilatation. This creates the need for expansion chambers.

They can be built with oversized pipe covering protected from mechanical pressure by a concrete chamber. The free space between the pipe and the insulation is calculated to allow the calculated movement when the pipe and the chamber reach their operating temperatures. The pipe is supported by concrete pads applied on FOAMGLAS® cellular glass coated with PC® HIGH TEMPERATURE ANTI-ABRASIVE. The size and the spacing of the pads are determined to limit the compressive strength on the cellular glass to an acceptable value (Fig. III.51).

Particularly for very large diameters, when the application of very large oversized pipe coverings or segments may become difficult, the "hot box" design may be chosen for the expansion chamber. In this case, FOAMGLAS® cellular glass slabs are used to internally line the concrete chamber. It is then very important to achieve perfect water and water vapour tightness for the joints of the concrete chamber cover (Fig. III.52).

Instead of an expansion chamber, which generally requires considerable space, an expansion bellow chamber can be used. Thanks to the possibility of bellows to accommodate substantial movement in a limited volume, these chambers can remain fairly small; since flanges and bellows may need inspection and eventually service, the concrete covers should be removable to allow openings and visits. Here again, it is of great importance to achieve water tightness by selecting a permanent seal, such as PITTSEAL® 444 (Fig. III.53).

A similar design can be adopted for accessible valve chambers (Fig. III.54).

If the pipeline has to be given the possibility to expand freely, its movements should only be axial and well controlled. This is the purpose of the line anchors, which should be placed at regular intervals and when necessary near angular points. They should have the required dimensions and weight, to be real fixed points.

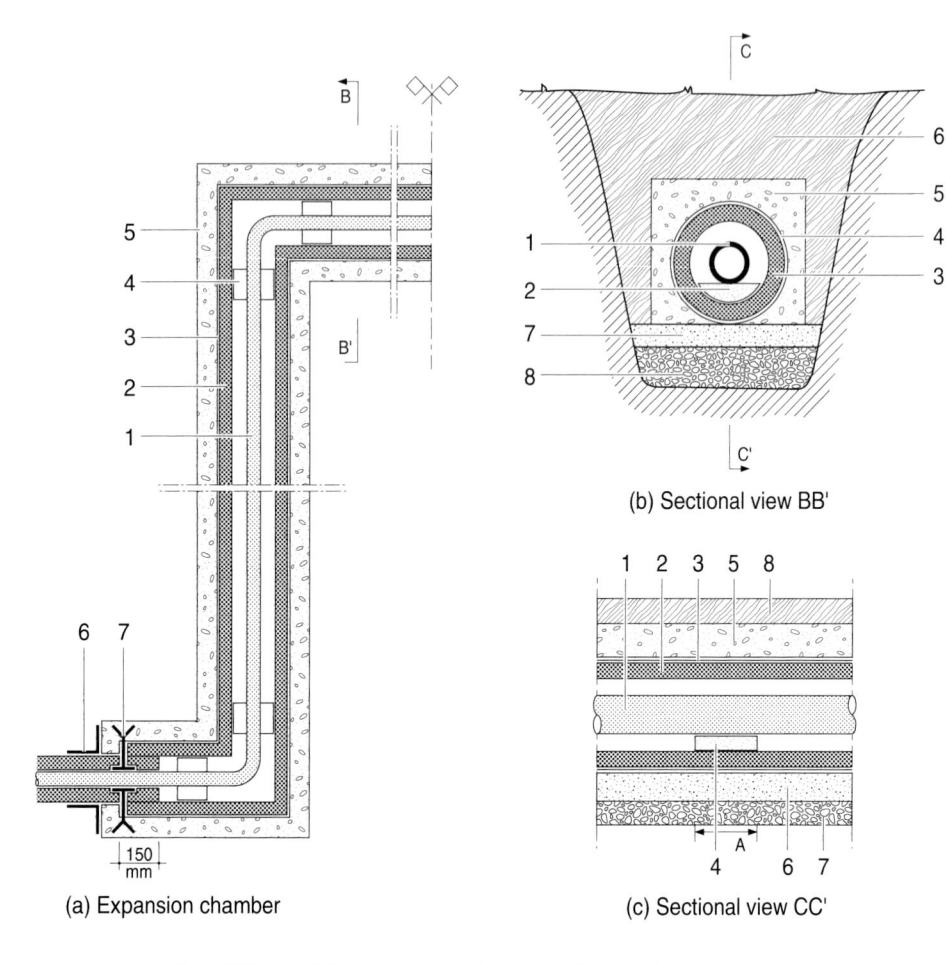

Fig. III.51. Underground insulated pipeline expansion chamber.

(a) **Expansion chamber** (oversized pipe covering)
1. Pipe.
2. Oversized FOAMGLAS® pipe covering.
3. PITTWRAP® finish.
4. Concrete support pad.
5. Concrete chamber.
6. Flashing.
7. Steel guide.

(b) **Sectional view BB'**
1. Pipe.
2. Concrete support pad.
3. Oversized FOAMGLAS® pipe covering.
4. PITTWRAP® finish.
5. Concrete chamber.
6. Backfill.
7. Sand backfill.
8. Gravel.

(c) **Sectional view CC'**
1. Pipe.
2. Oversized FOAMGLAS® pipe covering.
3. PITTWRAP® finish.
4. Concrete support pad.
5. Concrete chamber.
6. Sand backfill.
7. Gravel.
8. Backfill.

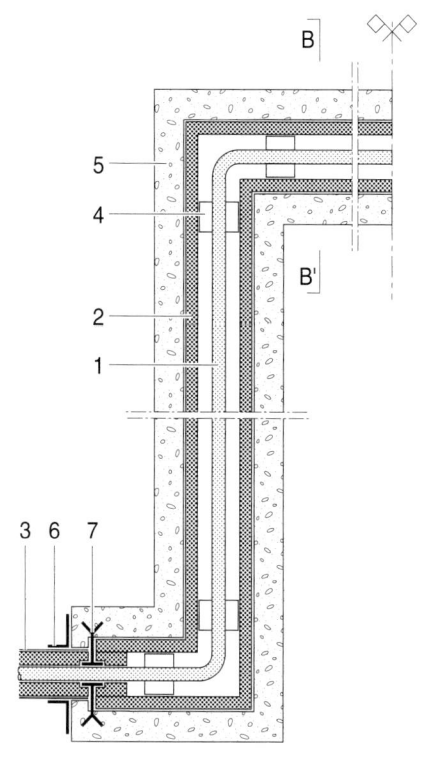

Fig. III.52. Underground insulated pipeline with "hot box" expansion chamber.

(a) **Expansion chamber**
1. Pipe.
2. FOAMGLAS® slabs adhered with PC® 56 ADHESIVE.
3. PITTWRAP® finish.
4. Concrete support pad.
5. Concrete chamber.
6. Flashing.
7. Steel guide.

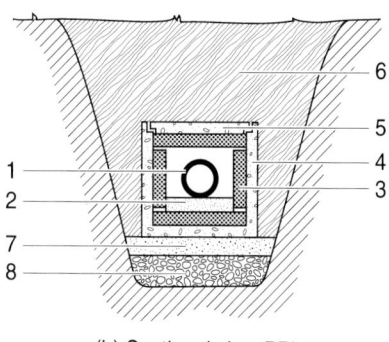

(b) **Sectional view BB'**
1. Pipe.
2. Concrete support pad.
3. FOAMGLAS® slabs adhered with PC® 56 ADHESIVE.
4. Concrete chamber.
5. Concrete chamber cover.
6. Backfill.
7. Sand backfill.
8. Gravel.

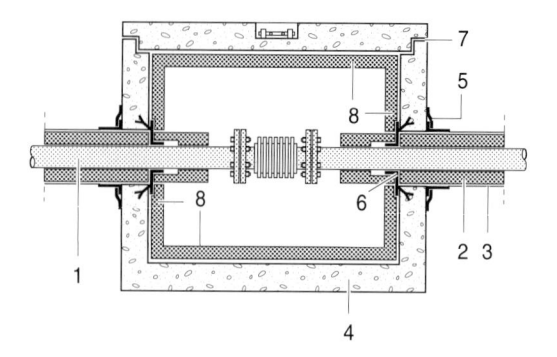

Fig. III.53. Expansion bellow chamber.

1. Pipe.
2. FOAMGLAS® pipe covering.
3. PITTWRAP® finish.
4. Concrete chamber.
5. Flashed and counter flashed covered with glass fabric reinforced PITTCOTE® 300.
6. Steel guide.
7. Watertight seal.
8. FOAMGLAS® slabs adhered with PC® 56 ADHESIVE.

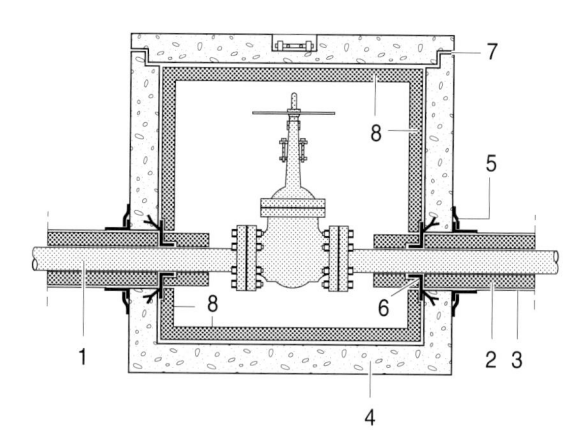

Fig. III.54. Valve chamber.

1. Pipe.
2. FOAMGLAS® pipe covering.
3. PITTWRAP® finish.
4. Concrete chamber.
5. Flashed and counter flashed covered with glass fabric reinforced PITTCOTE® 300.
6. Steel guide.
7. Watertight seal.
8. FOAMGLAS® slabs adhered with PC® 56 ADHESIVE.

Figures III.55 and III.56 show how fix points can be designed.

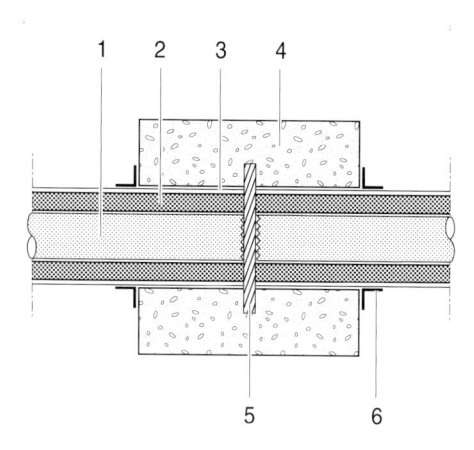

Fig. III.55. Line anchor with plate at right angle to the pipe.

1. Pipe.
2. FOAMGLAS® pipe covering.
3. PITTWRAP® finish.
4. Concrete block designed large enough for firm anchorage.
5. Steel plate welded to pipe.
6. Flashing.

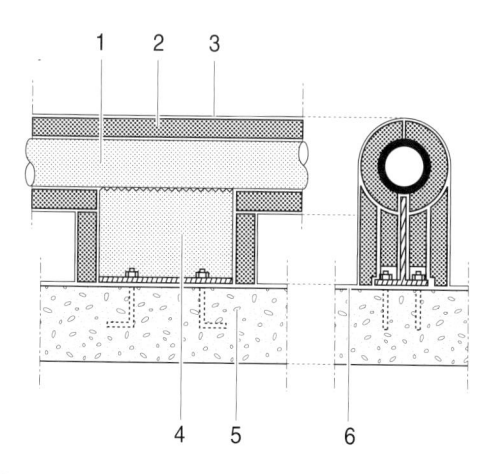

Fig. III.56. Line anchor with plate parallel to the pipe.

1. Pipe.
2. FOAMGLAS® pipe covering.
3. PITTWRAP® finish.
4. Steel plate or I beam section.
5. Concrete block designed large enough for firm anchorage.
6. Finish flashed to concrete.

Due to their dimensions and thermal resistance, the concrete blocks are not insulated. In case of need, they could be encapsulated with a layer of FOAMGLAS® cellular glass slabs covered with PITTWRAP®.

III.4.20 TABLE OF STANDARDISATION OF FABRICATED PRODUCTS, INSIDE, JOINTS AND OUTSIDE AREA

Flat Slabs (Fig. III.57 and Table III.13)

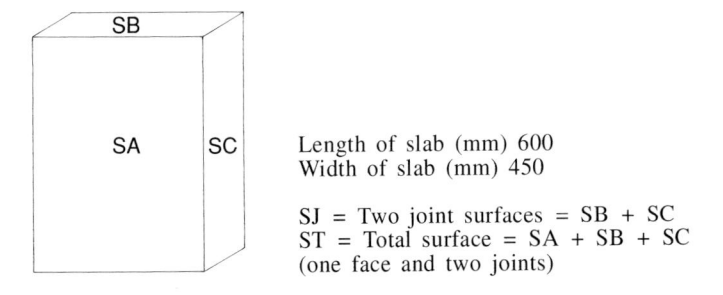

Length of slab (mm) 600
Width of slab (mm) 450

SJ = Two joint surfaces = SB + SC
ST = Total surface = SA + SB + SC
(one face and two joints)

Fig. III.57. Flat slab surfaces.

TABLE III.13

Thicknesses (mm)	Partial surface of slab (m^2)					Corresponding surface per m^2 of insulated surface (m^2)				
	SA	SB	SC	SJ	ST	SA	SB	SC	SJ	ST
40	0.270	0.018	0.024	0.042	0.312	1.000	0.067	0.089	0.156	1.156
50	0.270	0.023	0.030	0.053	0.323	1.000	0.083	0.111	0.194	1.194
60	0.270	0.027	0.036	0.063	0.333	1.000	0.100	0.133	0.233	1.233
70	0.270	0.032	0.042	0.074	0.344	1.000	0.117	0.156	0.272	1.272
80	0.270	0.036	0.048	0.084	0.354	1.000	0.133	0.178	0.311	1.311
90	0.270	0.041	0.054	0.095	0.365	1.000	0.150	0.200	0.350	1.350
100	0.270	0.045	0.060	0.105	0.375	1.000	0.167	0.222	0.389	1.389
110	0.270	0.050	0.066	0.116	0.386	1.000	0.183	0.244	0.428	1.428
120	0.270	0.054	0.072	0.126	0.396	1.000	0.200	0.267	0.467	1.467
130	0.270	0.059	0.078	0.137	0.407	1.000	0.217	0.289	0.506	1.506
140	0.270	0.063	0.084	0.147	0.417	1.000	0.233	0.311	0.544	1.544
150	0.270	0.068	0.090	0.158	0.428	1.000	0.250	0.333	0.583	1.583

Pipe Shells

These shells are fabricated following the ASTM method that means that the outer diameter of a shell always corresponds to the standard outer diameter of a standard steel pipe (Table III.14).

TABLE III.14
Standardisation of pipe shells & inside, joint and outside surfaces per linear meter of piping

Pipe size DN	inch	mm	Inside m²/m	1 FG th. mm	Joint m²/m	Outside m²/m	1 1/2 FG th. mm	Joint m²/m	Outside m²/m	2 FG th. mm	Joint m²/m	Outside m²/m	2 1/2 FG th. mm	Joint m²/m	Outside m²/m	3 FG th. mm	Joint m²/m	Outside m²/m	3 1/2 FG th. mm	Joint m²/m	Outside m²/m	4 FG th. mm	Joint m²/m	Outside m²/m
8	1/4	13.5	0.042	31	0.069	0.237	37	0.084	0.275	50	0.117	0.357	63	0.151	0.438	77	0.190	0.526	90	0.229	0.608	102	0.266	0.683
10	3/8	17.2	0.054	29	0.065	0.236	42	0.097	0.318	55	0.131	0.400	61	0.147	0.437	75	0.186	0.525	88	0.224	0.607	100	0.261	0.682
15	1/2	21.3	0.067	27	0.061	0.237	40	0.093	0.318	52	0.124	0.394	68	0.168	0.494	85	0.217	0.601	98	0.257	0.683	111	0.299	0.764
20	3/4	26.9	0.085	24	0.054	0.235	37	0.086	0.317	50	0.120	0.399	65	0.161	0.493	83	0.214	0.606	96	0.254	0.688	108	0.292	0.763
		30.0	0.094	29	0.067	0.276	42	0.100	0.358	54	0.132	0.434	69	0.174	0.528	81	0.209	0.603	94	0.249	0.685	107	0.291	0.767
25	1	33.7	0.106	27	0.063	0.276	40	0.095	0.357	53	0.130	0.439	67	0.169	0.527	80	0.208	0.609	92	0.245	0.684	105	0.286	0.766
		38.0	0.119	25	0.058	0.276	38	0.091	0.358	50	0.123	0.434	65	0.165	0.528	77	0.200	0.603	90	0.240	0.685	103	0.282	0.767
32	1 1/4	42.4	0.133	23	0.054	0.278	42	0.103	0.397	48	0.119	0.435	62	0.158	0.523	75	0.196	0.604	88	0.236	0.686	101	0.278	0.768
		44.5	0.140	28	0.067	0.316	41	0.100	0.397	47	0.117	0.435	61	0.156	0.523	74	0.194	0.605	87	0.234	0.686	99	0.272	0.762
40	1 1/2	48.3	0.152	26	0.062	0.315	39	0.096	0.397	54	0.137	0.491	72	0.189	0.604	85	0.229	0.686	97	0.268	0.761	111	0.315	0.849
		51.0	0.160	25	0.060	0.317	38	0.094	0.399	53	0.135	0.493	70	0.184	0.600	83	0.224	0.682	96	0.266	0.763	110	0.313	0.851
		57.0	0.179	28	0.068	0.355	41	0.103	0.437	50	0.128	0.493	68	0.181	0.606	81	0.221	0.688	93	0.259	0.763	108	0.309	0.858
50	2	60.3	0.189	27	0.066	0.359	39	0.098	0.434	54	0.140	0.529	66	0.176	0.604	79	0.216	0.686	92	0.257	0.767	106	0.304	0.855
		70.0	0.220	28	0.070	0.396	44	0.114	0.496	49	0.129	0.528	61	0.164	0.603	74	0.204	0.685	87	0.246	0.767	101	0.292	0.855
65	2 1/2	76.1	0.239	25	0.063	0.396	41	0.107	0.497	58	0.157	0.603	71	0.197	0.685	83	0.235	0.761	98	0.285	0.855	110	0.327	0.930
80	3	88.9	0.279	25	0.065	0.436	39	0.104	0.524	52	0.142	0.606	65	0.182	0.688	77	0.221	0.763	92	0.271	0.857	104	0.313	0.933
	3 1/2	101.6	0.319	33	0.089	0.527	46	0.128	0.608	57	0.161	0.677	71	0.206	0.765	85	0.253	0.853	98	0.298	0.935	111	0.346	1.017
		108.0	0.339	30	0.082	0.528	42	0.117	0.603	55	0.157	0.685	68	0.199	0.767	82	0.246	0.855	94	0.287	0.930	107	0.334	1.012
100	4	114.3	0.359	27	0.074	0.529	39	0.109	0.604	52	0.149	0.686	65	0.191	0.767	79	0.238	0.855	92	0.283	0.937	104	0.327	1.013
	4 1/2	127.0	0.399	33	0.094	0.606	46	0.134	0.688	58	0.172	0.763	73	0.222	0.858	85	0.264	0.933	98	0.311	1.015	114	0.372	1.115
		133.0	0.418	30	0.086	0.606	42	0.122	0.682	55	0.164	0.763	69	0.211	0.851	82	0.256	0.933	95	0.303	1.015	111	0.364	1.115
125	5	139.7	0.439	26	0.075	0.602	39	0.114	0.684	52	0.156	0.766	66	0.203	0.854	78	0.245	0.929	91	0.292	1.011	107	0.352	1.111
		159.0	0.500	29	0.087	0.682	42	0.128	0.763	56	0.175	0.851	69	0.220	0.933	82	0.267	1.015	98	0.328	1.115	110	0.375	1.191
150	6	168.3	0.529	25	0.075	0.686	37	0.114	0.761	52	0.164	0.855	64	0.206	0.931	77	0.253	1.013	93	0.313	1.113	106	0.364	1.195
	7	193.7	0.609				39	0.126	0.854	51	0.167	0.929	64	0.214	1.011	80	0.275	1.111	93	0.326	1.193	106	0.378	1.275
200	8	219.1	0.688				39	0.131	0.933	52	0.178	1.015	67	0.234	1.109	80	0.285	1.191	93	0.338	1.273	106	0.392	1.354
	9	244.5	0.768				39	0.136	1.013	55	0.196	1.114	67	0.243	1.189	80	0.296	1.271	93	0.350	1.352	100	0.380	1.396
250	10	273.0	0.858				40	0.146	1.109	53	0.196	1.191	66	0.249	1.272	79	0.304	1.354	88	0.342	1.411			
	11	298.5	0.938				41	0.155	1.195	53	0.204	1.271	66	0.258	1.352									
300	12	323.9	1.018				41	0.160	1.275	53	0.211	1.351												
350	14	355.6	1.117				37	0.150	1.350	47	0.193	1.412												
		368.0	1.156				40	0.165	1.407															

FG th. = FOAMGLAS® cellular glass thickness in mm.
Joint = joint surface in m² per meter (both longitudinal and radial joints are included).
Inside = inside surface in m² per meter.
Outside = outer surface in m² per meter.
Note: the ASTM standardisation method is followed by *Pittsburgh Corning Europe* fabrication shop.

Elbows

The insulation thicknesses are the same as for the pipe shells, which guarantees a perfect connection between both (Table III.15).

TABLE III.15
Standardisation of elbows & inside, joint and outside surfaces per piece (pc)

Nominal thickness (inch)				1	1 1/2		2		2 1/2		3		3 1/2		4	
Pipe size			Inside	Joint	Outside	Joint	Outside	Joint	Outside	Joint	Outside	Joint	Outside	Joint	Outside	Joint	Outside
DN	inch	mm	m²/pc	m²/pc	m²/pc	m²/pc	m²/pc	m²/pc	m²/pc	m²/pc	m²/pc	m²/pc	m²/pc	m²/pc	m²/pc	m²/pc	m²/pc
15	1/2	21.3	0.0034	0.0069	0.0119	0.0118	0.0160	0.0175	0.0198	0.0263	0.0248	0.0376	0.0302	0.0472	0.0343	0.0577	0.0384
20	3/4	26.9	0.0054	0.0070	0.0149	0.0122	0.0201	0.0184	0.0253	0.0275	0.0312	0.0394	0.0384	0.0493	0.0436	0.0601	0.0484
		30.0	0.0067	0.0096	0.0195	0.0155	0.0253	0.0223	0.0306	0.0313	0.0373	0.0402	0.0426	0.0503	0.0484	0.0614	0.0542
25	1	33.7	0.0084	0.0096	0.0219	0.0157	0.0284	0.0229	0.0348	0.0320	0.0418	0.0413	0.0483	0.0514	0.0543	0.0627	0.0608
		38.0	0.0107	0.0095	0.0248	0.0159	0.0321	0.0231	0.0388	0.0328	0.0473	0.0421	0.0540	0.0527	0.0613	0.0643	0.0686
32	1 1/4	42.4	0.0133	0.0094	0.0277	0.0196	0.0397	0.0235	0.0434	0.0332	0.0522	0.0430	0.0604	0.0539	0.0685	0.0657	0.0767
		44.5	0.0147	0.0124	0.0331	0.0197	0.0417	0.0236	0.0456	0.0335	0.0548	0.0434	0.0634	0.0544	0.0720	0.0662	0.0799
40	1 1/2	48.3	0.0173	0.0122	0.0359	0.0197	0.0452	0.0303	0.0559	0.0440	0.0688	0.0552	0.0780	0.0672	0.0866	0.0820	0.0966
		51.0	0.0193	0.0121	0.0381	0.0198	0.0479	0.0306	0.0593	0.0442	0.0721	0.0556	0.0819	0.0680	0.0917	0.0829	0.1023
		57.0	0.0240	0.0152	0.0477	0.0238	0.0586	0.0307	0.0662	0.0452	0.0814	0.0569	0.0924	0.0694	0.1025	0.0850	0.1152
50	2	60.3	0.0269	0.0151	0.0510	0.0236	0.0617	0.0347	0.0751	0.0454	0.0858	0.0573	0.0974	0.0702	0.1090	0.0858	0.1215
		70.0	0.0363	0.0181	0.0653	0.0303	0.0819	0.0346	0.0870	0.0457	0.0995	0.0583	0.1130	0.0718	0.1264	0.0880	0.1409
65	2 1/2	76.1	0.0429	0.0171	0.0710	0.0298	0.0891	0.0457	0.1082	0.0586	0.1229	0.0722	0.1364	0.0891	0.1533	0.1049	0.1668
80	3	88.9	0.0585	0.0196	0.0914	0.0324	0.1098	0.0450	0.1269	0.0587	0.1440	0.0730	0.1598	0.0909	0.1796	0.1073	0.1954
	3 1/2	101.6	0.0764	0.0299	0.1260	0.0434	0.1456	0.0569	0.1621	0.0728	0.1832	0.0911	0.2043	0.1088	0.2238	0.1274	0.2434
		108.0	0.0863	0.0284	0.1343	0.0417	0.1535	0.0565	0.1743	0.0724	0.1951	0.0911	0.2174	0.1087	0.2366	0.1277	0.2574
100	4	114.3	0.0967	0.0265	0.1424	0.0402	0.1627	0.0555	0.1847	0.0717	0.2067	0.0908	0.2304	0.1093	0.2524	0.1282	0.2727
	4 1/2	127.0	0.1194	0.0365	0.1814	0.0526	0.2059	0.0690	0.2284	0.0896	0.2566	0.1082	0.2792	0.1284	0.3036	0.1549	0.3337
		133.0	0.1309	0.0344	0.1900	0.0501	0.2136	0.0675	0.2392	0.0879	0.2668	0.1075	0.2924	0.1280	0.3180	0.1550	0.3495
125	5	139.7	0.1445	0.0313	0.1982	0.0480	0.2251	0.0659	0.2520	0.0867	0.2810	0.1060	0.3058	0.1270	0.3327	0.1544	0.3658
		159.0	0.1871	0.0396	0.2554	0.0586	0.2860	0.0806	0.3190	0.1018	0.3496	0.1240	0.3802	0.1529	0.4178	0.1766	0.4461
150	6	168.3	0.2097	0.0353	0.2720	0.0540	0.3019	0.0775	0.3392	0.0985	0.3691	0.1212	0.4015	0.1508	0.4414	0.1758	0.4738
	7	193.7	0.2777			0.0647	0.3896	0.0871	0.4240	0.1113	0.4613	0.1429	0.5071	0.1694	0.5444	0.1970	0.5817
200	8	219.1	0.3553			0.0725	0.4818	0.0984	0.5240	0.1308	0.5727	0.1589	0.6148	0.1880	0.6570	0.2182	0.6992
	9	244.5	0.4425			0.0804	0.5837	0.1157	0.6416	0.1443	0.6850	0.1749	0.7321	0.2066	0.7791	0.2238	0.8045
250	10	273.0	0.5517			0.0922	0.7133	0.1237	0.7659	0.1561	0.8184	0.1895	0.8710	0.2137	0.9073		
	11	298.5	0.6596			0.1017	0.8407	0.1343	0.8938	0.1693	0.9512						
300	12	323.9	0.7766			0.1099	0.9732	0.1449	1.0307								
350	14	355.6	0.9360			0.1091	1.1308	0.1385	1.1834								
		368.0	1.0024			0.1213	1.2204										

Joint = joint surface in m² per elbow (both longitudinal and radial joints are included).
Inside = inside surface in m² per elbow.
Outside = outer surface in m² per elbow.

Pipe Segments

These segments are fabricated with thicknesses varying from 40 mm up to 100 mm with steps of 10 mm. The number of segments needed to cover the circumference on a pipe depends of the pipe diameter (Table III.16).

TABLE III.16
Standardisation of pipe segments & inside, joint and outside surfaces per linear meter

Thickness (mm)				40		50		60		70		80		90		100	
Pipe size			Inside	Joint	Outside	Joint	Outside	Joint	Outside	Joint	Outside	Joint	Outside	Joint	Outside	Joint	Outside
DN	inch	mm	m²/m	m²/m	m²/m	m²/m	m²/m	m²/m	m²/m	m²/m	m²/m	m²/m	m²/m	m²/m	m²/m	m²/m	m²/m
250	10	273.0	0.858													0.895	1.486
	11	298.5	0.938									0.679	1.440	0.813	1.503	0.909	1.566
300	12	323.9	1.018							0.599	1.457	0.729	1.520	0.825	1.583	0.972	1.646
350	14	355.6	1.117					0.551	1.494	0.646	1.557	0.782	1.620	0.885	1.683	1.039	1.745
		368.0	1.156			0.434	1.470	0.554	1.533	0.686	1.596	0.788	1.659	0.936	1.722	1.045	1.784
	15	381.0	1.197	0.348	1.448	0.463	1.511	0.559	1.574	0.690	1.637	0.833	1.700	0.942	1.762	1.102	1.825
400	16	406.4	1.277	0.373	1.528	0.494	1.591	0.597	1.654	0.735	1.717	0.844	1.779	0.999	1.842	1.115	1.905
		419.0	1.316	0.376	1.568	0.498	1.630	0.600	1.693	0.739	1.756	0.889	1.819	1.005	1.882	1.172	1.945
	17	431.8	1.357	0.399	1.608	0.501	1.671	0.635	1.734	0.744	1.796	0.894	1.859	1.056	1.922	1.178	1.985
450	18	457.2	1.436	0.404	1.688	0.533	1.750	0.672	1.813	0.788	1.876	0.945	1.939	1.068	2.002	1.242	2.065
		470.0	1.477	0.427	1.728	0.536	1.791	0.677	1.854	0.793	1.916	0.950	1.979	1.119	2.042	1.248	2.105
	19	482.6	1.516	0.429	1.767	0.564	1.830	0.680	1.893	0.833	1.956	0.956	2.019	1.125	2.082	1.255	2.144
500	20	508.0	1.596	0.455	1.847	0.571	1.910	0.718	1.973	0.877	2.036	1.006	2.099	1.182	2.161	1.318	2.224
		521.0	1.637	0.457	1.888	0.599	1.951	0.723	2.014	0.882	2.077	1.012	2.139	1.188	2.202	1.375	2.265
	21	533.4	1.676	0.480	1.927	0.603	1.990	0.756	2.053	0.886	2.116	1.057	2.178	1.194	2.241	1.382	2.304
550	22	558.8	1.756	0.485	2.007	0.634	2.070	0.764	2.133	0.930	2.195	1.068	2.258	1.251	2.321	1.445	2.384
		570.0	1.791	0.508	2.042	0.637	2.105	0.798	2.168	0.935	2.231	1.112	2.293	1.256	2.356	1.451	2.419
	23	584.2	1.835	0.511	2.087	0.666	2.149	0.802	2.212	0.975	2.275	1.118	2.338	1.308	2.401	1.458	2.464
600	24	609.6	1.915	0.536	2.166	0.673	2.229	0.840	2.292	0.984	2.355	1.169	2.418	1.320	2.481	1.522	2.543
		622.0	1.954	0.539	2.205	0.701	2.268	0.844	2.331	1.024	2.394	1.174	2.457	1.371	2.520	1.528	2.582
	25	635.0	1.995	0.541	2.246	0.704	2.309	0.878	2.372	1.028	2.435	1.219	2.498	1.377	2.560	1.585	2.623
650	26	660.4	2.075	0.567	2.326	0.736	2.389	0.886	2.452	1.073	2.515	1.230	2.577	1.434	2.640	1.598	2.703
	27	685.8	2.155	0.592	2.406	0.743	2.469	0.924	2.531	1.082	2.594	1.281	2.657	1.491	2.720	1.661	2.783
700	28	711.2	2.234	0.617	2.486	0.774	2.548	0.962	2.611	1.126	2.674	1.331	2.737	1.503	2.800	1.725	2.863
		720.0	2.262	0.619	2.513	0.777	2.576	0.965	2.639	1.130	2.702	1.335	2.765	1.552	2.827	1.729	2.890
	29	736.6	2.314	0.623	2.565	0.806	2.628	0.970	2.691	1.171	2.754	1.342	2.817	1.560	2.880	1.788	2.942
750	30	762.0	2.394	0.648	2.645	0.813	2.708	1.008	2.771	1.215	2.834	1.393	2.897	1.616	2.959	1.801	3.022
	31	787.4	2.474	0.673	2.725	0.844	2.788	1.046	2.851	1.224	2.914	1.443	2.976	1.628	3.039	1.865	3.102
800	32	812.8	2.553	0.679	2.805	0.876	2.868	1.054	2.930	1.269	2.993	1.494	3.056	1.685	3.119	1.928	3.182
		820.0	2.576	0.700	2.827	0.878	2.890	1.086	2.953	1.271	3.016	1.497	3.079	1.689	3.142	1.932	3.204
	33	838.2	2.633	0.704	2.885	0.908	2.947	1.092	3.010	1.313	3.073	1.505	3.136	1.742	3.199	1.941	3.262
	34	863.6	2.713	0.729	2.964	0.914	3.027	1.130	3.090	1.322	3.153	1.555	3.216	1.799	3.279	2.005	3.341
	35	889.0	2.793	0.735	3.044	0.946	3.107	1.168	3.170	1.366	3.233	1.606	3.296	1.811	3.358	2.068	3.421
900	36	914.4	2.873	0.760	3.124	0.977	3.187	1.176	3.250	1.411	3.312	1.617	3.375	1.868	3.438	2.131	3.501
		920.0	2.890	0.761	3.142	0.979	3.204	1.178	3.267	1.413	3.330	1.659	3.393	1.871	3.456	2.134	3.519

Joint = joint surface in m² per linear meter (both longitudinal and radial joints are included).
Inside = inside surface in m² per meter.
Outside = outer surface in m² per meter.

The number of segments per circumference vary from 6 segments for the smallest diameter up to 16 segments for the largest.

277

III.4.21 APPLICATION GUIDE

In this section, a number of recommendations for the installation of FOAMGLAS® insulation will be given. Although they are mainly aimed at insulation contractors, particularly those who have not yet installed FOAMGLAS® insulation, they probably present also some interest for installation engineers, jobsite inspectors and designers.

The fact that quite a few suggestions and recommendations are provided should not be understood to mean that FOAMGLAS® installation is difficult. Actually, experienced contractors have often indicated that it is easier than other insulating materials, because:

- FOAMGLAS® cellular glass insulation can be easily shaped on the jobsite
- Its geometry, its flatness can be corrected even when it is already in place by abrading it with another piece of FOAMGLAS® cellular glass. Possible "steps" between the slabs can be eliminated in a matter of minutes, if not seconds
- It does not absorb water or water vapour and does not need any special precautions except that its surface should be dry to obtain good adherence when adhesives and sealers are used
- Cellular glass as such is not affected by variations of temperature and relative humidity
- No special precautions have to be taken against fire. Welding near it or sparks due to other activities on the jobsite do not constitute a particular danger for FOAMGLAS® cellular glass, as it is the case for some other insulation materials.

It can be further demonstrated that FOAMGLAS® cellular glass installation is not very difficult by the fact that many insulation contractors, without experience of this specific product, have done excellent work after a short training session. This does not mean that training and experience should not be considered as useful for providing a good and efficient work.

The purpose of the recommendations concerns the quality of the final job but it is also aimed at achieving the expected results with a limited amount of labour, for instance through good equipment and well-planned site organisation.

To make the application guide rather complete, some points already indicated in the beginning of this chapter under the title "Preliminary Conditions before

Insulation Application" will be mentioned again but focusing on the insulation contractors rather than on the designers. Some explanations will not be repeated.

Training

The very first step before even transporting and storing the products on the jobsite should be personnel training for the foreseen application. The specification and the foreseen method of application have to be well known by those who will actually have to apply them, from the site management to the worker level.

Pittsburgh Corning organises training sessions for contractors when necessary, before any large project.

If the team in charge of the job has not already attended such a training session, it is highly advisable to request one. It may cost an evening for travel and generally one working day but it will pay off substantially later on and help to make a good job in a shorter period of time.

Training in *Pittsburgh Corning Training Centre* is often supplemented by training on the jobsite to analyse specific problems.

Jobsite Preparation

FOAMGLAS® cellular glass insulation should be transported and stored vertically and the packages should be protected from weather during transport and storage.

Although negative or very high ambient temperatures do not affect FOAMGLAS® cellular glass itself, some accessory products may be damaged by negative temperatures. PC® 56 ADHESIVE that is a bitumen emulsion would be an example. PC® 88 ADHESIVE is not damaged by negative temperatures but its workability will only be satisfactory when it has been brought back at 20 to 25°C.

In the same way, FOAMGLAS® cellular glass does not need any protection against fire, the only element that can burn being the polyethylene packing. It does not contain any combustible material to present a serious fire hazard.

For accessory products like PC® 88 ADHESIVE or PITTCOTE® 300 that support negative temperatures without degradation but need a temperature of 20 to 25°C to be easily applied, a classic method consists in having intermediate storage in which the next day consumption is kept at the required temperature, for instance by thermostaticly regulated electrical heating.

For large jobsites, planning of the transport and of the storage position is important to minimise disturbance of the installation process and limit physical movements in the field. To achieve this result, good identification of the various shipments is necessary.

Before installation starts, a random check that the surface dimensions and the insulating product dimensions correspond to the design specifications is useful, particularly in the case of piping, equipment and fabricated ware such as pipe coverings, elbows, valves, etc. It should be remembered that insulation should be slightly oversized for hot work to allow for thermal expansion.

On fabricated elements, anti-abrasive coatings should have been taken into account for the determination of the inside dimensions.

Equipment

Scaffolding is expensive, but a safe and stable one is necessary if the work has to be carried out in a correct and efficient manner.

For the same reason, tarpaulins may be necessary depending on the weather conditions. They should protect the applications and the work from rain but allow the drying of coatings, when these are used.

If required on the jobsite, grinding equipment such as the "push box" (Fig. III.7) should be made available. Especially for large jobsites, mobile pipe covering cutting machines are useful to fabricate special elements for complicated items. They also have the merit of eliminating the danger of possible mistakes when transmitting complicated dimensions and geometry from the site to the fabrication shop, possibly located some distance away.

When applying the adhered system, PC® 88 ADHESIVE mixing equipment has to be available on the site. If for small jobsites, simple equipment such as a drilling machine with a mixing blade clamped on it may be used (Fig. III.58), more elaborate equipment is necessary as soon as the number of drums to mix becomes considerable.

A typical machine has a framework that can be dismantled, on which an electric motor is fitted with the mixing blade and on which the drum of PC® 88 ADHESIVE is firmly held. It also has a timer to program the mixing time at the desired value in function of the local conditions such as the temperature.

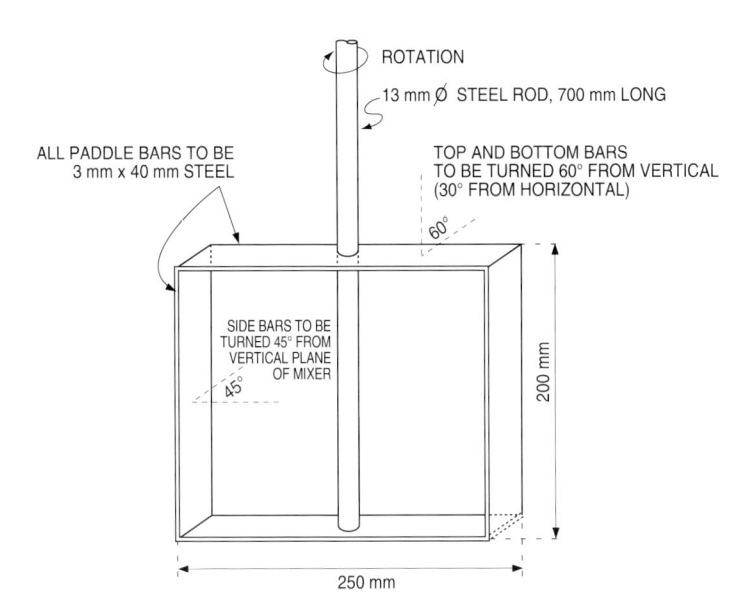

Fig. III.58. Mixing blade.

When this equipment is well designed, the worker has only to position and fix the drum, start the motor and come back when the mixing is done to change the drum. He can use the mixing time to carry another drum to the jobsite or to do other work. The worker time saving and the constant quality of the mixing justifies easily the cost of the mixing equipment (Fig. III.59).

Pittsburgh Corning has detailed descriptions of efficient mobile pipe covering cutting machines and PC® 88 ADHESIVE mixer. A few machines are regularly leased to insulation contractors or Approved Fabricator Contractors (AFC).

For the application of sealers, the use of special guns is recommended since these products are generally difficult to apply with a trowel because of their physical nature. Several special guns have been evaluated for the application of PITTSEAL® 444:

- One corresponds to small jobsites and is manually operated
- Two others are operated with compressed air.

In one case, a certain amount of compressed air is stored in a small vessel, which the operator carries with him to give him access to remote places.

Fig. III.59. Automatic mixer.

In another case, the compressed air comes through a flexible tube directly from the jobsite compressor.

PITTSEAL® 444 is available in drums but also in cartridges adapted to the guns.

Regarding the spraying of PITTCOTE® 300 and PITTCOTE® 404, *Pittsburgh Corning* has tested several machines and can provide reports indicating which one is the most suitable in given conditions.

Depending on the site conditions, the electrically driven equipment may have to be of the explosion proof type.

Questions related to jobsite equipment to install FOAMGLAS® insulation in the best technical and economic conditions are reviewed during the training session given at *Pittsburgh Corning Training Centre*.

Insulated Surfaces

Hydrostatic, radiographic and other tests on the surface shall have been completed before the insulation is applied.

As far as possible, the surface should be regular and smooth to avoid FOAMGLAS® cellular glass elements having to bridge discontinuities, steps, etc. When this is not possible, for instance when steel plates overlap, the FOAMGLAS® cellular glass elements should be locally fitted to the support by abrasion on a jobsite form covered with carborundum or equivalent simulating the discontinuity to insulate.

The surface should be clean and free of any traces of grease, rust or dust. If the surface is corroded or otherwise unsuitable, it shall be sandblasted to grey metal and primed with an acceptable primer.

If necessary it shall be cleaned. The cleaning of the surface is of special importance in the case of stainless steel on which chlorides or fluorides contained in the atmosphere could lead to stress corrosion. Only the part that can be insulated immediately will be cleaned to limit the danger of chloride deposit to a minimum.

Obviously traces of moisture will be eliminated. Shuttering oil, if any, particularly near concrete constructions shall also be eliminated.

Piping and other elements bridging long spaces shall be adequately supported to avoid undesired movements or stresses in the insulation.

Expansion joints, elbows, etc. shall be positioned to avoid undue stress in the insulation.

FOAMGLAS® Cellular Glass Installation

The application specification shall be followed. If this proves difficult, it should be discussed with all concerned before proceeding.

Before adhering or fixing, the fabricated elements shall be dry presented to the surface and the already fixed elements to make sure that they fit with narrow and regular joints. Whenever specified, joints shall be filled with the indicated product; any excess of it shall be troweled flat with the insulation surface.

Voids and cracked slabs or fabricated elements shall be repaired by making a well-filled joint where the crack has appeared. If this is not possible, the total slab or fabricated element shall be replaced. The application of the first layer shall be made quickly after the surface preparation and inspection.

The first layer, which becomes the surface for the second layer, shall be made smooth and even by abrasion if necessary. Any steps shall be levelled. Dust shall be eliminated, particularly in the case of the adhered system, with a vacuum cleaner if necessary.

Whenever weather conditions impose it, the second and the possible third layer will be quickly installed to allow the weather protection to be applied.

If the application specification does not give adequate details, reference can be made to the guidelines for industrial applications or other specialised documents prepared by *Pittsburgh Corning Europe,* although their general character should be recognised.

In the case of the adhered systems, notched trowels for applying the adhesive should have teeth with width, spacing and depth of about 5 mm, although they may range from 3 to 6 mm in function of the surface irregularities.

Two sets of each tool shall be available for each worker to enable him to clean them regularly and avoid the adhesive or sealer clagging the teeth of the trowel.

The mixing time shall be respected.

The potlife of the adhesive shall be respected.

Sealer

Before using a sealer or an adhesive in the case of an adhered specification, it should be verified that it is compatible with the coating if any.

The sealer should be applied with a notched trowel or with a gun.

In the case of PITTSEAL® 444, the use of a gun is strongly recommended.

No open flame should be approached to a sealer drum. With some of them, it could be very dangerous due to the vapour contained in the top part of the drum.

Reinforced Coatings

It should be verified that the fabric corresponds to the coating and can be correctly embedded in it. This requires that the mesh opening, the thickness of the fabric should be adapted to the type of coating. The specified elongation and tensile strength should be respected.

If necessary, the FOAMGLAS® cellular glass surface to be coated should be made smooth by abrasion. Any steps should be levelled. Dust should be eliminated if necessary with a vacuum cleaner. Before applying the coating, the cellular glass surface should be clean, free from dust or any other contamination.

Interior corners of the insulation should be filleted to limit mechanical stresses in the coatings.

Coatings applied on performing insulation materials experience considerable surface temperature variations resulting in mechanical stresses. Other stresses should not be added.

When coatings are foreseen, wire to fix insulation should be avoided as much as possible. Although not so bad as wires, bands also tend to cut coatings because they have a different coefficient of thermal expansion and surface temperatures change considerably.

Expansion springs for bands should also be avoided as much as possible although the length of bands, if any, should not exceed 12 to 15 metres.

Coatings can be applied with trowels or even gloves for a small area. For a large area, airless spray equipment should be used.

The specified coverage should be applied for each layer. It should be clearly defined either after drying or in quantity of wet material, taking into consideration that the water or solvent will evaporate.

Consumption is normally 10 to 20% higher than the actual wet coverage. Before starting and when applying coatings, attention should be paid to the temperatures, the relative humidity and the position of the sun.

The first layer is sprayed through the fabric that is then embedded in it with a squeegee making sure that no air spaces remain behind the coating and that the FOAMGLAS® cellular glass surface open cells are well-filled. Slight pressure with the squeegee is generally sufficient to achieve this result. Alternatively the coating can be applied by a trowel.

The first layer has to dry almost completely before the second layer can be applied.

The first layer has to be inspected to make sure that pinholes do not exist before the second layer can be applied. It should be remembered that a coating with pinholes is generally worse than no coating at all. The overlapping of the fabric should range from 7 to 10 cm. The work should process from the top to the bottom, applying the first coat on the total surface before starting the second coat. The fabric should be completely covered with the second coat but its patterns should remain slightly visible when the coating is fully cured.

Excessive thickness, considerably greater than the specified value, is not an advantage: it takes a very long time to dry and can create stresses in the thinner part of the coating. For the same reason the spray gun should not be stopped on local points.

If a top coat of a different product (for instance Chemglaze No A-276 on PITTCOTE® 404) is applied for aesthetical or protection reasons, the second coat of the main coating should be dry before application of the top coat.

Although this requirement may imply retaining of the scaffolding, it should be taken into account to avoid blisters and problems in the future.

The liquid part of the coating will not be absorbed by FOAMGLAS® insulation and can only evaporate on the free surface.

If blisters (water sacs) develop, the source of water entry should be traced and the problem corrected by repair as soon as possible.

Metal Jacket

As previously indicated in III.2.6, metal jackets are not water vapour barriers but only weather barriers. And even to achieve a good weather barrier, some rules should be followed on the jobsite, the basic one being to avoid water entry but to let water escape if it has penetrated the insulation system.

The overlaps in the longitudinal and in the circumferential directions should be made in the right direction to eliminate the danger of water entry.

When they are specified, sealers between metal jackets should be applied, although they do not always perform as well as expected.

On horizontal lines, particularly at low points, weep-holes at the bottom part of the metal jacket should be considered to drain the water out, if any.

On elbows at the bottom of vertical piping, provisions for drainage should also be considered. Finally, it is important to choose the same metal for all the elements of the metal jacket including small pieces, straps, parker screws, etc. if used, to avoid any electrolytic corrosion.

Details and Singular Points

Here also, avoiding water entry should be the main consideration. Flashing and generous use of high quality sealers are good investments and avoid that, as often, the deterioration of a good insulation system starts too early at singular points such as inlets, outlets, protrusions, supports or connections.

It should be kept in mind that relative movements may take place and open entry for water if flashing and sealing have not been carried out with sufficient care.

Special attention should be paid to the top part of the job where gravity helps water to penetrate in it if all the details have not been made actually tight.

In the case of large size tanks, it should be made sure that water that would have penetrated the tank roof insulation does not find a way to get into the tank wall insulation and from there to the tank base insulation. This problem is of a special importance when permeable insulation is mixed with FOAMGLAS® insulation on the same job: here again good flashing will pay.

At pipe supports, it should be verified that the relative positioning of the pipes and the supports are such that the bearing surfaces actually present the foreseen areas.

III.4.22 MAINTENANCE

It is well known that FOAMGLAS® cellular glass does not age and that its properties are unchanged over many years when it has been correctly installed and used. Many investigations of jobsites as old as 38 years have proven this point.

But other reasons call for regular inspection, maintenance and repair.

Reasons of Maintenance

1. Firstly, although the life of coatings like PITTCOTE® 404 on FOAMGLAS® insulation may exceed 10 years, particularly when a layer of Chemglaze has been applied on it, it is still limited and could be further limited at local points due to particular conditions. In general terms, sealers, coatings and metal jackets tend to age. If this phenomenon is specially slow for sealers, particularly when they are applied between two FOAMGLAS® cellular glass elements, it is more rapid for coatings and for metal jackets. For instance, cases of metal jackets badly deteriorated by sand and wind in large terminals within a few years are known.

2. Deterioration of insulation systems may be caused by storms, hails or other natural causes. Coatings, jackets and flashings, etc. can be badly damaged.

3. Personnel can also create damages: people tend to use large size insulated pipelines as scaffoldings, walk on it, support ladder on it, let fall tools and products on it. These actions often result in local deterioration. Even when platforms and walkways have been installed, workers can use insulated lines as short cuts.

4. Operating procedures may be changed: for instance the operating temperature is sometimes exceeded or unforeseen operations are carried out.

5. The initial insulation design or installation may have been faulty and fail after a few years.

6. Often the insulated equipment has to be repaired or modified and the insulation has to be opened or removed with corresponding damages.

Purpose

In general terms, the purpose of maintenance of an insulation system consists in detecting and correcting defects or damages before extensive replacements or repairs are necessary. As it has been often experienced, a small repair made early can avoid a major problem later on.

Economies of Maintenance

Although it is not always correctly appreciated, maintenance properly carried out in due time should improve the profitability of a plant since it reduces the unplanned shut down and helps to keep the production at the maximum level. Moreover since insulation has generally an excellent return on investment and a short pay back time and since maintenance and repair done in due time are substantially less expensive than a new installation, their return on investment is often higher than for new insulation. Independently of the increased danger of shut down, a deteriorated insulation on a pipe or a vessel means considerable energy loss and the corresponding cost.

Limit of Maintenance

Maintenance is obviously limited to deterioration that can be investigated by energy analysis or by visual inspection carried out without dismantling and with only moderate removal of the insulation system. From this standpoint, it should be pointed out that defects are generally easier to locate and identify under a reinforced coating than under a metal jacket.

When Should Inspection and Maintenance Operations Be Carried out?

Inspection and maintenance operations can be periodic or linked to a special event. On a periodic base, it is advisable to have a plant inspection before summer and before winter, thus twice a year. Moreover a plant inspection should certainly be done after a storm, a fire, an explosion or a serious production incident, etc. It should also be part of a shut down procedure before re-commissioning the plant.

Finally, the analysis of the energy consumption, obviously related to production and other parameters that may influence it, may show an anomaly that suggests that an inspection should take place.

It should also be mentioned that the period of the day should be chosen in relation to the type of plant to be inspected. If it has modest influence in many cases, it is critical for search of condensation places on locations where joints may show through coatings. Early morning after a clear night allows to see more condensation problems than a warm day.

Particular Points to Investigate during Inspection for Maintenance

The points that are mentioned in the following paragraphs should be understood as examples more than as an exhaustive list. It would obviously be very difficult and very lengthy to present a general list containing all possible points to control during an inspection. But such a list, or something close to it, could and should be developed for a specific plant by those who are in charge of its maintenance; it can be based on engineering evaluation, know how and experience with a given plant.

As already mentioned, energy consumption is one of the parameters that should be analysed on a regular basis to find the possible need for maintenance and repair. If a global analysis already offers interest, modern technology allows the inspector to go to a more detailed investigation. It is possible to measure with portable equipment the heat transfer and the surface temperature of a plant in operation. For instance, should the surface temperature at a certain location be very different from the calculated value, it would be wise to study more in depth the real situation. We quote this possibility although in the case of a FOAMGLAS® insulation system, the increase of energy transfer can only result from excessive joints, missing elements in complicated fabricated pieces or similar causes, but not from water absorption by the FOAMGLAS® cellular glass itself. Consequently, the reasons for higher energy transfer or high surface temperature should mainly be found in systems utilising other insulations. This type of phenomenon appears more often near particular points than on the main lines.

The first thing that is generally noted during a plant inspection concerns the possible deterioration of reinforced coatings or metal jackets.

Surface deteriorations such as cracks, blisters in the coatings, trace of impacts, deformations, particularly on the top of horizontal surfaces where standing water may occur, should be noted for quick action.

Joints between the elements can be marked through the coating, particularly when the inspections take place during cold and humid weather. Even general condensation can take place.

In the same way metal jackets can be deteriorated, twisted, or even removed; fixing straps may be cracked or getting loose; overlaps may be placed in the wrong direction letting water enter instead of directing it to the outside; weepholes may be closed by dirt or whatever. Expansion/contraction joints should also be inspected to see if the coating has been damaged locally by these movements. Fittings deserve special attention since breakdowns occur here more often than on straight lines: for instance flashings that can have been removed, broken or twisted.

Renewable insulations, applied with soft sealers and mechanical fixings constitute another point, worth checking with great care to make sure that reinstallation does not allow water entry in the system.

Finally, insulated underground line is one of the most difficult applications to inspect without major openings: here, one has generally to rely on a comparison between measured and calculated temperature drops if the ground does not reveal problems by signs such as burnt grass, evaporating rain water or melting snow.

How to Maintain or Repair Insulation Systems

Even more than for the description for normal application, only general concepts can be given for the methods of maintenance and repair. A detailed specification can only be proposed for a specific problem. But these general concepts can be useful when working on a particular case.

The first question to rise is as follows: should the repair be carried out with the initial specification or should it be modified to improve its performance. Obviously the answer depends on the causes of the identified problems.

An example will illustrate the situation. If a coating has some cracks that run at random without particular pattern and do not penetrate the insulation, ageing could be the cause and the repair may consist in brushing the coating to come back to the sound part of it and applying a further layer of the same coating with a suitable reinforcement. But if a coating shows a clear cut crack, with a straight pattern such as for instance a circumferential one and if the crack penetrates the insulation material, it can be the sign that an expansion joint is either not there or it not operating correctly.

In this case, a repair similar to the previously described one will not last long and the problem will appear again very soon, similar causes providing similar results.

When the damage to the coating clearly indicates frequent foot traffic, the solution generally consists in repairing the coating but also in taking appropriate measures to avoid further foot traffic at the same place. These measures may include an efficient interdiction (difficult solution) or a mechanical solution such as the construction of a foot-bridge.

When traces of spillage or of water coming from the regular testing of a deluge system are seen, the problem can usually be solved by the installation of a well-designed flashing leading the water or the chemicals in another direction... A well-positioned gutter can also solve the problem. Correct flashing is often the solution to cure water entry problems at inlets, outlets, protrusions, etc.

Record

Last but not least, it is always very useful to keep short but accurate records of the inspection, the location of problems and the method that was used to repair them.

This point has even more importance for FOAMGLAS® insulation systems than for others, since FOAMGLAS® insulation systems are by nature designed to perform successfully during a long time.

ANNEX
CALCULATION METHOD FOR ABOVE GROUND INSULATED PIPES

Insulation Thickness Determination

The method of calculation consists in writing that the amount of heat passing through a material by conduction is equal to the sum of the amount of heat coming out of the material by convection and radiation.

Since these three amounts of heat are expressed by rather involved formulas as a function of the surface temperature of the insulation thickness which is unknown, the calculation should be made by numerical method (Fig. III.60).

- Radiation in W/m²:

$$Q_r = 5.67 \times \varepsilon \times \left(\left(\frac{T_a}{100}\right)^4 - \left(\frac{T_s}{100}\right)^4 \right)$$

- Convection in W/m²:

$$Q_{cv} = 1.94 \times (T_a - T_s)^{1.25} \times \sqrt{\frac{w + 1.26}{1.26}}$$

- Conduction in W/m²:

$$Q_{cd} = \frac{2 \times k \times (T_s - T)}{d_2 \times \ln \frac{d_2}{d_1}}$$

T_a = ambient temperature in K
T_s = surface temperature in K
T = service temperature in K
ε = emissivity of outer finish
w = wind speed in km/h
k = thermal conductivity at mean temperature in W/(m.K)
d_1 = external pipe diameter (without insulation) in m
d_2 = external diameter (pipe and insulation) in m
ln = natural logarithm.

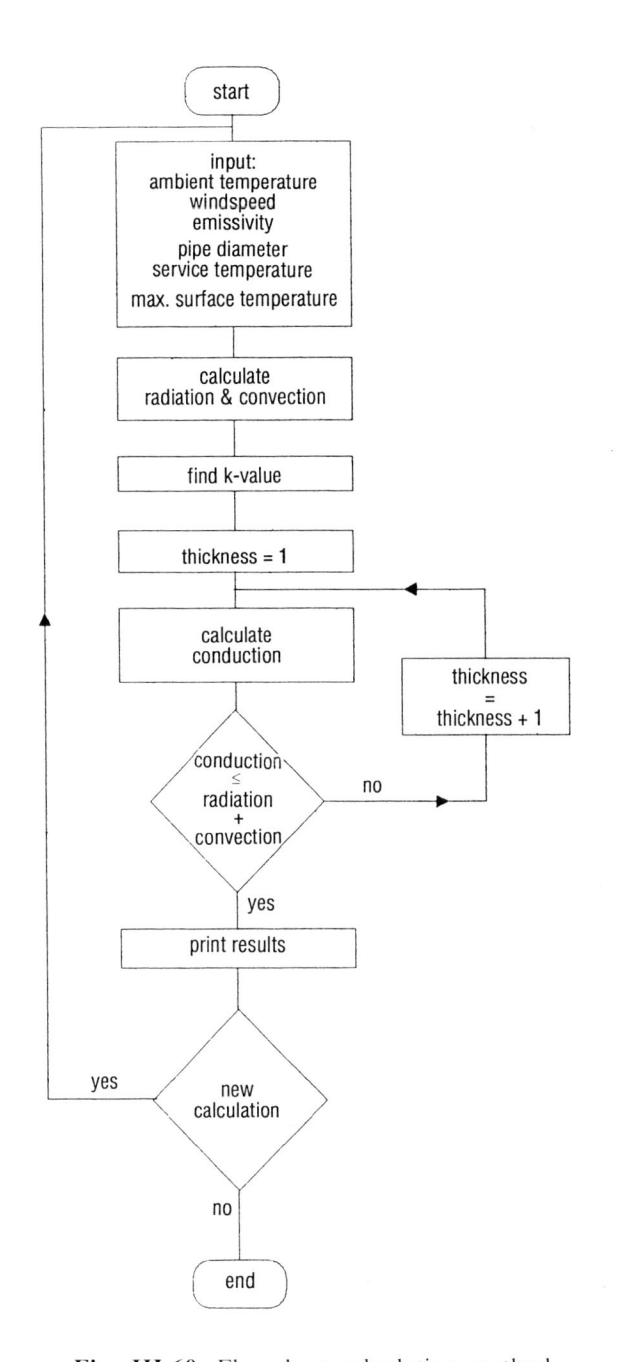

Fig. III.60. Flowchart calculation method.

IV

CASE STUDIES

IV.1

COOL IT!

Construction of *Courage*'s *Berkshire Brewery*, now one of the largest in Europe, began 13 years ago.

The first brew was produced in 1979 and since then two expansion schemes have been carried out.

The most recent of these was completed in 1988 when eight additional fermentation vessels fabricated by *APV Burnett and Rolfe Limited* went on stream, bringing the total number of vessels to 40 and increasing the brewery's capacity by 20 per cent. As a result, 1.8 million barrels (518,400,000 pints or 295 million litres) of beer a year can now be brewed. Ten brands of ale and four lagers are produced.

Fermentation, which is a biochemical reaction between yeast and starch, takes up to 10 days and generates considerable heat. Unless the beer's temperature is carefully controlled its flavour will be spoiled.

Every beer has its own unique fermentation 'temperature profile' within a scale ranging from 4.4 to 15.6°C; the temperature must be held to within one degree of a specific datum on this scale for predetermined lengths of time.

Lagers ferment at lower temperatures, and over a longer period, than other beers and consequently require a larger investment in fermentation vessels and cooling plant. *Courage*'s recent expansion scheme cost about £3 million and was carried out mainly to increase lager output.

To cool the 21 m high, 4.5 m diameter, stainless steel fermentation vessels – whose missile-like profiles are a familiar sight to motorists passing the brewery on the M4 motor way near Reading – propylene glycol is circulated at –2.8°C through stainless steel coils surrounding the vessels.

Each vessel's cooling system is divided into five zones, and each zone has its own independent coil assembly with separate inlets and outlets. A computer constantly monitors the vessels and controls the temperature of each zone by regulating the flow of coolant through the coils.

Efficient insulation is vital to the fermenting process. The bases of the vessels are enclosed within the fermentation plant, and because the ambient temperature is usually higher than that of the fermenting beer, insulation is necessary to prevent heat gain in the vessels. The upper sections protrude through the roof of the plant and have to be also protected from solar gain.

Selecting a suitable insulant for the vessels was quite a challenge for *Courage*'s engineers. Whilst the brewery was still at the design stage there was a series of disastrous fires at food factories and two other breweries, caused or exacerbated by flammable insulation.

At the insistence of *Courage*'s insurers, only non combustible insulants were allowed in application within *Berkshire Brewery*, and therefore FOAMGLAS® cellular glass insulation was specified for the vessels and for the 1500 m of coolant pipework.

Because cellular glass cannot absorb moisture it retains its original insulating value for all time, whereas most other insulants are liable to become damp in service and thus lose much of their efficiency.

Another benefit of insulating brewing plant with cellular glass, is that unlike some organic insulants the material does not contain free chloride.

When materials containing free chloride are placed in contact with stainless steel in highly humid and damp environments, such as breweries, the material is liable to absorb moisture, causing chloride to leach out, attack the steelwork, and induce stress corrosion cracking.

Installation of the FOAMGLAS® cellular glass on most of *Berkshire Brewery*'s fermentation vessels was carried out by *Kitsons Insulation Contractors*, who erected totally enclosed scaffolding around the vessels to allow insulation work to proceed in all weathers and protect the vessels from rain prior to being insulated.

The vessels were first cleaned with Deoxidine derusting compound to remove surface corrosion or contamination, and then washed with cold clean water. Although a chloride-free insulant was being used, the vessels were coated with Foster anti-stress corrosion paint to protect them from any chloride that may leach from ancillary materials applied with the insulation.

Flat 300 × 450 mm FOAMGLAS® cellular glass slabs with a thermal conductivity value of 0.044 W/(m.K) at 10°C were applied vertically, their edges being bevelled in a simple jig to form neat butt joints and make the slabs fit closely to the sides of the vessels. Flexseal, a viscous vapour barrier sealer for low temperature applications, was applied to the butt joints. Overall thickness of the insulation was approximately 90 mm.

The insulation was held in place with stainless steel and polypropylene straps. Aluminium-faced Sisalcraft paper, lap-bonded with Howstik rubber-based adhesive, was then applied between the insulation and the profiled aluminium cladding, made by *Precision Metal Forming Limited*, that protects the vessels from the weather and impact damage.

The cladding was fixed temporarily with rubber straps whilst its overlapping edges were riveted. Giving the cladding independent fixings ensures that if necessary it can be removed without disturbing the insulation.

The domed tops of the vessels required special detailing. Small curved elements of FOAMGLAS® cellular glass were applied and covered with a coating of PC® 88 ADHESIVE reinforced with a glass fabric. Solar reflective glassfibre reinforced polyester (GRP) preformed elements are dry applied over the FOAMGLAS® dome insulation. They can be removed independently and form a weather and mechanical barrier.

The 50, 100 and 165 mm diameter glycol pipes serving the vessels' cooling systems were insulated with preformed FOAMGLAS® insulation sheathed in aluminium.

Use of FOAMGLAS® cellular glass at *Courage* over a period of seven years has helped establish the material in the UK and Ireland as a reliable fireproof and waterproof insulant for fermenting plant.

Several other brewers including *Guinness, Harp Lager* and *Allied-Lyons* have now used the material to benefit from its impermeability, stable *k*-value and ability to protect vessels from stress corrosion cracking.

In one scheme, *Hastie Insulation* was cladding 18 bright beer vessels 9 to 14 m high and 4 to 6 m in diameter with 100 to 130 mm of FOAMGLAS® cellular glass. The cooling and insulation systems installed on some of these vessels are required to ensure that the temperature of the beer does not increase by more than 0.25°C over a 24-hour period.

The exceptionally thick insulation is formed by applying two layers of FOAMGLAS® cellular glass, the first fully embedded in PC® 88 ADHESIVE, the second bonded and sealed with PITTSEAL® 111 [1].

FOAMGLAS® cellular glass has also been used in numerous plant and structural insulation schemes in other breweries, including *Heineken* (Netherlands), *Feldschlösschen* and *Arkina* (Switzerland), *Mao Brewery* and *El Aguila* (Spain), *Maes, Stella Artois* and *Vieux Temps* (Belgium), *Gosser Brauerei* (Austria), *Le Pélican* and *SCA* (France), *Whürer* and *Peroni* (Italy), *Carlsberg* and *Tuborg* (Scandinavia) and *Union Brauerei, Mohr* and *Hofbrauhaus* (Germany).

(See colour photos 17, 18 and 19 between pages 320 and 321).

---

[1] PITTSEAL® 111 is now replaced by PITTSEAL® 444.

IV.2

THERMAL INSULATION ON CHILLED WATER PIPES AT THE LURGI GmbH COOLING SYSTEM IN FRANKFURT

Lurgi, the world-famous engineering company, established their engineering offices during 1987 in outstanding buildings located in the countryside near Frankfurt.

The premises have a large fenestration area, which apart from self-ventilation, is also provided with a temperature regulating system that, whilst not air-conditioning as such, can, in summertime, keep the ambient temperature lower than 30°C.

Moreover, additional cooling systems are placed in the computer rooms and CAD pools, to meet the severe temperature conditions laid down.

The complex being star shaped, each of the seven wings of the building has its own regulating system.

Piping networks run from each technical station fed by three cooling electric sets and water temperature is kept between 6 and 12°C.

As these pipes operate at low temperature, it is essential to cover them with an effective thermal insulation to avoid any substantial loss of cooling efficiency, also to avoid the risk of condensation on their external side and prevent corrosion.

Application

Because of its very nature and its structure, FOAMGLAS® cellular glass is totally impervious to water and water vapour and thus meets both of the above-mentioned objectives without the need for a vapour barrier.

FOAMGLAS® insulation, which has been applied in various thicknesses depending on the pipe diameter, also meets the requirements of those authorities who make it compulsory to use non combustible insulation materials in clearance areas and emergency exits.

The iron pipes, of 33.7 to 355.6 mm diameter, have been sanded to remove any oxidisation and covered with zinc phosphate based rust preventive paint immediately after manufacture.

After being fitted on site, the welded pipe joints and valve/gauge connections of the feeders and measuring devices have also been covered with the same paint; a second layer of paint was then applied over the whole system, according to the DIN 18 364 requirements.

FOAMGLAS® cellular glass pipe shells and elbows, 27 to 63 mm thick were then applied and bonded to the pipes with PC® 56 ADHESIVE, a two-component bituminous adhesive. It has a high resistance to water vapour diffusion ($\mu = 40,000$) and all sectional joints were also fully coated with it, FOAMGLAS® insulation provided the required imperviousness to water vapour.

Afterwards, two galvanised steel straps mechanically secured each 600 mm half shell.

In order to prevent the insulation from being damaged, a 1 mm thick galvanised steel plate was placed around the insulation at every support collar.

Most of the piping network is covered and therefore cannot be seen; the FOAMGLAS® insulation remains uncovered, as neither coating, membrane nor metal jacket are needed as a vapour barrier.

Only those parts that are exposed to mechanical shock or require an aesthetic finish are covered with aluminium jackets.

(See colour photos 20, 21 and 22 between pages 320 and 321).

IV.3

WRAPPING UP AT MOSSMORRAN

Shell brought their Far North Liquid and Associated Gas System (FLAGS) on stream in November 1984 at Mossmorran. Part of its modern specification called for high performance thermal insulation, the subject we consider here with a general description of the plant and its operation.

Correct temperatures in pressure vessels, storage tanks and pipelines at the extraction and fractionation plants served by *Shell Expro*'s £1500M FLAGS project are maintained with the aid of cellular glass insulation slabs and fabricated pipe coverings.

The material was chosen for parts of the plant where thermal cycling outside the temperature ranges of conventional cold or hot insulation is required as part of the process. Cellular glass was also selected for combined load bearing and insulation service.

Temperatures at which the natural gas liquids have to be processed and stored range from –46 to –175°C. This is within the operating limits of FOAMGLAS® cellular glass, which remains efficient when subjected to thermal cycling and also has the benefit of being fireproof and water resistant.

FLAGS collects gas from *Shell*'s Brent oil-fields complex, which has the highest gas/oil ratio of all the major North Sea oil fields. The gas is delivered by a 447 km pipeline on the sea-bed, at a rate of 1100 M standard cubic feet a day, to an extraction plant operated by *Shell* at St Fergus, north of Aberdeen. Here it is separated into methane and natural gas liquids (NGLs).

Methane is delivered to a nearby *British Gas* plant and fed into the natural gas grid, whilst the NGLs are pumped south through a 222 km underground pipeline to *Shell Expro*'s Mossmorran, Fife, fractionation plant.

Before they begin their journey to Mossmorran the NGLs are processed in six dual-temperature vessels, up to 3.1 m in diameter and 6.5 m long overall. These form part of a plant engineered and constructed by *Ralph M. Parsons Co. Ltd* to *Shell* designs and specifications. The vessels and their associated 19-1220 mm diameter pipelines have been insulated with 40-140 mm FOAMGLAS® cellular glass T2 insulation, which has a k-value of 0.044 W/(m.K) at 10°C.

The insulation was fabricated in single, double and triple layers with PC® 85 POWDER used as an adhesive. FOAMGLAS® cellular glass has also been specified for an extension to the plant that went on stream in 1985.

Up to 80,000 barrels of NGLs can be pumped every day from St Fergus to Mossmorran, and fractionated into ethane, propane, butane and natural gasoline. The ethane is piped 'over the fence' to *Esso Chemicals'* Fife Ethylene Plant.

Shell's plant at Mossmorran includes two propane and two butane double containment storage tanks, 35 m high and 45-55 m in diameter, partly embedded in a hillside and surrounded by rock and earth embankments. The bottoms of the tanks are supported by and insulated with 1154 m³ of FOAMGLAS® cellular glass HLB 107 [1] (High Load Bearing) insulation slabs, which have a compressive strength of 0.75 N/mm² and a k-value of 0.045 W/(m.K) at 10°C.

The load bearing concrete ring beams on which the inner walls of the tanks rest are supported and insulated by 115 m³ of FOAMGLAS® cellular glass HLB 125 [2] slabs, which have a compressive strength of 0.88 N/mm² and a k-value of 0.046 W/(m.K) at 10°C.

Cold and dual temperature pipe runs, up to 300 m long and 19 to 915 mm in diameter, are insulated with single and multiple layers of preformed FOAMGLAS® cellular glass T2 coverings up to 150 mm thick. Altogether some 10,000 linear metres of FOAMGLAS® cellular glass coverings, including 1500 preformed valve boxes and 45 and 90 degree bends, were supplied for Mossmorran's pipework.

---

[1] FOAMGLAS® HLB 107 has been recently replaced by HLB 115.
[2] FOAMGLAS® HLB 125 has been recently replaced by HLB 135.

From Mossmorran 38,000 barrels of liquid propane, 15,000 barrels of liquid butane and 10,500 barrels of gasoline can be pumped daily to a deep water terminal at Braefoot Bay, 4 1/2 miles away, and despatched in tankers to customers in the UK and overseas.

It is 669 km from the Brent oil field to Braefoot Bay. Thanks to FLAGS and its associated plants it is possible for gas to leave its reservoir 3048 m below the North Sea at, say, 6 a.m. on a Monday morning and arrive at Braefoot Bay at 10 a.m. the following Saturday, having been converted into valuable fuels and chemicals without being seen or heard – and leaving very few people aware that it has passed their way!

(See colour photo 23 between pages 320 and 321).

IV.4

DURABLE INSULATION FOR LIQUID GAS STORAGE TANKS

In 1969 a double-walled LNG tank with a capacity of 5080 tonnes was built for the *East Midlands Gas Board* at Ambergate, Derbyshire, by *Whessoe Heavy Engineering Ltd*. It was the first to be constructed at a strategic inland location as an extension to Britain's liquefied natural gas storage and handling complex at Canvey Island, in the Thames Estuary.

The gas, imported from Arzew in Algeria, was transported by refrigerated road tankers, stored at Ambergate at –162°C, and supplied to the national gas grid at times of peak demand.

The tank consisted of an aluminum inner vessel 28.95 m in diameter and 18.89 m high, and a mild steel outer vessel 30.63 m in diameter and 20.82 m high. The interspace was filled with powdered mineral insulant and pressurised with dry nitrogen gas, creating an inert and moisture free environment for the inner vessel.

The base of the tank rested on several layers of FOAMGLAS® Type S3S [1] cellular glass insulation. This material was chosen because it provided the standard of low temperature insulation demanded by the tank's designers, and was capable of withstanding high dead-loads. Its fireproof and waterproof properties were other factors that influenced the tank's designers.

The development of the North Sea gas fields eventually made the tank redundant, and in late 1986 it was decommissioned and dismantled by *British Gas*. This gave *Pittsburgh Corning*, who supplied the FOAMGLAS® cellular glass, an opportunity to assess the condition of the material after it had been exposed to severe conditions for 17 years.

---

[1] FOAMGLAS® S3S has been replaced by HLB 115.

PART IV. CASE STUDIES

The insulation for the tank base contained several layers of FOAMGLAS® cellular glass slabs, laid with staggered joints. From ground level upwards the structure consisted of a layer of bedding sand, two layers of bituminous felt, a 50 mm concrete screed, one layer of hessian felt, 50 mm paving slabs, four layers of FOAMGLAS® cellular glass slabs, (each covered with specially selected hessian felt) and a final layer of FOAMGLAS® cellular glass covered with bituminous felt and 50 mm of concrete.

The two lower layers of FOAMGLAS® cellular glass were 57.15 mm thick, the other 114.30 mm thick.

The insulation supported the weight of the tank at all times. The imposed load varied according to the amount of LNG in the tank, with a maximum when the gas was at 18.9 m. During water tests carried out at the time the tank was being commissioned, the insulation had to support a load equal to 125 per cent of the maximum compressive strength expected in normal service.

Once in service, the slabs close to the inner vessel were subjected to cryogenic temperatures and were liable to be stressed in the event of thermal contraction. The temperature of the faces of the coldest slabs was between −160 and −90°C.

When the tank base was dismantled, stresses and shocks were imposed on the insulation. For instance, a pneumatic drill was used to remove the concrete screed. The general state of the slabs was found to be good, although inevitably some were broken. Slabs had been installed under the shell of the inner vessel and in the annular space between the two vessels.

Samples were obtained from beneath the inner vessel (close to its wall), under the wall, and in the interspace. They were then placed in sealed boxes, shipped to *Pittsburgh Corning*'s factory, and opened in the presence of laboratory staff from *Liège University*.

A test programme, aimed at comparing the main properties of the samples with their original values 17 years ago, was then carried out. This measured the material's compressive strength, thermal conductivity (using the 'guarded hot plate' method) and percentage of open cells (determined by using a helium pycnometer). These values are the best indication of the condition of an insulating material.

The compressive strength was measured according to ASTM C 240-85, a method that entails coating the two bearing faces of a sample with bitumen at a ratio of 1200 g/m^2 and then applying a bituminous felt weighing 700 g/m^2. This method follows a widely accepted standard and simulates many actual applications. Coating cellular glass with bitumen is a common practice when insulating the bottoms of LNG tanks. Other factors, like the speed of deformation, also closely follow the ASTM procedure.

For the thermal conductivity tests, the method adopted was that set out in NBN B 62-201 ("Determination in dry state of the thermal conductivity or the thermal permeance of building materials by the method of the guarded hot plate"). The samples were prepared according to the procedure advised in ASTM C 240-85. Their dimensions were 326 × 326 × 25 mm.

The decision to measure the material's percentage of open cells does not suggest that cellular glass is basically an open cell material. On the contrary, it is a closed cell material but the cells cut when the material is sawn to the required dimensions are open, and are called "surface open cells".

Results of the Tests

- **Compressive strength**

Two series of tests were conducted. The first series of 18 samples gave an average compressive strength of 0.9 N/mm^2 with a standard deviation of 0.1 N/mm^2. The absolute minimum was 0.78 N/mm^2 and the highest value was 1.18 N/mm^2.

The second series of eight samples gave an average compressive strength of 1.15 N/mm^2 with a standard deviation of 0.25 N/mm^2. The absolute minimum value was 0.75 N/mm^2 and the highest value was 1.39 N/mm^2.

At the time of manufacture FOAMGLAS® cellular glass S3S had a published average compressive strength of 0.75 N/mm^2 when tested in ASTM conditions, although the average actual compressive strength was generally about 0.9 N/mm^2.

The test therefore showed that the compressive strength of the FOAMGLAS® cellular glass used in this application was not reduced after 17 years of service.

- **Thermal conductivity**

Four measurements were carried out, giving the following readings: 0.054 W/(m.K) at an average temperature of 21°C, 0.0542 W/(m.K) at 20.64°C, 0.0551 W/(m.K) at 20.44°C and 0.0541 W/(m.K) at 20.96°C. If one selects the typical temperature coefficient of 0.0015 W/(m.K) per Kelvin, these values at 10°C respectively read 0.0525, 0.0537, 0.0543 and 0.0540 W/(m.K), giving an average value of 0.0536 W/(m.K).

The published value of FOAMGLAS® cellular glass S3S used 17 years ago was 0.0535 W/(m.K).

Here again the values are very close and indicate that the material's thermal conductivity was unchanged after 17 years.

- **Percentage of open cells**

Four tests were conducted, giving the following results: 15.94, 13.53, 14.03 and 14.20%.

These readings appear to be somewhat high, but due to the rather small size of the samples (which were cylindrical in shape, 4 cm in diameter and 7 cm high) they correspond to a penetration depth of open cells of 0.13, 0.11, 0.115 and 0.115 cm (an average of 0.1175 cm).
This value should be compared with the cell diameters of 0.5 to 2 mm.

The average depth of penetration for FOAMGLAS® cellular glass S3S 17 years ago was 0.11 to 0.12 cm, so here also no significant change in the material characteristics was found.

To summarise, the results of the tests carried out for compressive strength, thermal conductivity and percentage of open cells were similar if not identical to the values of FOAMGLAS® cellular glass S3S in 1969.

(See colour photos 24 and 25 between pages 320 and 321).

IV.5

LAGGING THE DRAGON

Silicones have been manufactured at the *Dow Corning Corporation*'s plant at Barry, South Wales, since 1952. In 1985 the completion of a £60 million modernisation programme called the Dragon Project doubled the plant's capacity and it is now Europe's sole producer of intermediate fluids for most of the 1500 silicone products made by the corporation.

The basic raw materials used in the production of silicones are hyper-pure silicone metal, which is ground to a fine powder, and methanol, which is reacted with hydrochloric acid to make methyl chloride. The powdered silicon is mixed with a copper catalyst and reacted with the methyl chloride to give a mixture of methyl chlorosilanes in which dimethyldichlorosilane predominates.

The chlorosilanes are then distilled to produce pure compounds that are processed to provide intermediates for conversion into a variety of fluids, resins and elastomers. These have applications in a multitude of industrial processes, and in thousands of technical and consumer products.

Although the silicon and methyl chloride are reacted at moderate pressures, high temperatures have to be maintained in the two reactors, which are heated by gas- and oil-fired steam generators.

One of Barry's products is a polydimethyl siloxane heat transfer fluid, Syltherm 800, for chemical processes and this is used to heat the reactors, and return waste heat from the reactors to the steam generators.

Several other processes are also heated or cooled with Syltherm 800.

PART IV. CASE STUDIES

About 27,000 litres of heat transfer fluid are in constant circulation through a complex of alloy steel and stainless steel pipelines, valves and heat exchangers. Most of this equipment is exposed to the weather and some of the pipe runs are 150 m long, crossing roadways on overhead gantries.

In view of the energy-intensive nature of *Dow Corning*'s processes, efficient plant insulation is essential. Exposure to Barry's damp climate of the equipment carrying the heat transfer fluid is only one of the factors that have to be considered when specifying insulation systems.

Also pertinent are the high temperature (about 380°C) of the heat transfer fluid at the beginning of a heating cycle, the rapid thermal cycling of the fluid (from 380 to 180°C and vice versa within a period of a few minutes), and the high flammability of the fluid circulating in adjacent pipelines.

Until 1974, calcium silicate insulation was used extensively at Barry but this proved to be permeable to silicone. Leakages are, of course, rare and are dealt with immediately they are reported, but if they do occur they must not be able to soak into pipework insulation and create a risk of auto-ignition.

Some cases of spontaneous combustion of silicone fluid that had seeped into calcium silicate lagging did occur, and this was the main reason why it was decided progressively to replace it with FOAMGLAS® cellular glass insulation – a decision taken only after many types of insulation materials had been subjected to stringent tests at *Dow Corning*'s headquarters in Midland, Michigan.

In the 13 years that have elapsed since that decision was made, Barry has accumulated considerable experience of the capabilities of FOAMGLAS® cellular glass in all operating conditions.

FOAMGLAS® sealed glass cells make it completely impermeable, and this has been confirmed by its performance at Barry on the rare occasions when it has been exposed to silicones and other fluids.

The possibility of auto-ignition occurring within pipe insulation has been obviated, and no reaction of any kind between FOAMGLAS® cellular glass and the chemicals used and produced at Barry has been noticed. Nor have there been any instances of stress corrosion cracking of stainless steel pipework, a problem that had occurred when insulation materials containing free chlorides – which are liable to leach out and corrode the steel – were used.

The non combustibility of FOAMGLAS® cellular glass, and its inability to act as a wick for any flammable liquids spilled on or near it, are other factors that help endear FOAMGLAS® cellular glass to processing plant engineers and safety officers.

The material impermeability has an everyday advantage, as well as a safety benefit. Insulation materials that are permeable to chemicals will also absorb rain and ambient water vapour and consequently lose most of their thermal efficiency. The loss will become permanent if the insulation is in an exposed area and unable to dry-out.

This defect makes nonsense of theoretical k-values, and the adoption of a type of insulation material that retains its original k-value in all conditions and throughout its service life has made a significant if unquantifiable contribution towards Barry's energy conservation programme.

Resistance to degradation due to thermal cycling has also been proven; the 180 to 380ºC operating range of Barry's insulated pipes and vessels is well within the range for which FOAMGLAS® cellular glass is designed.

The only instances of deterioration that have occurred have been where vibration has caused abrasion of abutting faces of pipework insulation, revealed as 'hot spots' during routine infrared scans of the plant, and rectified accordingly.

Encouraging correct application methods is important to the reliability of any insulation system, and Barry's method in the case of preformed pipe coverings is to use support rings to prevent the insulation slipping down on the pipe flanges. The flanges are then lagged separately. On vessels, too, the FOAMGLAS® cellular glass is secured with support rings – neither impact fixings or adhesives are used.

Generally, insulation with an overall thickness of 150-200 mm is applied, built up from two or three equal layers covered with colour-coated galvanised steel cladding. Computer studies are made to calculate the most cost-effective thickness, taking into account such considerations as the price of the insulation and current and predicted energy costs.

Valves have been insulated by building boxes around them and filling them with FOAMGLAS® cellular glass chips. Shaping FOAMGLAS® cellular glass slabs to make them fit around awkwardly shaped valves had proved difficult, and as the heat losses from valves are minimal the tendency now is simply to box them in with perforated steel guards and where necessary inject some of *Dow Corning*'s Q3-6548 Fire Stop silicone foam into the box.

However, FOAMGLAS® custom-made prefabricated valve boxes are now available in the UK to make this job easier.

(See colour photo 26 between pages 320 and 321).

IV.6

FOAMGLAS® INSULATION 32-YEAR PERFORMANCE

Petro Wax PA, Inc is a well known manufacturer of high quality lube oils. From this plant in Emlenton, PA, they ship various grades of motor oil, lubricants, greases and paraffin. Many of the tanks at this location are insulated with FOAMGLAS® cellular glass.

One of the tanks insulated with FOAMGLAS® insulation is a vacuum bottom storage tank holding a combination of wax and oil not yet separated having the following characteristics:

Operating temperature	88°C
Tank dimensions	10.7 m diameter by 10.7 m high
Size of the slab	300 mm × 450 mm
Thickness	50 mm
Application	Banded in place with 19 mm metal strapping, applied dry with no sealing at joints

The tank must be heated and insulated to retain the temperature above ambient to prevent the wax in the mixture from setting up. Heat is supplied by steam coils located inside the bottom of the tank.

Built in 1958, this tank was originally insulated with FOAMGLAS® cellular glass insulation. Although a rare practice today, the FOAMGLAS® cellular glass insulation was installed without jacket. The insulation had completed a 32-year service when it was removed and tested in 1990.

Although the FOAMGLAS® insulation was completely exposed to the environment and extremes of temperature during all that time, the properties measured in 1990 compare favourably with the same properties at the time of installation. The table IV.1 shows those physical properties of FOAMGLAS® insulation published in 1958 along with the test results obtained for the 32-year old insulation removed from the tank.

TABLE IV.1

	Density		Thermal conductivity	
	(pcf)	(kg/m^3)	(BTU.inch/ (hr.ft^2.°F))	(W/(m.K))
Published 1958 ...	9.0	144	0.40	0.058
Test 1990 *..........	9.8	156	0.40	0.058
ASTM test...........	C 303		C 518	

* Tests conducted by Holometrix Inc, an independent testing laboratory located in Boston, Mass.

IV.7

THE NETHERLANDS: INSULATING THE BASE OF A TANK (110°C) AT THE SUIKERUNIE, DINTELOORD

Suikerunie is an agricultural cooperative, whose members produce roughly four million tonnes of sugar beet per year. During the autumn season, six factories transform the beet into sugar, pulp (food for livestock), and molasses (a raw material in alcohol production).

At Dinteloord, the Stampersgat factory uses high pressure steam for electricity production and the concentration of sugar juices. The condensates produced are stored in a large tank and are reused at the beginning of the season or in the case of an incident such as a lack or loss of pressure. At the end of the sugar season (December), the tank is filled with condensates at 110°C. As it is well insulated, there is no risk of the water freezing during the winter period.

The base of the tank is insulated with 100 mm thick HLB (High Load Bearing) FOAMGLAS® cellular glass. This material was chosen because of its high compressive strength (from 80 to 120 t/m^2 depending on type) and its perfect resistance to water and water vapour. High compressive strength is of primary importance as the tank and its content rest on the insulation.

What is more, as FOAMGLAS® cellular glass, by its nature, absorbs no moisture from the ground or from possible leaks from the tank, its ability to insulate remains constant. Slabs of cellular glass are bonded to the surface with a special mastic for tank bases developed by *Pittsburgh Corning*: PC® 86T.

To avoid the formation of rising damp along the tank, slabs of 80 mm thick FOAMGLAS® cellular glass have been applied to the lower part of the sides of the tank. Moreover, as the roofing of the tank needed to be accessible, access points, also insulated with FOAMGLAS® cellular glass, have been installed. Here as well, the high compressive strength of FOAMGLAS® cellular glass insulation was a decisive factor.

(See colour photo 27 between pages 320 and 321).

IV.8

SCHWEDENECK-SEE OFFSHORE PROJECT

Oil discovered by the German consortium *Texaco AG/Wintershall AG* in the Kiel Bay area a few kilometres off the Baltic Sea coast was scheduled to come on line commercially at the end of 1984. After extensive technical planning and discussion, the companies decided to go ahead with the necessary offshore installation to exploit the Schwedeneck-See oil field. The overall investment was estimated at 370 million DM.

The technology for this offshore project requires the very highest safety and environmental protection standards, even with the insulating material used, the latter having to withstand the most extreme conditions and be completely non combustible, hence the use of FOAMGLAS® cellular glass safety insulating material.

The project calls for two stationary drilling and conveying platforms to be prepared, each comprising a concrete core and steel superstructure. The contractors were *Bilfinger & Berger/Dyckerhoff & Widmann*, with *Howaldswerke Deutsche Werft AG* of Kiel main subcontractor to the consortium and supplier of the steel decks. Each platform was scheduled to operate seven producing wells that had to be sunk into the petroliferous sandstone layer at a depth of some 1500 metres and at a deflection of up to 55°, enabling the northern section of the deposit, with its expected exploitable reserves of 2.5 million tonnes, to be operated. Drilling took up to about a year and a half. The Schwedeneck-See A1 bore has already been successfully completed and measured, its end depth measuring as much as 1635 metres.

The platforms are interconnected by a pipeline system under the sea bed stretching as far as a land station three kilometres behind the coastline, from

PART IV. CASE STUDIES

where the conveyor plant is controlled and monitored by a process control computer and screen. The oil ready for processing at the land station is pumped to the Heide *Texaco* refinery via an 8 inch (200 mm) pipeline. The peak production figure of 400,000 tonnes a year was reached as early as 1986. FOAMGLAS® insulation was used on open drain, process and fire fighting ducts and on all vessels. The pipes are from 26.9 to 355 mm in diameter and the FOAMGLAS® insulation varies in thickness between 40 and 73 mm.

In all, a total of 3,800 running metres and 900 elbows of FOAMGLAS® cellular glass with anti-abrasive layer and groove for the parallel electric heating system were applied.

(See colour photo 28 between pages 320 and 321).

Photo 17. Berkshire Brewery's FOAMGLAS® cellular glass insulated fermentation vessels.

Photo 18. Detail of insulated pipes supplying glycol to the fermentation vessel's cooling systems.

Photo 19. The bases of the fermentation vessels, showing their aluminum-sheathed FOAMGLAS® insulation.

Photo 20. Overall view of the Lurgi headquarters where more than 7 km of chilled water pipes are insulated with FOAMGLAS® cellular glass shells.

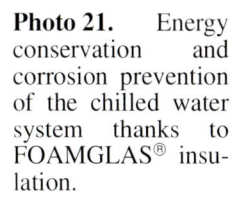

Photo 21. Energy conservation and corrosion prevention of the chilled water system thanks to FOAMGLAS® insulation.

Photo 22. Substation where FOAMGLAS® insulation has been covered with aluminium jacket.

Photo 23. Mossmorran's four double containment storage tanks, supported and insulated with FOAMGLAS® cellular glass type HLB.

Photo 24. The LNG tank at Ambergate, during construction. *Photo: Whessoe plc: British Gas.*

Photo 25. The LNG tank at Ambergate, after construction. *Photo: Whessoe plc: British Gas.*

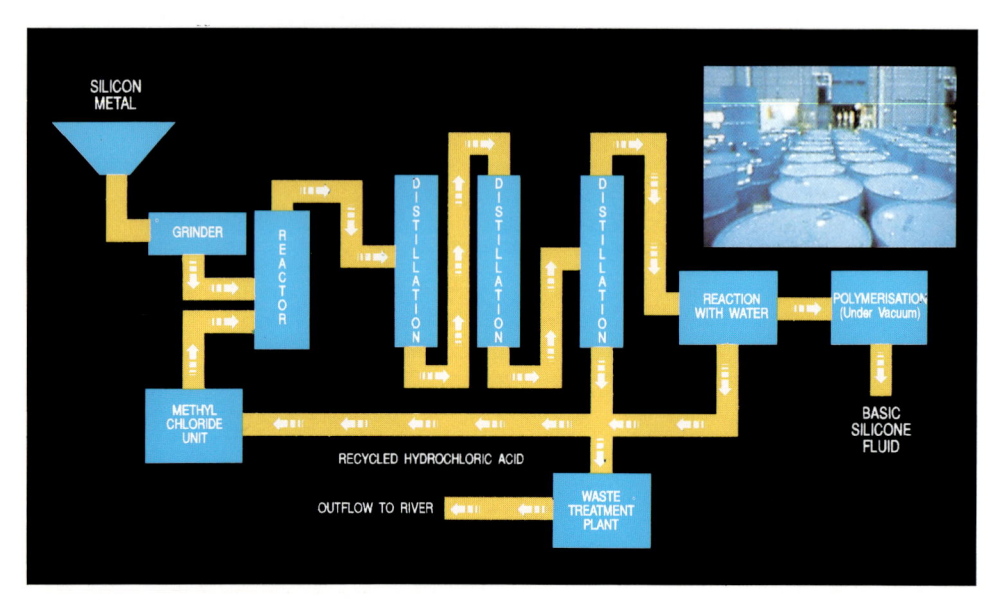

Photo 26. The silicone production process.

Photo 27. Hot tank base insulation.

Photo 28. Schwedenek-See offshore oil drilling platform.

V

WHY TO CHOOSE FOAMGLAS® INSULATION

V.I

FIRE AND SMOKE

V.1.1 THE PROBLEM: IMPORTANCE OF FIRES

V.1.1.1 Insulation Flammability and Wicking

Although the role of thermal insulation in industrial fires has only been recognised recently, it is now fully recognised as a major culprit.

With respect to metal deck roofs, A. Fuchs says [1]:

> "The analysis of fires showed that insulants have a considerable influence on fire development and fire propagation through
> - development of flammable gases
> - burning droplets
> - burning through the closed roof surface".

More generally, one should make a distinction between the insulating materials that are inherently combustible and those which are not combustible. But one should also consider that some non combustible insulating materials may become combustible if they absorb combustible liquids. In other words, two properties have to be evaluated to judge the fire safety of a material: the non combustibility and the non absorptivity. Insulating materials that are non combustible and non absorbent will offer a fire protection to the structure they insulate.

Insulating materials that are combustible or may absorb combustible liquids will increase the combustibility of the system.

---

[1] "Neue Brandschutz-Erkenntnisse bei Flachdächern", A. Fuchs, *Vorbeugender Brandschutz,* August (1983), p. 30.

Moreover, as indicated by Thomas Castino, Manager, *Fire Protection Department Underwriter's Laboratory* [2]:

> "The ignition of material itself is generally not a problem. More often, it serves as a wick for something that's being carried in the pipe, and the spread of it can be a problem. Penetrations of fire rated walls present problems. Smoke generation, heat release, and toxicity are complex problems".

V.1.1.2 Consequences of Fire

Although we do not want to establish an exhaustive list of all the drama caused by fire, it is worth pointing out that in a rather small country like Belgium, the ANPI mentions 102 deaths due to fire in 1987 when considering routine articles, both in normal life and in industrial settings [3].

In an article published by *Modern Plastics* in August 1973 [4], one can read:

> "Let's start with two basic facts:
> 1. The annual toll of human lives destroyed by fire stands at 12,000, year in and year out, and plastics are frequently blamed.
> 2. The annual cost in property losses runs at about 2.5 billions ($)".

Presumably *Modern Plastics* speaks for the USA only.

In *Techniques Nouvelles*, February, 1984, Lieutnant-Colonel R. Lion mentions [5]:

> "In France, fire is the cause of death of 3,000 persons per year..."

Fig V.1 from the *Verband der Sachversicherer* (Union of the Insurances in West Germany) indicates the cost of damages caused by industrial fire.

It should be added that the statistics generally do not include subsequent damages like shut-down costs, loss of production capacity, loss of customers, etc. which can prove very substantial and are frequently much higher than the immediate repair costs.

---

[2] "Thermal Insulation and Energy Conservation", Part II, *Heating/Piping/Air Conditioning*, July (1979), p. 17.
[3] "Dossier Sécurité", *Entreprendre*, June (1988), p. 24.
[4] "Fire !", Reprint from *Modern Plastics*, August (1973), pp. 1, 2, 5, 6.
[5] "Effects Toxiques des Fumées", *Techniques Nouvelles*, February (1984), pp. 11-12.

THOUSAND MILLIONS OF DM

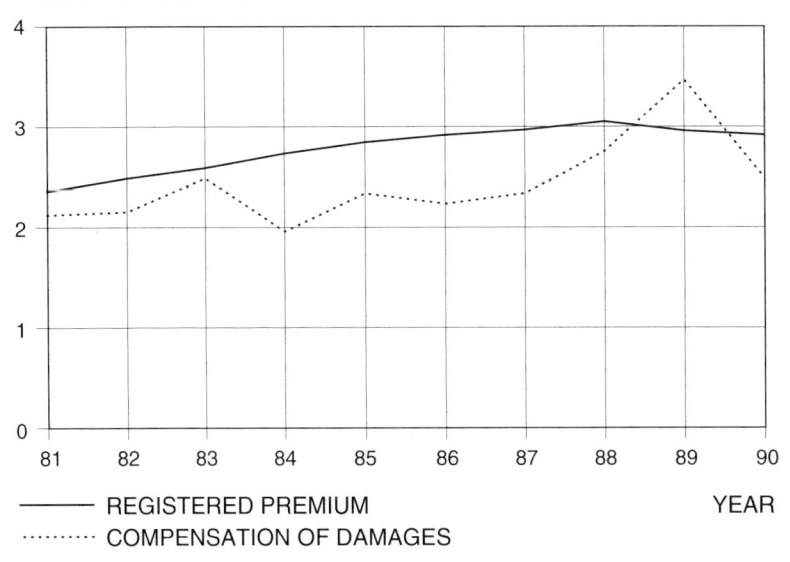

Year	Registered premium (thousand DM)	Compensation of damages (thousand DM)	Ratio of damages to premium (% of the premiums)
81	2,362,212	2,131,865	91.1
82	2,509,062	2,167,371	86.8
83	2,600,632	2,508,381	96.1
84	2,729,501	1,959,353	72.1
85	2,849,481	2,321,052	81.8
86	2,906,820	2,235,597	76.9
87	2,957,496	2,289,372	77.1
88	3,039,340	2,663,301	89.0
89	2,949,169	3,452,123	117.3
90	2,900,064	2,513,141	86.5
81-90			87.5

Fig. V.1. Costs for damages through industrial fires in West Germany *(Verband der Sachversicherer)*.

V.1.2 MISLEADING TERMINOLOGY

V.1.2.1 Historical Terminology

One of the reasons for the confusion that often characterises fire safety probably comes from the misleading terminology that has been used in the past and is still sometimes used. Moreover, the number of different test methods leading to different classifications has also created some misunderstandings. As example of the misleading terminology we can quote the *Federal Trade Commission* in the USA that, as early as in 1975, already states [6]:

> "The Federal Trade Commission believes that terms such as 'nonburning', 'self-extinguishing' or 'noncombustible' do not accurately reflect the hazards that may be presented by such products since in fire such products are not selfextinguishing and will burn rapidly if not properly protected".

Or *Modern Plastics*, August 1973, issue [7] that reads:

> "A roaring fire is no respecter of 'SE-O', 'NB' or '25 flame spread'. Realistic descriptions of plastics 'behaviour in actual fires would be a big step in solving the problem of plastics' fire hazards".

Or Mr. John Rhodes, vice-president and chief operating officer, *Factory Mutual Research Corporation*, who said [8]:

> "small scale tests, however, sometimes give misleading and potentially tragic results.... White pine, for instance, is 'self-extinguishing' by this test".

V.1.2.2 The Development of Current Terminology

Three elements should be available at the same place and in adequate quantities to start a fire: a combustible material, an oxidiser (or supporter of combustion) and finally a source of ignition. This is the so-called fire triangle (Fig. V.2).

A very important distinction should be made between the reaction to fire of a material and the fire resistance of a structure. According to ISO/TAG 5, the proposed definitions read as follows [9]:

> "**Reaction to fire**: the response of a **material** under specified test conditions in contributing by its own decomposition to a fire to which it is exposed".

and the

> "**Fire resistance**: the ability of an **element of building construction, component or structure**, to fulfill for a stated period of time the required stability,

---

[6] "Important Notice Concerning Certain Cellular Plastic Products", SPI, *Time,* January (1975), p. 52.
[7] "Fire !", Reprint from *Modern Plastics,* August (1973), pp. 1, 2, 5, 6.
[8] "The Building-Fire Problem", John M. Rhodes, *Record, 49,* September-October (1972), pp. 2-3.
[9] "ISO/TAG 5: United Glossary of Terms – Fire Test", Nederlands Normalisatie-instituut, March (1987), p. 1.

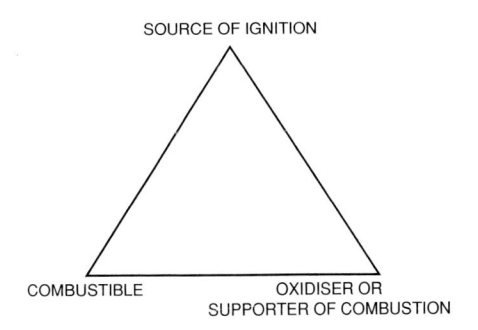

Fig. V.2. Fire triangle.

integrity, thermal insulation and/or other expected duty, specified in a standard fire resistance test".

V.1.3 TEST METHODS

The first purpose of this paragraph is to point out the misleading conclusions that can be drawn from many small scale tests. Expert opinions support this position.

For example, *Factory Mutual* in the *Loss Prevention Data 1- 57* of December 1978 writes [10]:

> "Some additives can markedly reduce the flame spread rating of polyurethane as measured by the ASTM E 84 test. Large scale corner testing has indicated however that the performance of expanded polyurethane under actual fire conditions is not significantly affected by the use of these additives.
> Consequently, claims for fire retardancy based on ASTM E 84 tunnel test should be disregarded for foam polyurethane".

---

[10] "Loss Prevention Data, Urethane in Building", *Factory Mutual Systems, 1-57,* December (1978), p. 1.

Similarly, G. Klose, in a paper published in *Forschung und Technik im Brandschutz*, 4/84, describes an ad hoc test on an insulated metal deck and offers among others the following comments [11]:

> "The polystyrene foam shows on the steel deck roof a completely different behaviour as suggested by the 'difficult to ignite' B1 classification following DIN 4102 First Part".

These remarks should not be interpreted as meaning that all the small scale tests are useless. For instance ISO 1182, ASTM E 136, BS 476 Part 4 and other rather similar test methods appear to be severe enough and meaningful as a way to determine if a material is truly "non combustible".

About fire resistance of systems, the difference of temperature increase versus time in a typical building fire and in a petrochemical fire has been recently emphasised.

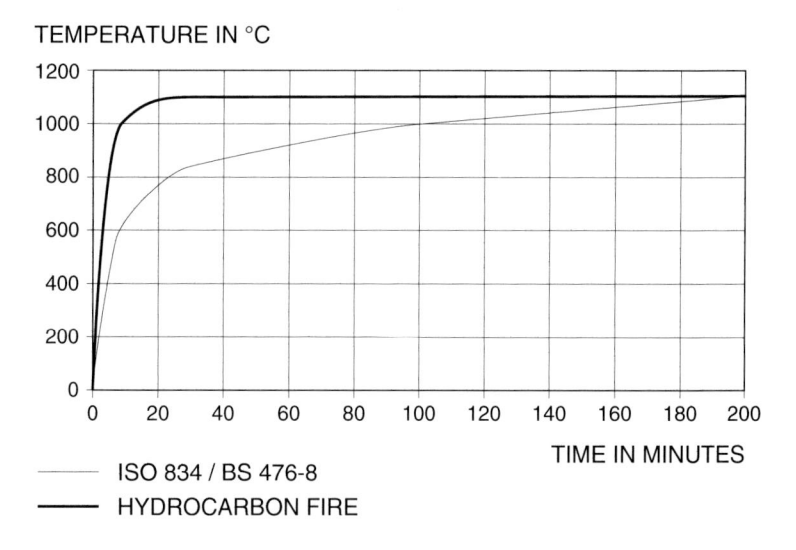

Fig. V.3. Graph comparison between building fire test temperature curve and typical hydrocarbon fire test curve.

---

[11] "Vergleichende Brandversuche mit nichtbrennbaren und schwerentflammbaren Dämmstoffen auf Stahlprofilblech-Dächern", G.R. Klose, *Forschung und Technik im Brandschutz*, April (1984), p. 181-184.

The temperature increase in a building fire can be adequately simulated by the ISO curve that calls for a temperature of 925°C after one hour. But this curve is considered by many experts [12] as too slow to represent a petrochemical fire that is characterised by very rapid heat release and temperature increase. Although there is no ISO or universally accepted petrochemical temperature increase curve, the proposed values by *Factory Mutual* and several major petrochemical companies imply a temperature of 982 to 1093°C after only 5 minutes. One can see on Fig. V.3 the harshness of this test.

V.1.4 INFLUENCE OF INSULATING MATERIALS

V.1.4.1 Generalities

The effect of fires on human beings has been described in details in the Report of the *National Commission of Fire Protection and Control* entitled *America Burning* [13] that lists the five ways in which fire can kill, presenting them in order of declining importance.

They work as follows:

- Asphyxiation or reduction of oxygen concentration in the air that can cause death in a few minutes
- Attack by superheated air or gases which causes loss of conscience or death in several minutes
- Smoke carrying toxic products and irritant products. The irritants attack the mucous membranes already in the early stage of a fire
- Toxic products
- Flames.

Although it may seem surprising, the relatively modest role of flames in human casualties has been confirmed by several other experts.

---

[12] "Thermal Insulation and Energy Conservation", Part II, *Heating/Piping/Air Conditioning,* July (1979), p. 17.
[13] "The Hazards Created Through Materials", National Commission on Fire Prevention, *America Burning,* May (1973), pp. 61-63.

Various independent and competent organisations have emphasised the influence of insulation materials.

The *Federal Trade Commission* has issued a complaint which among other reads as follows [14]:

> "A proposed Federal Trade Commission class action complaint alleges that certain plastic products used in the construction and furnishing of buildings and homes may constitute serious fire hazards. The plastics involved are cellular (or foamed) polyurethane and all forms of polystyrene and its copolymers".

and later:

> "These plastics are organic materials and, as such, are combustible, the complaint continues. Once ignited, they frequently produce greater fire hazards than the traditional materials.

And Shaw and Gillette write [15]:

> "Urethane is inherently flammable. Fire investigators call it 'solid gasoline'. Although its flammability varies from one formulation to another, most formulations burn hotter than wood and twice as fast. Urethane melts and flows as a flaming liquid and generates dense smoke and toxic gases".

Lieutnant-Colonel R. Lion [16] points out:

> "But, with modern fires, which force firemen to automatically use breathing masks, the majority of the victims are not burned to death but asphyxiated".

About this danger, one can provide the information of the gases produced by pyrolysis and combustion [17] of plastic materials as indicated in an article published in the *Revue Belge du Feu* [18].

---

[14] "Commission Proposes a Complaint Challenging the Knowing Marketing of Plastics Presenting a Serious Fire Hazard", *Federal Trade Commission News,* 30 May (1973) p. 1.
[15] "Urethane: Hazardous to your Health – A Deadly and Pervasive Peril", G. Shaw, R. Gillette, Reprint from *Los Angeles Times,* 21 January (1979), p. 2.
[16] "Effets Toxiques des Fumées", *Techniques Nouvelles,* February (1984), pp. 11-12.
[17] "Combustion is the burning action in which flames are involved whereas pyrolysis is the breaking down of a product by heat without any flames being present", definition given in "Some Aspects for Consideration in the Insulation of Chilled Water Pipes" J.G. Smit, R.I. Kirkman, *Journal of the Mine Ventilation Society of South Africa, 32, No. 4,* April (1979), p. 69.
[18] "Products Generated by Combustion of Plastics", *Revue Belge du Feu,* (1982) pp. 18-20.

Substance	Pyrolysis products	Combustion products

Polystyrene	Mainly:
	  Styrene (monomer, dimer, trimer), CO, CO_2
	Secondly:
	  H_2, aliphatics, aromatic hydrocarbons
		CO_2, CO
		
		H_2, aliphatics, aromatic hydrocarbons

PVC	Mainly:
	  HCl, CO_2, CO
	Secondly:
	  Aliphatics, aromatic hydrocarbons, aldehydes
		HCl, CO_2, CO
		
		Aliphatics, aromatic hydrocarbons

Phenolic resins	Mainly:
	  CO_2, CO, phenol, aliphatics, ketone, alcohols
	Secondly:
	  H_2, aromatic hydrocarbons, aldehydes
		CO_2, CO, formic acid
		
		Aliphatics, aromatic hydrocarbons, H_2

Polyisocyanurate based foams	Acrylonitrile, acetonitrile, pyride, benzene, cumene, toluene, styrene, aniline, p-toluidine, tolunitrile, HCN, CO, CO_2, pyrrolidine, phenyl isocyanurate	Mainly:
		  CO_2, CO, HCN
		Secondly:
		  Nitrogen composites, aromatic hydrocarbons

V.1.4.2 Examples of Large Fires

A number of large fires have confirmed in full scale what has already been said about the influence of insulating materials on fire. Without listing a long series of catastrophic situations, it may suffice to mention a few of them, eventually with some expert comments.

On February 10th, 1973, a fire took place in the *Texas Eastern Transmission Corporation's* 95,000 m³ capacity LNG storage tank on Staten Island resulting in the death of 40 workmen who were repairing the inner liner of the empty tank. The tank was above ground made of reinforced and/or poststressed concrete and insulated with polyurethane foam covered with a liner on the inside.

PART V. WHY TO CHOOSE FOAMGLAS® INSULATION

The report of the *Federal Power Commission* issued on July 9th, 1973 [19] reads:

> "Some of the materials employed in the tank construction, i.e., the tank internals, were inappropriate for the functions proposed. The most serious of these deficiencies was the flammability of the urethane insulation and to a lesser extent, that of the laminated liner".

On March 22nd, 1975, at the *Tennessee Valley Authority's Browns Ferry* (Alabama) nuclear plant, an engineering aid tried to block up a hole through which many of the plant's 2,000 electrical cables passed beneath the control room and he checked the air tightness with a candle. The draft sucked the flame right into the hole, igniting the foam that caught fire. After several attempts, working several hours in difficult conditions, the control room being filled with smoke and fumes, the TVA workers managed to get the nuclear reactor under control and extinguish the fire.

In the *US Nuclear Regulatory Commission* report issued in February 1976 [20], one can read:

> "The immediate cause of the fire was the ignition of polyurethane foam which was being used to seal air leaks..."

On September 16th, 1986, a welding team was repairing a broken track in the Kinross underground gold mine at about 1.6 km depth.

In *Fire Prevention*, No 197, issued in March 1987, the following appears [21]:

> "At least 170 people died and 235 people were injured as a result of a fire 1.6 kilometres underground in a gold mine. An investigation into the cause and events surrounding the fire has been launched by the South African Government Mining Engineers. However, it is clear that polyurethane foam used underground as a sealing compound was ignited during a welding operation".

One should not have the impression that only unprotected and misused plastic foams are dangerous from a fire standpoint. The following case shows fuel oil saturated fibreglass insulation can also be dangerous.

---

[19] "Preliminary Staff Report on Investigation of Disaster of Texas Eastern Transmission Corporation LNG Storage Tank on Staten Island", Federal Power Commission, Bureau of Natural Gas, 9 July (1973), p. 13.
[20] "Recommendations Related to Browns Ferry Fire", C. Harold, L. Saul, M. Warren, *US Nuclear Regulatory Commission, NUREG-0050,* February (1976), p. 1.
[21] "Polyurethane Ignited During Welding Operation", *Fire Prevention, 197,* March (1987), p. 39.

In Bayonne, New Jersey, on June 1st, 1973, as described by the *Fire Protection Manual* [22]:

> "Arc welding ignited the 2" aluminium jacketed fibreglass pipe insulation which was heavily satured with No 6 fuel oil"

and later on

> "Seven of the tanks were destroyed or severely damaged".

As it can be seen, the analysis of these fires confirms the important role of the insulating materials and shows that not only combustible materials but also incombustible materials that have absorbed liquids constitute a fire hazard.

V.1.5 REVIEW OF INSULATING MATERIAL MANUFACTURERS' LITERATURE

About the described situation, it is interesting to review the literature published by insulation material manufacturers and see what are the reported fire tests and which type of warnings are given. Great differences of revealed information levels are noted, probably in relation with the countries, liability, legislation and corporations.

Dow Chemical take a very responsible attitude and write in their publication [23]:

> "Polyurethanes and polyisocyanurates produced from these materials may present a fire risk in certain applications if exposed to fire or excessive heat, e.g., welding and cutting torches. The use of polyurethane and polyisocyanurate foam in interior applications may present an unreasonable fire risk unless the foam is protected by an approved fire-resistance barrier such as one-half inch (1.27 cm) gypsum wallboard, or its equivalent. Polyurethane and polyisocyanurate foam used in exterior applications should be coated for protection from the elements and to lower the fire risk. Consultation with building code officials and insurance personnel before application is recommended".

---

[22] "100 Largest Losses: A 30-Year Review", *Fire Protection Manual for Hydrocarbon Processing Plants, 1,* (1984), p. 14.
[23] "Trymer Rigid Foam Insulation" The Dow Chemical Company, *109-764-86,* p. 3-4.

and in another brochure [24]:

> "Styrofoam, Roofmate and Wallmate contain a flame retardant additive to inhibit accidental ignition from a small fire source, but the boards are combustible and if exposed to an intense fire may burn rapidly. All fire classifications are based on small scale tests and may not reflect the reaction of the material under actual fire conditions".

In a similar manner, *Upjohn* write [25]:

> "One conclusion which can be drawn and emphasized is: Polyurethane and related materials are combustible and can create a fire risk. The degree of risk may range from very low to extremely high and depends upon the degree of exposure to heat. Recognition and avoidance of these factors which can create a high degree of risk is essential for safety".

And later on:

> "Furthermore, the gas diffusion process which occurs on these foams can create an additional risk under certain circumstances. Rigid polyurethane and polyisocyanurate foam can absorb combustible gases, such as low molecular weight, highly inflammable hydrocarbons. Therefore, if used in applications where such exposure may occur, the foam must be protected by a gas-impermeable membrane".

Owens-Corning Fiberglas [26] in their brochure about urethane give rather similar warnings under the title "fire safety".

Note: The liquefied oxygen case

Before finishing this chapter, it is worth pointing out that special precautions have to be taken when insulating liquid oxygen equipment, vessels and piping systems operating at lower temperature than liquid oxygen or systems operating in oxygen-enriched atmosphere. In these cases, the use of plastic foam is particularly dangerous. As quoted in a report of the *Donald S. Gilmore Research Laboratories* [27]:

> "Upjohn does not recommend that these cellular plastic materials be used at temperatures lower than –321°F (–196°C)".

---

[24] "Styrofoam Plan", Dow Construction Products, Dow EU, *4600-E-583,* April (1985), p. 23.
[25] "Precautions for the Proper Usage of Polyurethanes, Polyisocyanurates, and Related Materials", Upjohn Co., *Technical Bulletin, 107,* May (1980), p. 29.
[26] "Owens-Corning Urethane", Brochure, *Owens-Corning Fiberglas, 1-UF,* May (1975), p. 1.
[27] "Cryogenic Insulation", Donald S. Gilmore Research Laboratories, *Insulation Outlook,* November (1981) p. 23.

In some countries, like Germany, materials containing more than 0.5% organic content by weight are prohibited in the case of installations for the treatment or storage of liquefied oxygen and liquefied air.

V.1.6 THE FOAMGLAS® CELLULAR GLASS SOLUTION

After having seen the problems associated with the use of combustible insulation materials, it is worth considering how FOAMGLAS® cellular glass and FOAMGLAS® cellular glass insulation systems pass the fire tests and how they behave in actual full scale fire situations.

V.1.6.1 Combustibility

As far as FOAMGLAS® cellular glass fire reaction is concerned, many tests have been carried out following the ISO Recommendation 1182 method or the other methods that, without being identical, are very similar, i.e. ASTM E 136, BS 476 Part 4 and NEN 3881. This last method basically calls for a furnace preheated at 750°C in which samples having a length and a width of 40 mm and a depth of 50 mm are introduced and kept during 20 minutes. A product is classified non combustible when none of the three samples creates:

- A furnace temperature increase exceeding a specified value, often 30 or 50°C
- A sample temperature increase exceeding a specified value, often 30 or 50°C
- A flame lasting more than 10 seconds.

If one of the criteria (or more) is not satisfied, the sample is classified combustible. As one can easily appreciate, this test is very severe and only products that can withstand a temperature of 750°C without burning have a chance to pass it. Details of the test methods are given in [28, 29, 30].

---

[28] "International Standard ISO 1182 Fire tests – Building materials – Non-combustibility test", *ISO 1182-1983 (E)*, (1983), p. 9.
[29] "Fire Tests on Building Materials and Structures – Non-combustibility Test for Materials", B.S. 476: Part 4: 1970, pp. 10-11.
[30] "Standard Test Method of Behaviour of Materials in Vertical Tube Furnace at 750°C" ASTM E 136-82, p. 366.

FOAMGLAS® cellular glass has passed several types of these tests, either for passing the test as such or to obtain a certificate of approval of one of the internationally acknowledged authorities. One can quote [31, 32, 33, 34, 35, 36] (Fig. V.4).

V.1.6.2 Heat Evolution

FOAMGLAS® cellular glass also passed the epiradiateur test (which determines the degree of flammability as a result of thermal radiation) with the classification M0 [37] in France and A0 in Belgium. This is the best possible classification [38, 39], the gross calorific value was found at 0 kJ/kg that is certainly a very fine result.

Similar excellent results were found in Germany where FOAMGLAS® cellular glass is rated incombustible A1 [40, 41].

V.1.6.3 Smoke and Toxic Gases

Although the chemical glassy nature of cellular glass makes it a very remote possibility, the potential emission of smoke and toxic gases has been evaluated.

---

[31] "Combustibility Test of FOAMGLAS® Insulation for Pittsburgh Corning Corporation", *Factory Mutual Research,* 14 October (1986).

[32] "Test Report on Heated Vertical Tube Test for Nippon Pittsburgh Corning K.K.", *Japan Ship Machinery Quality Control, 81-116E,* 30 May (1981), p. 4.

[33] "Lloyd's Register of Shipping Certificate", *Certificate ICD/F85, 207,* 2 September (1985), p. 1.

[34] "Fire Test Report" University of Hong Kong, 29 September (1986), p. 1.

[35] "Noncombustibility Test on FOAMGLAS® Material", *Singapore Institute of Standards and Industrial Research,* June (1980), p. 3.

[36] "Report Concerning an Examination of FOAMGLAS® Type T4 on Non-combustibility According to IMO Resolution A 472", *Instituut TNO voor Bouwmaterialen en Bouwconstructies, B-86-302 (E),* August (1986), p. 2.

[37] "Essai de Réaction au Feu d'un Matériau", *Centre Scientifique et Technique du Bâtiment, P.V. 78.13991,* November (1978), p. 1.

[38] "Verslag van Proeven Nr. 5503 Insulation material FOAMGLAS® type T2", *Rijksuniversiteit Gent,* 7 July (1986), p. 3.

[39] "Verslag van Proeven Nr. 5486 Isolatiemateriaal FOAMGLAS® type T2", *Rijksuniversiteit Gent,* 26 September (1986), p. 2.

[40] "Prüfung nach DIN 4102, Teil 1, Klasse A1, FOAMGLAS® Typ T2", *Forschungs- und Materialprüfungsanstalt Baden-Württemberg, I.6-74628,* February (1987), p. 2.

[41] "Cellular Glass as Insulation Material for Industrial Equipment Applications", Arbeitsgemeinschaft Industriebau, *Work Sheet Q 137,* August (1985), p. 4.

CHAPTER V.1 FIRE AND SMOKE

Lloyd's Register of Shipping

71 Fenchurch Street, London, EC3M 4BS

Certificate No: SVG/F90/229

Date : 2nd September 1990

ITEM : "FOAMGLAS T2" (50mm THICK)

MANUFACTURER : PITTSBURGH CORNING (UNITED KINGDOM) LTD.,
SOUTHCOURT,
29, SOUTH STREET,
READING,
BERKSHIRE, RG1 4QU.

This is to certify that the above material, samples of which were tested at the Yarsley Technical Centre Ltd., Surrey, and reported in their Certificate No. C 66630/6 dated 11th July, 1979, will be accepted for compliance with this Society's Rules and Regulations and with the International Conventions for the Safety of Life at Sea 1960, 1974 and the 1981 and 1983 Amendments thereto, on ships classed with Lloyd's Register of Shipping and on ships for which the Society is responsible for the issue of the relevant certificate required by the above International Conventions as authorised by contracting Governments, and for use on Offshore Installations subject to the Society's Classification Rules and Offshore Installations for which this Society is authorised to issue Certificates of Fitness, and similar licences, permits, etc., as a:

MATERIAL HAVING LOW FLAME SPREAD CHARACTERISTICS.

This Certificate is valid for five years from the date of issue

Surveyor to Lloyd's Register of Shipping

NOTICE - This certificate is subject to the terms and conditions overleaf, which form part of this certificate

Fig. V.4. Lloyd's Register of Shipping Certificate.

When following NEN 3883, the *Instituut TNO* in Rijswijk (The Netherlands) found that [42]:

> "The measurement of smoke development during fire of FOAMGLAS® cellular glass type T2 was zero".

V.1.6.4 Fire Resistance Tests

A. Systems Involving Adhesives

Many tests have also been carried out regarding the resistance to fire of insulating systems incorporating cellular glass. The general opinion when considering a system made of FOAMGLAS® cellular glass and adhesives is to acknowledge that cellular glass is truly incombustible according to ISO 1182 but that organic adhesives can burn. A more thorough analysis indicates that even organic adhesives do not burn easily when they have been applied between two cellular glass slabs. In fact the temperature increase is substantially slowed by the insulating value of FOAMGLAS® cellular glass, because oxygen is not easily available in an adhesive layer surrounded by cellular glass slabs and because the amount of adhesive in volume is very limited compared to the volume of cellular glass, generally between 2 to 4%. Tests confirm this theoretical approach [43, 44].

Another test has been run on a complete insulation system simulating an adhered insulation system applied on a low temperature tank or a sphere wall. This system was basically made of

- A 6 mm thick mild steel plate
- Two layers of 50 mm thick FOAMGLAS® cellular glass T2 slabs adhered to the steel and between themselves with PC® 88 ADHESIVE, the joint being fully filled with PC® 88 ADHESIVE

---

[42] "Rapport Betreffende een Onderzoek Volgens NEN 3883 naar de Mate van Rookontwikkeling bij Brand van FOAMGLAS® Type T2", *Instituut TNO voor Bouwmaterialen en Bouwconstructies*, B-81-60, 3 February (1981), p. 4.

[43] "Prüfungsbericht: Brandschachtversuche an FOAMGLAS® mit geklebten Fugen für Dämmungen beim Bau von Kälteanlagen", *Forschungs- und Materialprüfungsanstalt Baden-Württemberg*, I.6-75301a, 19 December (1983), p. 4.

[44] "Prüfungsbericht: Brandschachtversuche an FOAMGLAS® mit geklebten Fugen und Beschichtung für Dämmungen beim Bau von Kälteanlagen", *Forschungs- und Materialprüfungsanstalt Baden-Württemberg*, I.6-75301B, 19 December (1983), pp. 2-3.

- A 20 gauge aluminium sheet strapped to the face of the specimen with 12.7 mm width stainless steel straps.

The test was run following the BS 476 – Part 8 temperature curve on the outside aluminium facing.

In this ad hoc test, the system satisfied the BS 476 – Part 8 criteria of Insulation, Integrity and Stability for 90 minutes. Reference [45] provides more details on this test. Obviously, only very few insulation systems can reach this level of performance.

B. Pipe Tests

Three different series of tests were carried out to determine how various insulation systems would protect the pipe, should an outside fire occur. The basic criteria were an outside temperature increase according to ASTM E 119 or BS 476 – Part 8 (about 925°C after one hour) and a fire resistance rating measured from the beginning of the tests to the time at which the pipe temperature would reach 538°C.

A first series of tests run in Japan [46] shows that two hours and even two hours and a half could be reached with a system made of two layers of FOAMGLAS® cellular glass held in place with stainless steel straps and covered with galvanised metal.

A second and a third series of tests compare a straight polyurethane pipe insulation, with a composite polyurethane/FOAMGLAS® cellular glass system and with a straight FOAMGLAS® cellular glass pipe insulation. As expected, the composite polyurethane/FOAMGLAS® cellular glass system offers far superior protection than the urethane system but without obtaining the results of a homogeneous FOAMGLAS® cellular glass system. Two hours fire rating are easily reached with adequately designed FOAMGLAS® cellular glass pipe insulation.

---

[45] "Ad Hoc Fire Resistance Test", *Yarsley Technical Centre, C 72756/2,* 13 December (1982), p. 2.
[46] "Fire Resistance Test on FOAMGLAS® Pipe Insulation Material as per JIS A 1304", Japan Testing Center of Construction Materials, 30 May (1980), p. 1.

C. Petrochemical Fire Tests

Two ad hoc tests were run according to a "petrochemical" temperature/time curve on an insulated steel pipe (see V.1.3). In spite of this harsh test, a system made of two 50 mm layers of laminated FOAMGLAS® cellular glass and stainless steel jacket kept the mean steel temperature underneath 256°C after a one hour exposure test [48] and a system of three 50 mm layers of laminated FOAMGLAS® cellular glass and stainless steel jacket, limited the mean steel temperature rise to 471°C after a two hour exposure [47] (Fig. V.5).

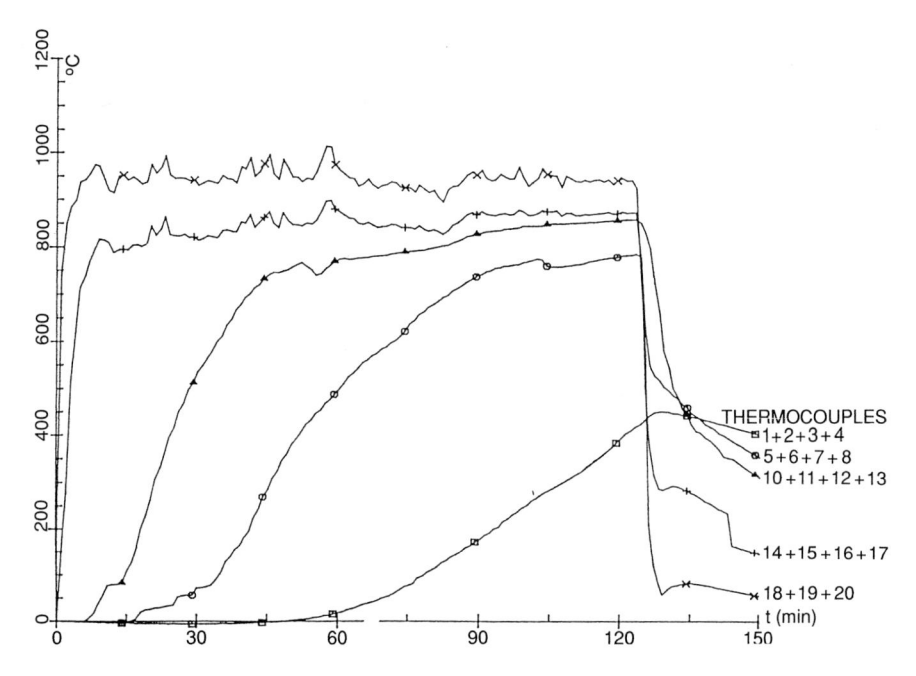

Fig. V.5. Fire test for evaluating fire protective coating of an insulated steel pipe with three-layers of FOAMGLAS® cellular glass.

As a conclusion, tests show that well-designed FOAMGLAS® insulation systems offer excellent fire protection for piping.

---

[47] "Verslag van proeven 5939, Hydrocarbon fire-3 layers", *Rijksuniversiteit Gent*, 10 November (1988).
[48] "Verslag van proeven 5939 A, Hydrocarbon fire-2 layers", *Rijksuniversiteit Gent*, 1 Februar (1989).

D. Penetrations

Another test that presents great interest addresses the problem of insulated pipes penetrating from one floor of a building to another one and being a potential way for fire to pass from a lower floor to a higher floor. Tests [49] were carried out with a polyurethane insulated pipe and a FOAMGLAS® cellular glass insulated pipe representing each typical thickness specified for chilled water lines. The conclusion reads:

> "Overall, the performance of the FOAMGLAS® cellular glass pipe insulation was better than urethane. When exposed to fire, the FOAMGLAS® cellular glass continually kept a steel pipe at least 200°F (111°C) cooler than a pipe covered with urethane and was not consumed by the flames as was urethane. Since the FOAMGLAS® cellular glass did not burn, no smoke or harmful cyanide gases were produced".

E. Panels

Before ending this section, it may be worth pointing out that FOAMGLAS® insulation as a core material has been tested for various types of sandwich panels achieving fire resistance of one hour and even two hours depending on the thickness, the adhesive and the selected facings [50, 51, 52].

Knowing to which extent design details can influence the results of a fire test, it is suggested that the reader reads the full reports to obtain complete information on the systems, the methods of testing, etc.

V.1.6.5 FOAMGLAS® Cellular Glass Fire Experience

Having briefly reviewed some of the most significant fire tests run on FOAMGLAS® insulation, it is most interesting to see how FOAMGLAS® cellular glass behaves in real life, what actually happens and how experts analyse the situations.

---

[49] "Fire Tests of FOAMGLAS® and Urethane Pipe Insulation", *Carnegie-Mellon University Pittsburgh,* PA, December (1976), p. 16.
[50] "Fire Resistance Rating FOAMGLAS® Porcelain Enamel Panels", *Ohio State University,* 25 May (1955), p. 1.
[51] "Partition Panel Units in a Non-bearing Wall Assembly", *Underwriters Laboratories Inc., R7759-3-4,* 13 July (1981) p. C1.
[52] "Partition Panel Units for Use in a Nonbearing Wall Assembly", *Underwriters Laboratories Inc., R7759-1-2,* 12 December (1975), p. C1.

PART V. WHY TO CHOOSE FOAMGLAS ® INSULATION

A. Ethylene Plant Fire

On July 13th, 1965, a fire with explosion took place in the ethylene plant in Lake Charles, Louisiana, causing loss of several million dollars each for property damage and business interruption.

At least two reports were prepared on this catastrophe. In [53], one reads:

> "The plant's fractionation towers came through relatively unscathed because of the low temperature (down to –200°F) (–129°C) of their contents and because they carried three layers of foamed-glass insulation".

and later on:

> "Using a critical path schedule for the master plan, and a workforce of 100 PCI personnel plus more than 700 employees of contractors, engineers had the plant back on stream on Oct. 18, the exact target date".

and in the second one [54], contrasting with the actual repair in three months:

> "First, we would have been down for better than a year, I imagine, if it had not been for foam glass or equivalent non-flammable insulation on all our towers and drums".

In another ethylene plant, an ethylene vapour cloud was released resulting in an explosion and fire that destroyed the entire facility in 1963. The plant was rebuilt with an insulation made of polyurethane foam protected with a layer of FOAMGLAS® cellular glass covered with a 0.25 mm thick aluminium jacket with a water spray coverage and a variety of loss prevention features. It was subject to a similar vapour cloud release in 1974 that created a fire, but as written in [55]:

> "After one hour all fires were out and sprinkler systems and monitor guns were shut down".

B. LNG System

In [56], S. Kastanas of *Colonia Gas* explained why after having a process fire including the release of a lubricating oil on LNG liquefaction facilities, he

---

[53] "Explosion Leads to a Safer Ethylene-Plant Desing", *Chemical Engineering,* 14 February (1966), pp. 96-97.
[54] "Flare Line Rupture in an Ethylene Plant", C.G. Kevil, Cities Service Oil Co., Lake Charles, Lousiana, *Safety in Air and Ammonia Plants, 9,* 1967, p. 39.
[55] "Plant Loss Prevention: Vapor Clouds and Fires in a Light Hydrocarbon Plant", S.A. Saia, *Chemical Engineering Progress,* November (1976), p. 61.
[56] "Updating Piping and Vessel Insulation with Composite Systems", S.T. Kastanas, *AGA 1987 Operating Section Proceedings,* p. 470.

decided to specify FOAMGLAS® cellular glass at least for the layer exposed to the outside. His conclusion reads:

> "After reviewing the insulation perspectives noted in this paper, consideration should be given to incorporating inorganic cellular foam glass (e.g. FOAMGLAS® cellular glass) for maintaining better thermal efficiency and for added fire protection of existing equipment and insulation systems".

C. Metal Decks

Reference [57] quotes the case of a large metal deck on which the waterproofing membrane adhered to FOAMGLAS® insulation applied on the steel profile was undamaged on the major part of the surface after an important fire coming from within the building.

In reference [58], explanation and experimental confirmation of large scale tests are given.

D. Cold Stores

The fire protection offered by FOAMGLAS® insulation in cold stores has been highlighted in [59].

E. Insurances

The fact that FOAMGLAS® insulation system provides good fire protection has been acknowledged by some insurance companies. As indicated in [60]:

> "Our insurance rates are considerably less with FOAMGLAS® insulation", Hoy admitted. "It's giving us a good return on our investment".

F. LNG Pool Fires

Finally, an unusual application of FOAMGLAS® cellular glass in fire situations is the control of LNG pool fires. FOAMGLAS® cellular glass elements

---

[57] "Ein Dämmstoff in der Bewährung", I. Sauerbrunn, Reprint from *112-Magazin der Feuerwehr*, (1986), pp. 2-3.
[58] "Brandverhalten und sichere Anwendung von anorganischen Dämmstoffen bei Flachdächern", I. Sauerbrunn, Lecture Paper, *4*, p. 4.10.
[59] "Miscellaneous Fire References".
[60] "Insulation Installation Cuts Both Ways", Reprint from *Quick Frozen Foods*, February (1984), p. 2.

10 to 15 mm are floating on the LNG pool making a 200 mm thick layer. As written in [61]:

> "Typically peak radiation with cellular glass in the bund [impounding area] was 12-17 percent of that recorded during the free-burning LNG fire in the same bund".

---

[61] "A Novel Method of Controlling LNG Pool Fires", Y. Lev, *Fire Technology,* p. 278.

V.2

COMBUSTIBLE LIQUIDS IN INSULATION

V.2.1 THE PROBLEM

The absorption of combustible liquids in insulation can create a very serious fire hazard. Absorbent insulations have "the ability to store large quantities of combustible liquids should a leak occur" [1]. This leakage and accumulation may often go undetected until ignition takes place. Obviously such a fire presents a serious threat to personnel, property and production. According to Bowes:

> "Fires associated with the leakage of combustible liquids into lagging on hot pipes and other hot surfaces are widely known. They have occurred, for example, following leaks of lubricating oil and hydraulic fluids in power stations and ship engine rooms, and, with leakage of a wider range of liquids, in chemical and allied industries" [2].

Such events can have very serious consequences. For instance, at a fuel oil storage facility in Bayonne N.J., a No. 6 fuel oil pump was found to be leaking badly. Arc welding during repair operations:

> "...ignited the 2" (50 mm) aluminum jacketed fiberglass pipe insulation which was heavily saturated with No. 6 fuel oil. The subsequent fire resulted in failure of piping and eventually destroyed or damaged seven tanks. The loss, expressed in 1983 dollars, was slightly of over $ 10 million" [3].

In other cases, ignition has been a result of self-heating, resulting in spontaneous combustion. In such instances, an external source of heat, such as welding operations, it not required.

---

[1] "Criteria for Installing Insulation Systems in Petrochemical Plants", W.C. Turner, *Chemical Engineering Progress,* 70, [8], August (1974), pp. 3-7.
[2] "Fires in Oil Soaked Lagging", P.C. Bowes, *Building Research Establishment Current Paper,* CP 35/74, February (1974), pp. 1-11.
[3] "100 Largest Losses: A 30-year Review", M & M Protection Consultants (March & McLennan), *Fire Protection Manual for Hydrocarbon Processing Plants,* 3rd edition, edited by Charles H. Vervalin (Houston, Texas: July; 1984), 1, p. 14.

V.2.1.1 Explanation of the Problem

Hucke presents the following explanation for fire in oil soaked insulation [4]:

> "An oil leak in a mineral wool insulation may cause, for example, an autocombustion, even if the leak is small. The volume occupied by the penetrating oil in the mineral wool enlarges a thousandfold, like a drop of gasoline that spreads through a pad of cotton wool. The oxidation process begins immediately and, according to the type of oil, the ignition point may be so reduced, that the mixture of oil, oxygen and insulating wool bursts into flames by itself at the usual operating temperature. A fire and a chain of explosions may eventually result from this.
> Even small quantities of oil are sufficient to bring about fires of this nature,..."

Heat transfer fluids are manufactured by several companies including *Monsanto* (Therminol®, Santotherm®), *Dow* (Dowtherm®) and *Mobil Oil Corporation* (Mobiltherm®). They all offer the same basic explanation and the same recommendation. For instance, in their brochure on these products [5], *Monsanto* tells us:

> "Organic heat transfer fluids such as Therminol exhibit a slow oxidation reaction with air in the presence of insulating materials when system temperatures are above 500°F (260°C). Porous insulation such as calcium silicate offers a large reaction surface in the face of poor heat dissipation conditions, and this, along with possible catalysts from the insulating material can cause a temperature buildup. This temperature buildup can result in ignition of the fluid when the satured insulation is exposed to air (i.e., should the insulation be opened for repair, etc.).
> This phenomenon is not fully understood, but appears not to occur with cellular glass, possibly because of its closed cell structure. Cellular glass should be used in all areas where leakage is a possibility and the system temperature is greater than 400°F (200°C)".

The fact that the ignition temperature of combustible liquids, when they are in insulating materials, is well below the normal autoignition temperature has been examined by Bowes [6] and the results are published [7]:

---

[4] "Transfert de Chaleur par Fluide Chauffé à l'Electricité", Hans Hucke, *Informations Chimie, 248*, Avril (1984), p. 157.
[5] "Therminol® Heat Transfer Fluid", Data Sheet; Monsanto Industrial Chemicals Co, MIC-4-081, pp. 1-24.
[6] "Fires in Oil Soaked Lagging", P.C. Bowes, *Building Research Establishment Current Paper*, CP 35/74, February (1974), pp. 1-11.
[7] "Piped Services", Fire Safety Data; *Industrial and Process Fire Safety, FS 6017* (London: Fire Protection Association, June, 1982), pp. 1-4.

"LAGGING OF PIPES
Oils and other flammable liquids impregnated on lagging, even if it is noncombustible, can ignite spontaneously. Spontaneous heating, which can continue until ignition occurs, can start at temperatures well below the normal autoignition temperature of a particular flammable liquid. Ignition has occurred when plant temperatures have been as low as 80°C:

Type of oil	Typical dangerous temperatures (°C)
Essential oils..	80 to 150
Coal tar distillates...............................	100 to 150
Mineral oils	200 to 300

Usually the thicker the lagging the lower the temperature required for spontaneous ignition".

According to manufacturer's literature, the heat transfer fluids, when they have penetrated insulation, can experience autoignition – once again without any other source of energy – at temperatures ranging from 200 to 400°C (Fig. V.6).

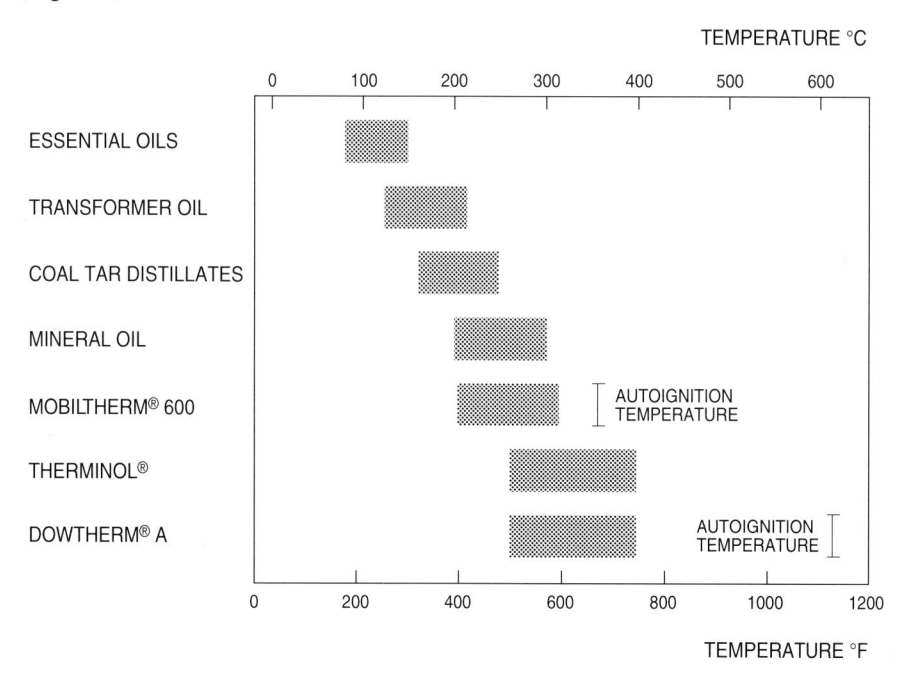

Fig. V.6. Ignition temperatures of combustible liquids in insulation.

A lowering of ignition temperatures is not limited to liquids, it can also occur in gases.

W.C. Turner [1], a well-known author and consultant in the field of thermal insulation, tells us that:

> "ethylene oxide has an ignition point of 571°C. In contact with calcium silicate, the ignition point is 298°C. The hazard is that 298°C is below process temperature, thus any leak can result in immediate burning".

V.2.1.2 Sources of Leaks

Dow [9] lists the sources of leaks as follows:

> "Principal trouble points occur around leaky valve packing glands, thermocouple connections, equipment flanges, and in places where insulation contacts horizontal flat surfaces such as floor plates. Leakage collects on the flat surface and is absorbed into the insulation by capillary or wicking action".

V.2.2 THE SOLUTION

As noted earlier, *Monsanto* [8], *Dow* [9] and *Mobil* [10], indicate that closed cell insulation material is preferred when leakage and contamination of organic material are likely. Bowes [11] agrees, stating that:

> "... the use of foamed glass lagging can be recommended... This material has a closed pore structure and being therefore impermeable to liquids and air, there is no risk of self-heating following a leak".

---

[8] "Therminol® Heat Transfer Fluid", Data Sheet; Monsanto Industrial Chemicals Co, *MIC-4-081*, pp. 1-24.

[9] "Accident Prevention in High Temperature Heat Transfer Fluid Systems", A.R. Albrecht, W.F. Seifert, Reference Paper presented at *Loss Prevention Symposium*, 67th Annual Meeting, American Institute of Chemical Engineers, Atlanta, Georgia, 15-18 February, 1970 (Dow Chemical Co., revised and updated May 1987), pp. 11-33.

[10] "Heating with Mobiltherm® Heat-Transfer-Oils", Mobil Oil Corporation, *Mobil Technical Bulletin, TBL6887101,* 3 September (1988), pp. 1-15.

[11] "Fires in Oil Soaked Lagging", P.C. Bowes, *Building Research Establishment Current Paper,* CP 35/74, February (1974), pp. 1-11.

Dow Corning [12] has tested their heat transfer fluid Syltherm 800 for ignition when absorbed in various insulation materials including calcium silicate, perlite, glass fibre and rigid blown glass. They report that in their tests:

> "the rigid blown glass was crushed (in all tests) to facilitate the absorption of the test fluid. Despite this pronounced bias, it matched or exceeded.. all other materials tested".

V.2.3 INSTALLATIONS USING FOAMGLAS® INSULATION

Many large size installations have been insulated with FOAMGLAS® cellular glass to avoid the danger associated with combustible liquids (Fig. V.7). We can quote:

Fig. V.7. Example of pipe insulation.

---

[12] "Method for Fire Hazard Assessment of Fluid-Soaked Thermal Insulation", Robert R. Buch, Daniel H. Filsinger, *Plant/Operations Progress, 4,* July (1985), pp. 176-180.

- Solar energy systems using heat transfer fluids
- Chemical plants
- Polypropylene oxide plants
- LNG plants
- Solvent plants like soybeans extraction
- Synfuel
- Synthetic resin
- Fatty acids
- Silicones
- Explosives
- ...

V.2.4 HIGH TEMPERATURE SYSTEMS

Heat transfer fluids are generally employed in the temperature range from 175 to 400°C. At these temperatures FOAMGLAS® insulation is subject to thermal cracking. This tendency is a function of insulation thickness and geometry. For instance, in pipe sections both radial and longitudinal cracks can develop. Since the early 1970's *Pittsburgh Corning* has developed a number of systems to control thermal shock. With these systems, thermal cracking is controlled and mechanical and thermal performances are unimpaired.

V.2.5 LOW TEMPERATURE CONSIDERATIONS

V.2.5.1 Hydrocarbons

At low temperatures, the risk of fire may involve the condensation of various gases in the insulation. The problem of absorption of hydrocarbon gases was discussed by *Upjohn* [13]:

> "... the gas diffusion process which occurs in these foams can create an additional risk under certain circumstances. Rigid polyurethane and polyisocyanu-

---

[13] "Precautions for the Proper Usage of Polyurethanes, Polyisocyanurates and Related Materials", Technical Bulletin; The Upjohn Chemical Division, *107*, May (1980), p. 44.

rate foam can absorb combustible gases, such as low molecular weight, highly flammable hydrocarbons. Therefore, if used in applications where such exposure may occur, the foam must be protected by a gas-impermeable membrane".

V.2.5.2 Oxygen

At cryogenic temperatures, we must also consider the possibility of condensation of oxygen. The *National Fire Protection Association* [14] in the United States tells us:

> "Liquid oxygen is the most concentrated common source of oxygen. Contamination of liquid oxygen with most organic substances often renders the mixture subject to violent explosion".

Thus a risk of liquid oxygen explosions exists when liquid oxygen, liquid nitrogen, liquid hydrogen and liquid helium insulation systems are in use, since liquid oxygen can be condensed from the air at such temperatures.

FOAMGLAS® insulation has been proven to be LOX (liquid oxygen) compatible. This has been stated explicitly in a NASA Technical Memorandum [15]:

> "Two varieties of cellular glass, Foamsil [16] and Foam Glass, have proven satisfactory when tested for LOX compatibility".

For such applications it is, of course, necessary to ensure that LOX compatible accessory materials are also used. The use of Hydrocal® B-11 adhesive for FOAMGLAS® insulation has been indicated by the White Sands Test Facility in the US [17].

PC® 85 POWDER is equivalent to Hydrocal® B-11.

As a conclusion, the fact that cellular glass does not absorb combustible vapour or liquid can prevent many serious accidents.

---

[14] "Fire Hazards in Oxygen-Enriched Atmosphere", National Fire Protection Association, *ANSI/NFPA 53M*, 7 December (1985), pp. 10, 17.
[15] "Compatibility of Materials with Liquid Oxygen", C.F. Key, W.A. Riehl, NASA Technical Memorandum, *NASA TM X-985*, August (1964), pp. 6, 25-26.
[16] Foamsil is another cellular glass produced on special order by *Pittsburgh Corning*. It is characterised by a very low coefficient of thermal expansion.
[17] "Cellular Glass Insulations for LOX Systems", *SAF-ALERT*, 30 July (1976).

V.3

LIQUID WATER ABSORPTION AND RETENTION IN INSULATION

V.3.1 THE PROBLEM

George Lang of *DuPont* has estimated that

"98% of the problems with insulation systems are due to moisture" [1].

Some of the consequences of water in insulation systems have been listed by Herb Moak, also from *DuPont*, who stated:

"Water is the worst enemy of thermal insulation systems because it increases heat loss, deteriorates insulation, can corrode carbon steel, and can cause chloride stress corrosion cracking of austenitic stainless steel" [2].

Increased heat loss (or gain for cold equipment) leads not only to increased energy consumption but may also cause such problems as increased boil-off from low temperature tanks and the risk of solidification of certain materials in piping designed to operate at high temperature.

Deterioration of insulation can range from complete destruction to dimensional changes that can compromise system integrity.

Corrosion under thermal insulation can give rise to major safety and economic consequences.

---

[1] "Common In-Plant Problems Encountered with Insulation", George Lang, *Chemical Processing,* January (1984), pp. 38-39.
[2] "An Owner's View of Industrial Thermal Insulation", Herbert A. Moak, *Insulation,* June (1983), pp. 9-11.

V.3.2. WATER IN LOW TEMPERATURE INSULATION

Low temperature insulations are generally cellular materials that serve to minimise the intrusion of liquid water. However, the predominant mode for water entry into cellular materials such as some plastic foams that are in low temperature service is generally not via the intrusion of liquid water. Rather it is by way of water vapour transported in response to the water vapour pressure gradient resulting from the temperature difference between ambient conditions and the low temperature equipment. In fact, manufacturers of polyurethane make statements in their literature that indicate that:

> "Polyurethanes and polyisocyanurates are not vapor barriers and must be protected from water vapor, liquid water and the elements..." [3].

In the case of phenolic foams, the permeability to both liquid water and water vapour is much higher than for polyurethane and polystyrene and so water intrusion from both liquid water and water vapour must be considered [4].

Insulations that are in low temperature service have no possibility of drying out. W.C. Turner tells us that in such cases:

> "... all moisture that enters the system migrates to the substrate low temperature. It continues to build up until the insulation is completely saturated with liquid or ice" [5].

Thus the retention of water or ice in low temperature service is permanent.

Dr. J. Achtziger of the *Forschungsinstitut für Wärmeschutz e.V.* in Munich presents the following diagrams [6] (Figs. V.8, V.9 and V10):

---

[3] "Trymer Rigid Foam Insulation", Brochure; Dow Chemical U.S.A., *109-764-86*.
[4] "Moisture Gains by Foam Plastic Roof Insulation under Controlled Temperature Gradients", C.P. Hedlin, *Journal of Cellular Plastics,* September/October (1977), pp. 313-326.
[5] "Criteria for Installing Insulation Systems in Petrochemical Plants", W.C. Turner, *Chemical Engineering Progress, 70,* August (1974), pp. 3-7.
[6] "Messung der Wärmeleitfähigkeit von Schaumkunststoffen mit beliebigem Feuchtigkeitsgehalt", J. Achtziger, Reprint from *Kunststoffe im Bau,* Themenheft *23,* pp. 3-6.

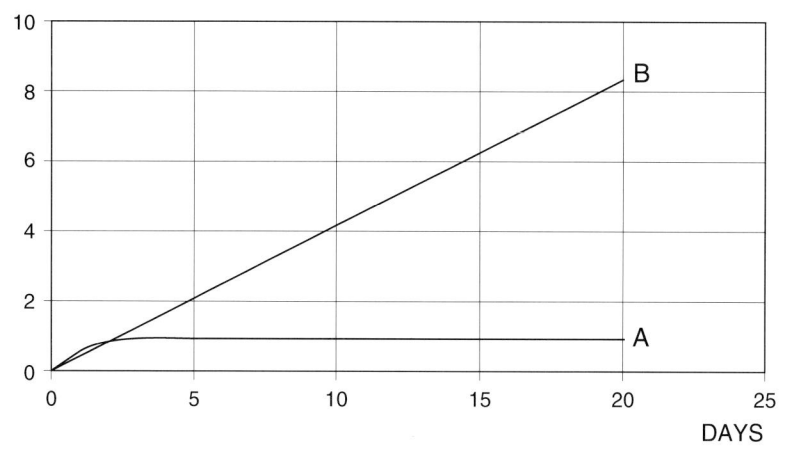

Fig. V.8. Water absorption of polyurethane rigid foams dependent on the time-density of the sample: 35 kg/m^3. (According to Dr. J. Achtziger, *Forschungsinstitut für Wärmeschutz*, Germany)

A. Immersion under water.
B. Absorption by vapour diffusion in the equipment.

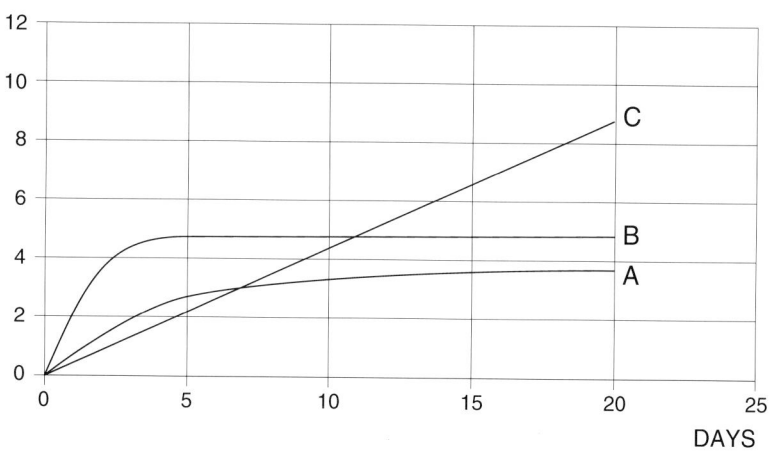

Fig. V.9. Water absorption of polystyrene rigid foam dependent on the time-density of the sample: 19 kg/m^3. (According to Dr. J. Achtziger, *Forschungsinstitut für Wärmeschutz*, Germany).

A. Immersion under water.
B. Immersion under water of the sample twice chilled at 10°C after being heated at 90°C.
C. Absorption by vapour diffusion in the equipment.

PART V. WHY TO CHOOSE FOAMGLAS ® INSULATION

MOISTURE CONTENT IN VOL %

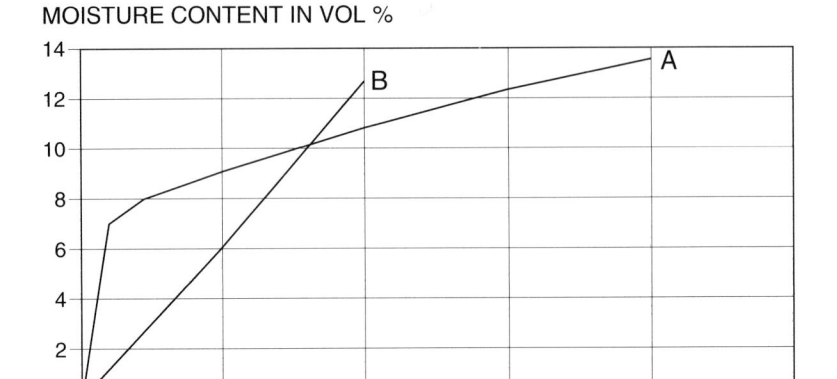

Fig. V.10. Water absorption of phenolic resin rigid foam dependent on the time-density of the sample: 49 kg/m³. (According to Dr. J. Achtziger, *Forschungsinstitut für Wärmeschutz*, Germany).

A. Immersion under water.   **B.** Absorption by vapour diffusion in the equipment.

V.3.3 WATER IN HIGH TEMPERATURE INSULATION

V.3.3.1 High Temperature Insulations

High temperature insulations are generally fibrous or particulate. Thus they are inherently absorptive to liquid water. V.M. Liss, an engineering consultant specialising in design and construction of chemical plants, has compiled water absorption data for various insulation materials [7]. For high temperature insulations he lists:

Insulation material	Water absorbed (submersion) (% by volume)
Calcium silicate	75
Mineral wool	70
Fibrous glass	65
Perlite	16

356

A. Calcium Silicate

Calcium silicate can absorb and retain vast quantities of water. Herb Moak of *DuPont* tells us [8]:

> "I'm not sure many people realise how much water calcium silicate will hold. It will hold from 300 to 400% of its weight without water literally dripping out. And to have insulation with 60 to 90% moisture is not unusual".

These quantities have been confirmed by Johns-Manville, a major manufacturer of calcium silicate. In their *Engineering and Technical Bulletin* Number 79-03-ISD, they indicate than their Thermo-12 calcium silicate can absorb 439% water by weight [9].

B. Mineral Wool

A very practical effect of water absorption on mineral wool occurred on the 300°C steam lines at *Shell Chemical*'s Carrington plant in the UK [10]. At this facility, it was found that:

> "If mineral wool insulation becomes water logged during a downpour, the boilers have to produce up to 60 tons of extra steam each hour to compensate... this can almost be worse than having no insulation at all".

As a result Carrington's:

> "... relagging programme [has used] stainless steel cladding over foam-glass insulation providing a rigid closed cell structure which cannot absorb moisture".

They conclude that:

> "... in certain circumstances, this relagging can give a return on your money in just half a year".

---

[7] "Selecting Thermal Insulation", V.M. Liss, *Chemical Engineering,* 26 May (1986), pp. 103-105.
[8] "Thermal Insulation and Energy Conservation", *Heating/Piping/Air Conditioning,* May (1979), pp. 3-12, July (1979), pp. 13-21.
[9] "Competitive Product Analysis, THERMO-12 Calcium Silicate vs. CELOTEMP & GOODTEMP Expanded Perlite, Water Absorption", *Engineering and Technical Bulletin;* Johns-Manville Sales Corporation, *79-03-ISD*.
[10] "Lagging Ahead", Mike Moss *Shell Times, 32,* pp. 14-15.

C. Water Repellent Insulations

As mentioned earlier, calcium silicate is not water repellent.

Other high temperature insulations are frequently treated with silicones to make them water repellent. Such water repellent additives are, to a certain degree, effective in preventing moisture intrusion during installation. However, we have found that such water repellent treatment may be destroyed by service conditions.

V.3.3.2 Water Entry

Water can enter insulation through jacket systems that have been rendered ineffective by design, workmanship, weathering, damage or simply removal. And water does not mean only rainwater, but also water coming from snow melting, deluge testing, sprinkler system, leaking equipment, etc.

V.3.3.3 Can Insulation Dry out?

A. Low Temperature

It is obvious that water or water vapour which enters an insulation system that is operating continuously at low temperature, will be permanently retained as liquid or ice.

It should never be forgotten that when insulation is applied on low temperature piping, equipment or vessels, the steel face of this equipment constitutes a perfect vapour barrier... but situated on the wrong side.

B. High Temperature

However, it is not as obvious that water will also be retained in insulation operating at high temperatures.

Win Irwin from *Certainteed*, a manufacturer of fibrous glass insulation, tells us that such insulation is difficult to dry out [11]:

---

[11] "Inspection Tips for Plant Managers", Win Irwin, *Insulation,* January (1982), p. 48.

"Remember that once moisture gets into insulation you can never completely remove it. Better to avoid moisture leaks in the first place than trying to overcome them later".

Herb Moak [2] from *DuPont* supports this position. He tells us:

"You can drive moisture back to a certain point, but you can't get rid of it totally. And if water got into it in the first place, you're probably going to get more in before you get what's already there out".

Walter Heinrich [15] of *Phillips Petroleum* and McChesney of *Fuel Save Associates* [12] appear to agree with Moak's position. Heinrich states:

"Insulation is never dry even on a steam line at 1000°F (538°C). It's never dry because when you use some economic thicknesses much of the insulation is below 212°F (100°C). Water, if it ever gets into an insulation system, is difficult to get out".

C. Mechanism of Water Accumulation in Permeable High Temperature Insulation

Since the thickness of high temperature insulation is often chosen to limit the heat transfer to a certain value or to limit the outside surface temperature for personnel protection to about 60°C, a substantial part of the insulation is actually at temperatures ranging from 60 to 100°C and only the part of the insulation close to the pipe or the vessel is at a temperature exceeding 100°C. In the part of insulation at temperatures lower than 100°C, and provided it is a permeable insulation, the water can be in liquid state or in water vapour state but not as superheated steam under considerable pressure. Observations have shown that the escape of this water is very slow. A period close to a month has been experimentally observed. In many locations, in fact almost everywhere, the rainfalls followed each other after much shorter periods and consequently new water coming from outside may penetrate the exterior part of the insulation when the weather barrier is not perfect before the water in the system may escape. This may result in water accumulation in the insulation.

When the water accumulates in the outside part of the insulation, its thermal conductivity is deteriorated and the isotherm 100°C is moved towards the hotter inside part. The part of the insulation situated on the exterior side where the

---

[12] "Preventing Burns from Insulated Pipes", M. McChesney. P. McChesney, *Chemical Engineering*, 27 July (1981), pp. 58-64.

temperature does not exceed 100°C becomes thicker and water accumulation increases further. The mechanism is explained in detail in V.4.4.

Experimental data and calculations have confirmed this explanation, corresponding to the previously quoted experts.

As a summary

- Substantial quantities of water can be retained in thermal insulation operating at high temperatures
- Water retention will increase heat loss, jacket temperature and insulation weight
- Insulation dry out is sufficiently slow that repeated rainfall intrusion will ensure that water retention will be permanent.

V.3.4 FOAMGLAS® INSULATION

V.3.4.1 Water Resistance

FOAMGLAS® insulation is composed of a network of non interconnecting glass cells. Therefore it has outstanding resistance to liquid water and even water vapour. This was initially demonstrated during World War II where large blocks of FOAMGLAS® insulation served as floats for steel anti-submarine nets.

Water absorption of FOAMGLAS® insulation is normally determined in accordance with ASTM C 240-91. This test, which requires two hours immersion, routinely gives a maximum sorption value of less than 0.5% by volume (with the specified test specimen dimensions). This corresponds to a theoretical water penetration of less than 0.1 mm. In fact, this is essentially surface absorption in the open cells in the exterior surface of the materials. Long term tests at *Pittsburgh Corning*, lasting many years, have resulted in negligible sorption which can be expressed as a theoretical water penetration of less than 1 mm.

V.3.4.2 Examples of Water Resistance

Perhaps the best practical demonstration of the non absorption characteristic of FOAMGLAS® insulation in a high temperature application was found during the inspection of a lube oil tank at a Petro-Wax plant in the US. Although this was an unjacketed system, which is not recommended, the FOAMGLAS® insulation on this tank remained dry over the course of 32 years [13]. We know of no other insulation material that could come close to demonstrating such performance.

For this reason, specifiers frequently recommend FOAMGLAS® insulation when there is a probability that the insulation may remain exposed during installation [14], although, this also is not a recommended practice.

V.3.4.3 Cautions

FOAMGLAS® insulation is unequalled in its ability to maintain its thermal effectiveness in the presence of liquid water, but to ensure optimum performance, certain precautions must be maintained at low and high temperatures.

A. Low Temperatures

At low temperatures which cycle around 0°C, the outside surface cells of FOAMGLAS® insulation, which have inevitably been cut, are susceptible to maintain liquid water and consequently it may be needed, in some cases, to fill them with a cell filler (adhesive, coating, sealer) to avoid damaging them by freeze and thaw action in wet conditions. The specifications edited by *Pittsburgh Corning* and major engineering offices for cold applications are taking care of that and recommend, when necessary:

- Sealer in the joints
- Adequate outside coating or weather protection.

The situation at low temperatures is best summed up by Walter Heinrich of *Phillips Petroleum* who states:

> "The best guide in low temperature work is really experience. After many years, I have gone to foamed glass with sealed joints and metal jacketing. We have systems that have been in service 20 years and are still serving adequately" [15].

---

[13] "FOAMGLAS® Insulation – 32 Years Performance", Case Study of Petro-Wax plant, Emlenton, Pennsylvania; *FI-165,* March (1992).
[14] "Solving the Problems of Heat Tracing Insulation: Procter & Gamble's Quincy, Massachusetts Plant", Edward Feit, *Insulation Outlook,* February (1987), pp. 16-19.
[15] "Thermal Insulation and Energy Conservation", *Heating/Piping/Air Conditioning,* May (1979), pp. 3-12 and July (1979), pp. 13-21.

B. High Temperatures

At room temperature, water absorption in FOAMGLAS® insulation is insignificant even after many years. However, at higher temperatures the prolonged continuous exposure to water may result in a more noticeable interaction between FOAMGLAS® insulation and water, and therefore caution should be observed. But its service life is still measured in long periods. In contrast we have seen that the useful service life of other high temperature insulations would be measured in much shorter periods of time in case of rapid absorption of water, if they are not protected.

For conventional high temperature application, continued insulation effectiveness can be assured by providing an efficient weather barrier to prevent water entry. In certain cases it may also be useful to locate weep holes at strategic locations as a means of rapidly venting any accidental water intrusion. Such weep holes will be uniquely effective with FOAMGLAS® insulation since it, in contrast to other high temperature insulations, does not permit the rapid absorption and retention of water.

V.3.5 FOAMGLAS® INSULATION SYSTEMS

V.3.5.1 Total FOAMGLAS® Insulation Systems

The resistance of cellular glass against water in both the liquid and vapour state presents a substantial advantage in many applications if not all of them. This advantage is probably even more decisive in the following:

- Low temperature pipe insulation
- Steam pipes [16]
- Heat traced pipes. In heat traced pipes used to transport such fluids as bitumen, high viscosity oils, waxes and sulphurs, the tracers often cannot generate sufficient heat to drive the water from the system. Prolonged contact between the tracers – especially electrical ones – and water is dangerous. And if water intrusion occurs and cannot be removed, solidification within the system becomes probable, since insulation efficiency has been destroyed. This may result in the total loss of the system. For this

---

[16] "Lagging Ahead", Mike Moss, *Shell Times, 32,* pp. 14-15.

reason, heat tracer pipes are often insulated with FOAMGLAS® cellular glass [17]
- Underground pipes
- Tanks
- Towers. FOAMGLAS® cellular glass has been used on towers operating at moderate temperatures, 80 to 100°C, to replace insulating materials affected by water absorption and retention. It illustrates that permeable insulation dry-out would be virtually impossible at such a low operating temperature
- Pulp and paper mills, textile mills and water treatment plants.
  The use of impermeable FOAMGLAS® insulation is particularly effective in industries where wet, high humidity conditions prevail. In these industries, FOAMGLAS® insulation should be considered not only for piping and equipment but also for buildings, particularly roofs
- Offshore
- Breweries and food processing applications.

V.3.5.2 FOAMGLAS® Cellular Glass/Mineral Fibre Composite Insulation System

For steam lines, the FOAMGLAS® cellular glass/mineral fibre composite insulation system offers a number of advantages. This system consists of an inner layer of high temperature fibrous insulation and an outer layer of FOAMGLAS® insulation. In this system, the outer FOAMGLAS® insulation layer will obviously stay dry because it is impermeable. Not so obviously, the mineral wool insulations will stay dry because, for most applications, they will be completely above 100°C. If water intrudes, the system becomes self-drying.

This system has not only been employed for steam lines, but also for equipment that experiences both temperature cyclings and vibrations. An example of such equipment are digesters used in the pulp and paper industry. In such cases the resilient fibrous inner layer effectively accommodates the vibrations and the dimensional changes resulting from cycling operations.

The system has also been used for underground steam lines, such as the 8 km, 412°C steam line for *Northern States Power Co.* in the US [18].

---

[17] "FOAMGLAS® Pijpschaalisolatie in Esso-Flexicoker", *Uit Europoort-kringen*, 25, [10] (1986), pp. 14-15.
[18] "Utility Supplies High-Pressure Steam to Industrial Plant through 5 mile Insulated Pipeline", Reprint from *Power Engineering*, June (1985).

V.3.5.3 Overfit/Retrofit

The overfit system merits serious consideration for upgrading the performance of previously insulated steam lines without the necessity of removing existing insulation or interrupting plant operations.

This system simply consists in an addition over the existing insulation of an outer layer of FOAMGLAS® insulation with open joints through to the outside to allow the drying out. The overfit will increase insulation thickness to reduce heat loss and, in addition, will dry out the original insulation material, if needed. The mechanism of drying out is explained in detail in V.4.4 and V.4.5.

As little as a 50 mm thickness of the FOAMGLAS® insulation layer is generally sufficient to increase the temperature of the original insulation to over 100°C to force water evaporation from the original layer. It is necessary, of course, to choose a completely impermeable material for the outer overfit layer to prevent the water evaporated from the inner layer from condensing in the overfit layer. Such condensation would, obviously, defeat the purpose of the overfit. At the Lubrizol facilities in Bayport and Deer Park, Texas, the pay back period for FOAMGLAS® overfit installations has been found to be less than one year [19]. As in the case of the FOAMGLAS® mineral fibre composite system, the overfit system will maintain its thermal performance due to its inherent self drying capability.

---

[19] "Steam Pipes Retrofitted On-Line with Cellular Glass Insulation to Avoid Plant Shutdown", Stanley Lee, *Chemical Processing,* April (1985), p. 28.

ANNEX
PATHS FOR WATER ENTRY

It is generally recognised that water is the primary enemy of insulation systems and therefore emphasis is placed on minimising the possibility of water entry into them. But if water somehow or other finds its way into an insulation system, those barriers to water intrusion now serve to retain the accumulated water. This will lead to a reduced thermal efficiency and possible metal corrosion of the insulated surface.

Sources of Water

The most common sources of water are external to the insulation system, rain or snow being the most obvious. However, in industrial facilities, there are additional water sources to be considered – some of which can be particularly damaging. These include [20]:

- Deluge testing
- Cleaning or washdown
- Cooling tower mists
- Melting and dripping from old equipment
- Steam traps.

In many of these cases the water source may contain chlorides and other chemicals that may accelerate corrosion [21].

Internal sources of water include water that has been absorbed in insulation during storage and installation. In fact in some cases, particularly light weight

---

[20] "Factors Affecting Corrosion of Carbon Steel Under Thermal Insulation", Peter Lazar, III, Paper presented at the symposium on *Corrosion of Metals under Thermal Insulation,* 11-13 Oct. 1983, edited by Warren I. Pollock, Jack M. Barnhart (Philadelphia, Pennsylvania: ASTM, August, 1985), pp. 11-26.

[21] "Factors Affecting the Stress Corrosion Cracking of Austenitic Stainless Steel Under Thermal Insulation", Dale McIntyre, Paper presented at the symposium on *Corrosion of Metals Under Thermal Insulation,* 11-13 Oct. 1983, edited by Warren I. Pollock, Jack M. Barnhart (Philadelphia, Pennsylvania: ASTM, August, 1985), pp. 27-41.

concrete, water is deliberately added to the system and may not dry out. Another internal source of water may be a leaking steam line or tracer.

The intrusion of water vapour into low temperature insulation systems will be covered in another chapter.

Water Entry

Water can enter an insulation through flaws in design, poor workmanship, damage, deterioration and poor operating and maintenance procedures.

In general more insulation failures have occurred on equipment in contrast to piping, due to the greater complexity of equipment that must be weather sealed [22]. The figure V.11 gives a drawing showing many of the most likely spots for moisture entry, retention and subsequent corrosion. These include flanges, nozzles, manways, vents, stiffening rings, lifting lugs and support brackets [23, 24]. All such protrusions are difficult to seal. Some may actually be designed in such a way as to channel water into the insulation. In other cases, insufficient clearance between pipes, equipment and supports may preclude installation of a complete thickness of insulation and weather barrier.

In piping, as in equipment, regions of complexity (e.g. valves, flanges, supports and changes in direction) offer the greatest problems to effective weather sealing.

Corrugated aluminium jackets, though widely used, have been described as almost completely unsealable [25]. In addition, deformation of jacket overlaps, which may occur by walking on a system with compressible insulation, or by expansion-contraction of the system, will provide a direct path for water entry. In some cases, insulation for personnel protection is stopped where accidental contact is unlikely. Such a termination may, in time, facilitate water entry.

---

[22] "Reducing the Need for Maintenance on Insulation Systems", John W. Kalis, *Insulation Specifiers' & Buyers' Guide* (1986), pp. 6-14.
[23] "Controlling Carbon Steel Corrosion under Insulation", Paul E. Krystow, Paper presented at the symposium on *Corrosion of Metals under Thermal Insulation*, 11-13 Oct. 1983, edited by Warren I. Pollock, Jack M. Barnhart (Philadelphia, Pennsylvania: ASTM, August, 1985), pp. 145-154.
[24] "Experience with Corrosion under Insulation", R.W. Gregory, Paper on a study by Esso Engineering (Europe) Ltd., pp. 1-5.
[25] "Methods for Keeping Thermal Insulation Dry to Preserve its Function for Energy Conservation", William C. Turner, John W. Johnson, *Journal of Thermal Insulation*, 2, October (1978), pp. 67-86.

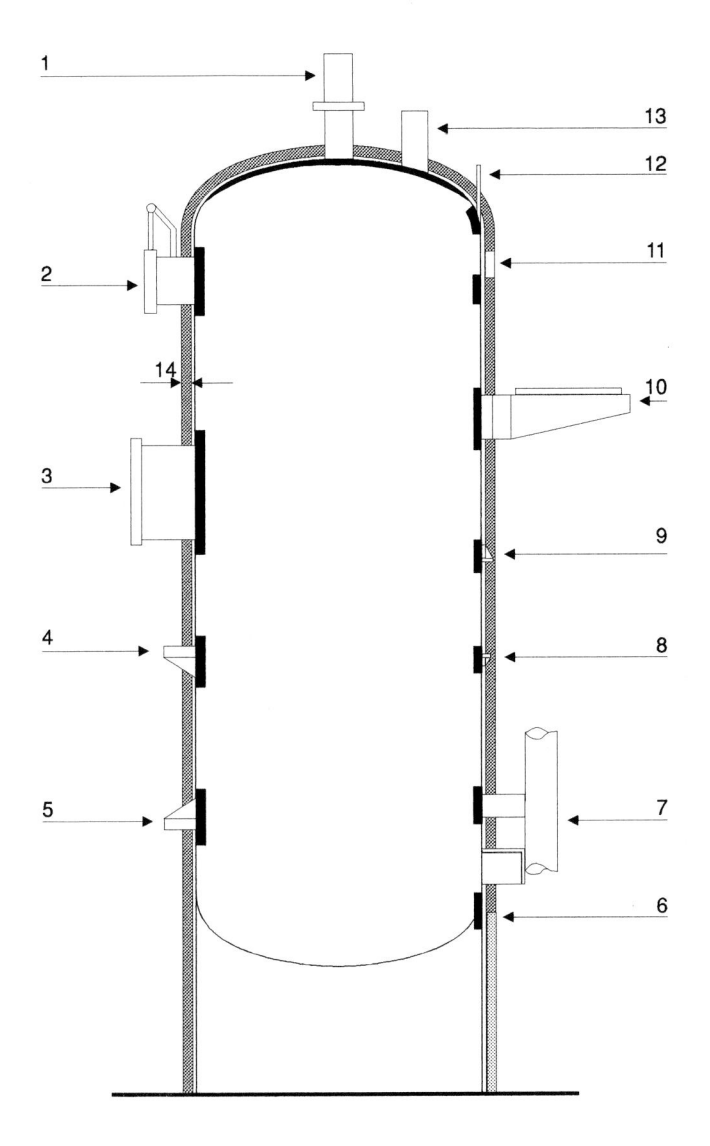

Fig. V.11. The most likely spots for moisture entry, retention and subsequent corrosion.

1. Head nozzles top.
2. Davits.
3. Large nozzle or manway.
4. Small gussetted connections.
5. Vessel support brackets.
6. Fireproofing, vessel support.
7. On-vessel pipe brackets, guides.
8. Stiffening or support ring, hidden.
9. Stiffening ring, exposed.
10. Platform, ladder brackets.
11. Inspection openings unrepaired.
12. Lifting lugs.
13. Top platform supports.
14. Insulation thickness.

PART V. WHY TO CHOOSE FOAMGLAS ® INSULATION

For both piping and equipment, the current interest in economic optimisation generally results in increased insulation thickness. This will have capacity to store additional water that, in turn, will result in both reduced efficiency and increased corrosion [26].

Deterioration of mastic coatings is also a well-known cause of insulation failure. Such deterioration frequently involves blisters and/or splitting. Fig. V.12 shows a badly blistered tank installation. Deterioration of sealers can result from weathering, extreme temperature, thermal cycling and chemical attack.

Fig. V.12. Example of badly blistered tank installation.

Maintenance of insulation systems is essential for long-term performance. However, operating problems that can result in water intrusion include opening insulation systems for inspection [27], maintenance or modification and not properly reclosing the system.

---

[26] "Pijpleidingen Aangetast", P. den Hollander, *AD Binnenland,* 17 juni (1986).
[27] "Factors Affecting Corrosion of Carbon Steel Under Thermal Insulation", Peter Lazar, III, Paper presented at the symposium on *Corrosion of Metals under Thermal Insulation,* 11-13 Oct. 1983, edited by Warren I. Pollock, Jack M. Barnhart (Philadelphia, Pennsylvania: ASTM, August, 1985), pp. 11-26.

368

V.4

WATER VAPOUR TRANSPORT AND CONDENSATION

V.4.1 THE PROBLEM

One of the most serious problems facing a low temperature insulation system is water vapour intrusion. It can intrude, condense and eventually turn to ice in low temperature facilities.

Although less known, problems due to water or water vapour may also occur in hot insulation systems, in the part of pipe coverings situated near the jacket when the temperatures are lower than 100°C.

The consequences of water vapour intrusion and its condensation may include degraded thermal performance, poor temperature control and increased corrosion hazard.

They have more often been pointed out for cold applications as for instance by W.C. Turner, Consultant, South Charleston, W. Va, USA [1], who states:

> "Low temperature pipe and equipment: the major problem in this case is prevention of moisture entry. In such installations, all moisture that enters the system migrates to the substrate low temperature. It continues to build up until the insulation is completely saturated with liquid water or ice".

In the medium temperature field, corrosion probably represents the most severe consequence of water vapour intrusion and condensation.

---

[1] "Criteria for Installing Insulation Systems in Petrochemical Plants", W.C. Turner, *Chemical Engineering Progress, 70* [8], August (1974), pp. 3-7.

V.4.2 EXPLANATION OF THE PROBLEM

V.4.2.1 Absolute Humidity

The air usually contains a certain amount of water in gaseous state, the water vapour. The possible amount increases with the temperature and for each temperature, there is a maximum amount that the air can contain, at saturation. Table V.2 (see Annex) indicates the corresponding quantities for saturated air in g/m^3 for various temperatures. If at a given temperature, this amount is exceeded, condensation takes place: a part of the water vapour turns to liquid (Fig. V.13).

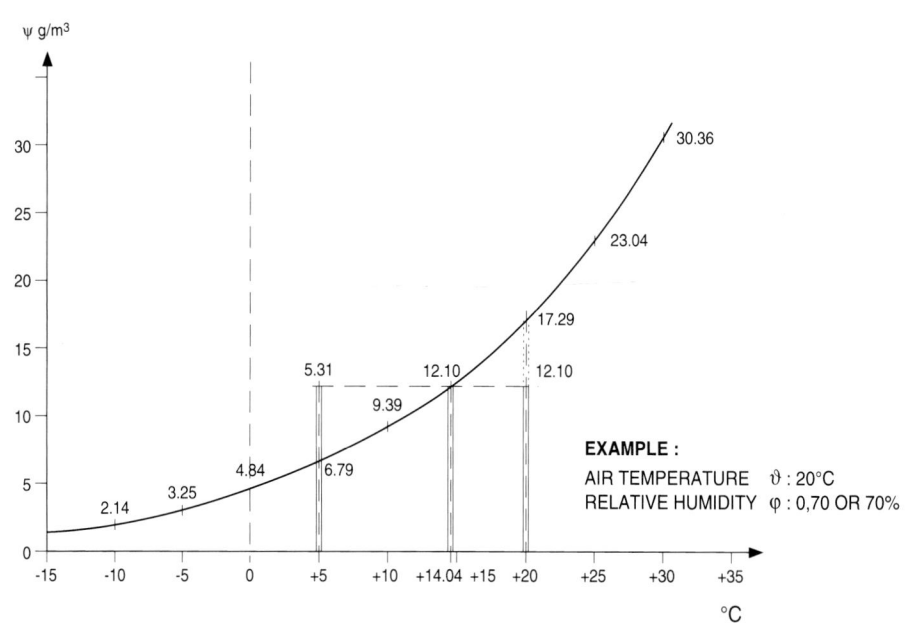

Fig. V.13. Absolute humidity in g/m^3 as a function of air temperature.

V.4.2.2 Relative Humidity

If the air is not saturated but contains a certain amount of water vapour, one can define the relative humidity φ as the ratio of the amount of water vapour

CHAPTER V.4 WATER VAPOUR TRANSPORT AND CONDENSATION

contained in the air to the maximum it can contain at saturation at a given temperature.

V.4.2.3 Partial Pressure of Water Vapour in the Air

Since each gas – for instance water vapour – present in the air exercises a given pressure, one can also refer to the partial pressure P_r of the water vapour and define the relative humidity φ by the ratio of the partial pressure of the water vapour present in the air to the partial pressure in saturated air at the same temperature. This definition is equivalent to the one given in the previous paragraph.

V.4.2.4 Condensation and Dew Point Temperature

If one reduces the temperature of a given volume of air containing a given amount of water vapour, one reaches a temperature at which the air becomes saturated and the water vapour condenses. This temperature is called the dew point temperature evoking the dew that one can see in the morning on an outside surface after a cold night.

Table V.3 (see Annex) indicates the dew point temperature for various conditions of temperature and relative humidity. For instance, the dew point temperature for air at 30°C and 80% relative humidity is 26.2°C.

Table V.4 (see Annex) provides the saturation water vapour pressure P_s in Pa (N/m^2) according to GOFF formulas. For instance, the saturation water vapour pressure P_s for air at 32.5°C is 4892 Pa.

Obviously the partial pressure of water vapour can be easily calculated when the relative humidity and the temperature are known.

V.4.2.5 Partial Pressure Difference

Let us consider two different volumes of air respectively characterised by

No. 1: t_1; φ$_1$; P_{r1}
No. 2: t_2; φ$_2$; P_{r2}

separated by a partition wall of a given material.

One immediately understands that a partial pressure difference exists between the two sides and it is defined by $\Delta p = P_{r1} - P_{r2}$. This difference of partial pressure constitutes the driving force to create a flow of water vapour from the high pressure side to the low pressure side.

V.4.2.6 Permeance and Permeability. Water Vapour Diffusion Resistance Number

Under the driving force of the partial pressure difference, the flow of water will depend on the reaction of the partition wall. If it is totally vapour tight such as metal or glass, the vapour flow will be zero. If it is not tight the vapour flow will be significant. Like in the electrical analogy, the reaction of the partition wall can be expressed by its resistance or its inverse value. In the metric system, a unit of permeance is called the metric perm and corresponds to the flow of 1 gram of water vapour through a material of given thickness per m^2 of surface per 24 hours under a pressure difference of 1 mm Hg.

This definition applies for thick products and for thin foil type material (roofing felt, membrane, vapour brake, etc.).

In the case of homogeneous materials, the flow is inversely proportional to the thickness. Consequently one can define a property which is material specific, independent of the material thickness : the permeability.

Due to the complexity of the units of permeance and permeability, one tends to use a ratio (without dimension) to express the resistance of a product to the flow of water vapour. The water vapour diffusion resistance number (or factor) or μ value is the ratio of the resistance of a layer of material (to water vapour diffusion) to the resistance of a layer of air of the same thickness, under the same conditions of temperature and atmospheric pressure. It does express how many times better the material resists to water vapour passage than air. As a matter of appreciation, it is often considered that an insulating material should have a μ value of at least several thousands to be satisfactory for most applications or that special means should be used to protect it from moisture penetration and transfer.

The μ value can also be defined in terms of water vapour diffusion. It is then the ratio of the water vapour diffusion through the air layer of equal thickness to the water vapour diffusion through the material, under the same conditions of temperature and atmospheric pressure.

To measure the permeance, permeability or water vapour diffusion number, the specimen is sealed to the top of a cup in which a very dry (dry cup method) or a very wet (wet cup method) atmosphere has been created. The cup is placed in a standardised controlled atmosphere. By weighing periodically the cup, it is possible to determine the amount of water vapour passing through the specimen. Without going into details, it is worth mentioning that the testing conditions have an impact on the results and that the measurements have to be done with great care to obtain valid results particularly when a water vapour tight material such as FOAMGLAS® cellular glass is tested.

It is interesting to compare the water vapour resistance number of different materials. Many tables exist. From DIN 4108 – Part 4 – Table 1, we can extract the following values.

The ranges are generally explained by different densities.

Material	μ value
Rendering	
Cement mortar	15/35
Building elements	
Reinforced concrete	70/150
Cellular concrete	5/10
Thermal insulation	
Wood wool	2/5
Polyurethane foam in situ	30/100
Ureumformaldehyde	1/3
Cork	5/10
Expanded polystyrene	20/100
Extruded polystyrene	80/300
Polyurethane foam board	30/100
Phenolic foam	30/50
Mineral and vegetal wool	1
Cellular glass	Practically water vapour tight

(to be continued)

(continued)

Material	μ value
Waterproofing layer and membranes	
Mastic asphalt > 7 mm...	Practically water vapour tight
Naked roofing felt...	2,000/20,000
Roofing felt applied with hot bitumen....................	20,000/60,000
PVC foil thickness > 0.1 mm	20,000/50,000

As it can be seen, FOAMGLAS® cellular glass and bitumen (asphalt) have a very high μ value, especially when comparing them with the other materials.

V.4.2.7 Specific Resistance against Water Vapour Diffusion

To calculate the water vapour flow through a given system, the specific resistance of each material is needed.

Obviously the resistance of homogeneous layers depends on the μ value and is proportional to the thickness s of the material [2]. It can be calculated by the formula:

$$\rho = \mu s N$$

where

    ρ is the specific resistance
    s is the thickness of the material
    N is an expression that depends of several parameters and can be found in Table V.5 (see Annex) [3].

---

[2] Here again, it may be interesting to make the analogy between water vapour flow, electrical current or heat transfer.

[3] More specific N is given by $N = \dfrac{R_D \times T_m}{\delta}$

where R_D is the gas constant of water vapour
        T_m is the mean absolute temperature
        δ is the diffusivity of water vapour in the air
(reference to *General Law of Gas* – Boyle-Mariotte and Gay Lussac).
Table V.5 (see Annex) provides the value of N from −50°C to +100°C in N·h/(kg·m) or J·h/(kg·m²).

Usually ρ is expressed in J·h/(kg·m), s in meter, N in J·h/(kg·m²) and μ is a number without dimension.

Since 1 J = 1 N·m, one can also express ρ in N·h/kg and N in N·h/(kg·m).

V.4.2.8 Water Vapour Flow

The water vapour flow G is proportional to the water vapour pressure difference (Δp: the "driving force") and inversely proportional to the specific resistance ρ so that

$$G = \frac{\Delta p}{\rho}$$

Δp being expressed in Pa or N/m²,
ρ   in J·h/(kg·m) or in N·h/kg and
G   in kg/(m²·h).

Based on these data and equations, one understands that it is possible to calculate the water vapour flow, the amount of water remaining at a given place in an insulation system, the dew point temperature and compare the result with the actual temperature at the same point, to see if condensation does occur.

V.4.3 EXAMPLES OF WATER VAPOUR TRANSMISSION THROUGH LOW TEMPERATURE INSULATION SYSTEMS

Without going through detailed calculations based on the previous equations, we can review some results and get a feeling for the order of magnitude of water vapour transmission in practical situations.

V.4.3.1 Water Vapour Transmission through an Insulating Material

As a first step we will compare the water vapour transmission passing through one square meter of a 10 cm thick insulation used to separate an ambience at +20°C and 80% relative humidity from an ambience at −25°C and 100% relative humidity, assuming that the movement lasts one year. To analyse the extreme conditions, we will suppose that no vapour brake has been applied (Fig. V.14).

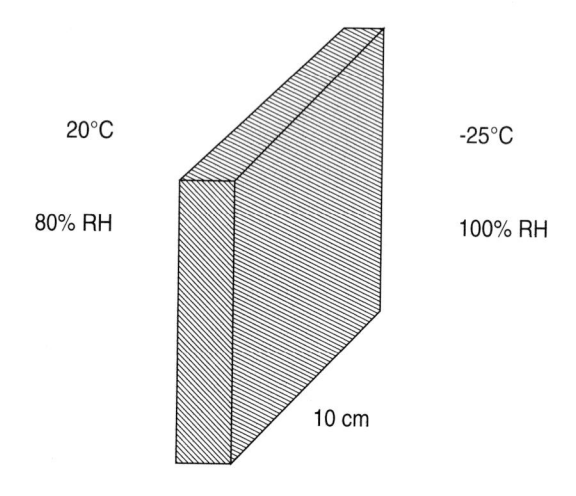

20°C   -25°C

80% RH   100% RH

10 cm

Fig. V.14. Water vapour transmission through an insulating material.

We will also consider several insulation materials having the following μ values, generally based on DIN 4108 – Part 4:

FOAMGLAS® cellular glass applied with hot bitumen.......... 70,000
Extruded polystyrene 250
Expanded polystyrene 100
Polyurethane foam 100
Mineral wool 1

As it can be seen below, the amount of water vapour passing through the insulating material is substantial, if its μ value is low.

μ value of the insulating material	Amount of moisture passing through in g/(m^2·year)
70,000 [4]	1.5
250	413
100 [5]	1033
1	103,333

V.4.3.2 Water Vapour Transmission through an Insulating System of a Low Temperature Application

Let us now consider the same situation but for industrial equipment like a vessel, a pipe operating at –25°C. The steel wall of the vessel or of the pipe is a perfect vapour barrier but applied on the cold side, the wrong side! And consequently there would be a considerable accumulation of water vapour condensing into water and then turning to ice due to the low operating temperature. How can we try to compensate for this perfect "steel" vapour barrier inevitably applied on the wrong side in a cold application? We should apply on the "warm" side another vapour barrier, as efficient as the first one, if the insulation is not vapour tight itself.

This practically means that we should use a steel casing on the outside with all the joints perfectly welded to achieve a tightness comparable to the vessel or the pipe. This is theoretically feasible and is sometimes done for double containment systems such as those used for some cryogenic applications but it remains a technical achievement, rather than a practical solution for the general use.

---

[4] The μ value given for FOAMGLAS® is a very conservative one. Actually, the limit of the μ value is more determined by the accuracy of the test method than the μ value of FOAMGLAS®. A value of 70,000 is often chosen because it corresponds to a thick layer of bitumen like the joint between two slabs.

[5] Although the purpose is not to go into calculations, we can detail one example. For instance for a polyurethane foam or an expanded polystyrene having a μ value of 100, one can write:

Water vapour pressure on the warm (20°C, 80%) side: $2338 \times 0.8 = 1870.4$ Pa
Water vapour pressure on the cold (–25°C, 100%) side: $63.24 \times 1.0 = 63.24$ Pa
Difference of water vapour pressure: $1870.4 - 63.24 = 1807.16$ Pa
N value (average temperature: –2.5°C): 1.532×10^6 N·h/(kg·m)

Resistance to water vapour flow:
$\rho = \mu \times s \times N$
$\rho = 100 \times 0.1 \times 1.532 \times 10^6$
$\rho = 15.32 \times 10^6$ N·h/kg

Water vapour flow: $G = \dfrac{\Delta p}{\rho}$

$G = \dfrac{1807.16}{15.32 \times 10^6}$ kg/(m²·h)   Pa/N·h/kg = N/m² × kg/(N·h) = kg/(m²·h)
$G = 117.96 \times 10^{-6}$ kg/(m²·h)
$G = 117.96 \times 10^{-6} \times 24 \times 365$ kg/(m²·year)
$G = 1.033330$ kg/(m²·year)
$G = 1033$ g/(m²·year)

In the majority of the cases the outside protection is made of a metal jacket that does not have tight joints and consequently is certainly not a vapour barrier, not even a vapour brake or is made of a layer of mastic – a few millimetre thick – having a high μ value but not comparable at all to steel or glass.

As an example, we can consider an insulation system used on an ammonia tank and make the following – realistic – assumptions:

Exterior temperature	20°C
Exterior relative humidity	80%
Operating temperature	–33°C
Insulation	100 mm of a material having a μ value of 100
Protection	2 mm of a coating having a μ value of 70,000
Tank diameter	30 m
Tank wall height	27 m

The calculation shows a water vapour flow of 8.4 mg/(m^2·h). Although this value seems rather low, it should be remembered that it represents a value of 21.5 g/h for the wall of the tank.

This is significant and would have after some time an influence on the thermal conductivity of the insulating material, particularly since this water vapour will mainly be located near the steel where the temperature is such that the water vapour not only turns to water but even to ice. Here again a layer of ice constitutes an excellent vapour brake, which like the steel plate, is situated on the cold side, the wrong side.

As a matter of comparison, the same tank wall having a 13 cm thick FOAMGLAS® insulation (μ practically infinite) with 2 mm thick joints between the slabs filled with an adhesive/sealer having a μ value of 22,000 and no vapour brake, will have a water vapour flow of 0.008 g/h for the wall of the tank. The difference with the previous case is several orders of magnitude.

An obvious conclusion that we should draw from the previous general analysis, is that perfect water vapour barriers are needed for low temperature applications made on steel equipment if the insulating material as such is not perfectly vapour tight.

Does this perfect vapour barrier exist?

V.4.3.3 The Water Vapour Barrier or Water Vapour Retarder

First of all, a strong difference should be made between a water barrier (also called weather barrier) and a water vapour barrier. Many products, provided that they remain undamaged, are excellent water barriers. Few are good water vapour barriers. The umbrella, which is water tight and protects from the rain, is neither water vapour tight nor gas tight. It is a classic example of this difference. It is well known that smoke can be blown through the umbrella, which indicates that the rain tight tissue is neither gas nor water vapour tight. The umbrella is a weather barrier but not a vapour barrier. In a much more scientific way, Hedlin of the *National Building Research Council*/London and Achtziger of the *Forschungsinstitut für Wärmeschutz*/Germany both found independently by laboratory experiments, that the penetration of water vapour is much more important and faster than the penetration of water [6, 7]. With the exception of glass and metals (in sufficient thickness), very few products are actually vapour barriers. This has been recognised by replacing the name vapour barrier by "vapour brake", "vapour retarder" or "vapour check".

If the efficiency of the large majority of the vapour retarders is already questionable from a theoretical stand-point, the situation is much worse when we consider the real facts in industrial life.

Vapour retarders, or weather barriers are subject to many detrimental influences and the calculation based on their theoretical characteristics is only an approach to their idealised performance in the very best conditions.

It should be remembered that vapour retarders are applied on the outside face of the insulation of low temperature equipment and consequently they experience considerable temperature changes, which may accelerate their ageing.

Condensation takes place on the vapour retarders when the exterior relative humidity increases too much, implying long term direct contact with water. Of

---

[6] "Moisture Gains by Foam Plastic Roof Insulation under Controlled Temperature Gradient", C.P. Hedlin, *Journal of Cellular Plastics,* September (1977), pp. 313-326.
[7] "Messung der Wärmeleitfähigkeit von Schaumkunststoffen mit beliebigem Feuchtigkeitsgehalt", J. Achtziger, Reprint from *Kunststoffe im Bau,* Themenheft *23,* pp. 3-6.

course rain, snow and hail can also deteriorate the vapour retarder. The same applies to the effects of ultraviolet and infrared rays.

External winds, often salt laden or sand grain laden in the case of terminals situated near the sea, also attack the vapour retarder.

In the case of many chemical plants, vapour retarders can be subject to the aggression of chemicals coming from spillages and damaging them.

With all the insulating materials that are not perfectly dimensionally stable, the vapour retarders are subject to local stress, when the insulation slab or pipe covering changes dimensions resulting in joint openings.

Moreover, although it should be forbidden, people tend to use pipelines, vessels and tank roofs as natural scaffoldings and the vapour retarders are aggressed by a number of mechanical abuse, such as abrasion, excessive point load, falling of sharp objects, even sparks or burning torches used for other activities, but placed on the insulation during preparation.

This damage of the vapour retarders has been acknowledged by some design engineers who apply a vapour retarder on the insulation and then a metal jacket to protect the vapour retarder from exterior negative influences. This solution possibly preferable when using insulating materials which are sensitive to water vapour represents of course a substantial cost increase and is not necessarily effective: galvanic corrosion may take place when pieces of metal of different electrical potential are in contact, directly or through an electrical connection, such as an insulating material soaked with water and protected by a damaged, inefficient moisture vapour retarder. Moreover screws are often used to fix the metal jacket and can penetrate the mastic water vapour retarders.

As a conclusion, vapour retarders are subject to a number of possible aggressions and are ineffective in many cases. This sad situation has been pointed out by several authors. For instance Smith and Kirkman state in the *Journal of Mine Ventilation* [8], under the heading "Vapour Barrier":

> "Of the many products on the market which people assume to be vapour barriers, relatively few are in fact impervious to water vapour. Other foils are excellent

---

[8] "Some Aspects for Consideration in the Insulation of Chilled Water Pipes", J.G. Smit, R.I. Kirkman, *Journal of Mine Ventilation, 32,* April (1979), pp. 61-70.

vapour barriers but unfortunately they are very susceptible to puncturing and corrosion unless excessively thick, whereupon the cost structure and mass render them unacceptable".

And W. O'Keefe, Associate Editor of *Power Magazine* [9], writes:

"Vapour barriers are the only practical answer to moisture problems. Completely effective vapour barriers do not exist in practice... Since it is practically certain that some moisture will pass the vapour barrier, other provisions must be made to keep insulation dry and prevent its eventual destruction".

V.4.4 WATER ABSORPTION IN MEDIUM AND HIGH TEMPERATURE INSULATION

If water or water vapour absorption in low temperature insulating systems is generally recognised as detrimental, some people still believe that a similar situation does not practically occur for insulation systems operating at medium or high temperatures. In fact, when the insulation material does absorb water that can turn to vapour or steam, it affects energy consumption, surface temperature, personnel safety, operating costs and constitutes one of the major factors leading to steel corrosion.

Sources of water are very common around industrial plants. Water can penetrate industrial systems if they are not made with water tight insulation materials.

The large majority of the metal jackets used to protect medium and high temperature insulation have no vapour tight joints but generally overlapped joints. These theoretically prevent rain and snow from penetrating into the systems. In fact, these systems are often far from being always tight and consequently rain and snow can run into the insulation materials when they are not water tight as such. The same applies to condensation water formed when the plant is idling. In this last case, water vapour can also play a major role.

Moreover, many insulated industrial plants are equipped with deluge systems that are regularly tested to control their efficiency. Finally, it is not unusual to have a leaking valve or flange which will allow water or other liquid to get into the insulation.

---

[9] "Thermal Insulation", W. O'Keefe, *Power,* August (1974), pp. 1-24.

This may occur when corroded pipes eventually become perforated and allow water to soak the insulation.

And a large part of this water turns to vapour or steam, dependant upon the temperature.

It should be remembered that at 100°C, when water boils, the saturation water vapour pressure reaches one atmosphere or one bar or 10^5 Pa.

The developed pressures, when the temperature exceeds 100°C are such that the water vapour can escape. When the temperatures are below 100°C and especially if they are well below, the water vapour pressure substantially decreases and the possibility of water to escape is very reduced (See Annex, Table V.4).

For instance, we can consider a pipe of a 168.3 mm diameter operating at 300°C, the ambient conditions being 25°C and no wind. We can also assume that the emissivity of the metal jacket is 0.1 and that a not water tight insulating material having an average thermal conductivity of 0.066 W/(m·K) is applied. For simplification we take the average thermal conductivity at the average temperature [10].

- It can be calculated that with a thickness of 70 mm of this insulating material the surface temperature is 57.21°C which is slightly under the generally accepted value of 60°C for personnel protection and the heat loss is 171.62 W/m² or 166.22 W/m.

  The local temperature is lower than 100°C in the outer 12.34 mm layer of insulation (ranging from 57.21 to 100°C) (see first line on the table V.1) of the pipe covering.

- As a theoretical model, we can now assume that a 10 mm layer near the outside face of the pipe covering is full of water having a thermal conductivity of 0.6 W/(m.K). It is possible since the outer 12.34 mm has a temperature lower than 100°C.

Since the insulating value of the system is reduced, we find logically that the surface temperature increases to 59.87°C, the heat loss to 189.32 W/m² or 183.37 W/m and the local temperature is reduced under 100°C in a 19.34 mm

---

[10] It results in a linear instead of a multilinear graph if we would take a thermal conductivity which varies with the temperature.

layer near the outside of the pipe covering (see line 2 on the Table V.1). Actually at this stage, the possibility exists that the water layer increases in thickness to 19.34 mm, which would result in a further increase of the surface temperature, heat loss and displacement of the 100°C point from the outside towards the steel pipe.

• As a further step with our theoretical model, we can consider a layer of 20 mm water near the outside surface. The surface temperature will increase to 63.31°C, the heat loss to 212 W/m² or 206.10 W/m, and the temperature will be reduced to under 100°C in a 26.35 mm layer near the outside of the pipe covering (see line 3 on the table). The cumulative character of the situation is easy to appreciate. It is developed in Table V.1 and Fig. V.15:

TABLE V.1

Item	Assumed thickness of the water layer (mm)	Surface temperature (°C)	Heat loss (W/m²)	Heat loss (W/m)	Zone in which the temperature is < 100°C (mm)
1	0	57.31	171.62	166.22	12.34
2	10	59.87	189.32	183.37	19.34
3	20	63.31	212.00	206.10	26.35
4	30	67.98	245.51	237.79	33.36
5	40	74.75	294.60	285.33	40.37

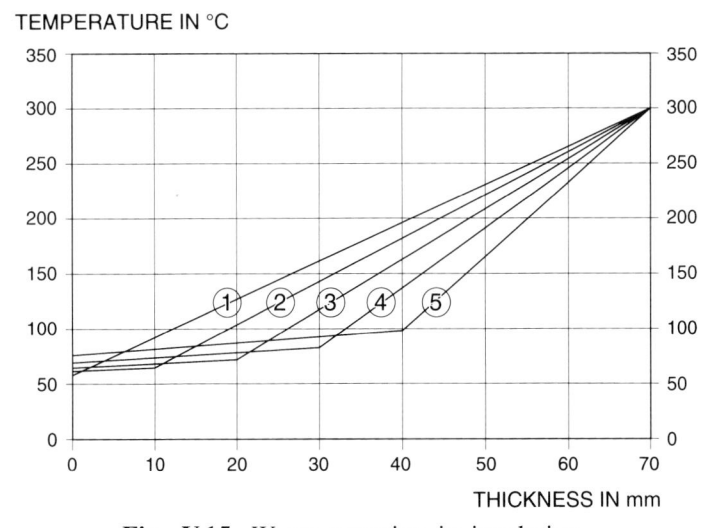

Fig. V.15. Water retention in insulation.

For each of the assumed thicknesses of the water, the zone in which the temperature is lower than 100°C is thicker and consequently the phenomenon may develop until the exterior layer of insulation gets saturated to such a thickness that the temperature locally exceeds 100°C, the water turning to steam and developing pressure high enough to force it to escape [11]. This point is usually reached when a very substantial part of the insulation is full of water and when the insulation is not performing according to the initial design.

We have based our analysis on the limitation of the surface temperature for personnel protection. Actually when using the criteria of economic thickness, the required thickness is generally substantially higher, the surface temperature lower and the part of the pipe covering operating under 100°C thicker. Thus in this case, the water can remain in a larger part of the insulation.

This does not mean that the water vapour cannot escape but the speed of drying, in the part where the temperature is lower than 100°C, is so low that new water entry exceeds, in many climatic conditions, the amount of water escaping as water vapour.

As a general conclusion, water can stay in permeable high temperature insulation resulting in a higher heat loss and a higher surface temperature thus creating a risk for personnel and increased possibilities of corrosion danger.

V.4.5 THE OVERFIT/RETROFIT SYSTEM

If we modify the temperature profile in a pipe insulation which is not water-tight applied on a hot line to the point that all the insulation would be at a temperature exceeding 100°C, the water eventually present in the system would boil into steam and its pressure would be high enough to rapidly force it out of the system resulting in an increase of the insulation efficiency.

---

[11] A further limitation would be when the insulation value of the pipe covering has decreased to the point that the surface temperature exceeds 100°C. This criterion is easier to understand but the previous one should be considered first.

We can achieve this by adding one layer of insulation on the outside of the applied one. But it must be made of a water and water vapour proof insulating material. Should it not be the case, we would have only displaced the problem without solving it since the new added layer will have a temperature lower than 100°C and could absorb water. The additional layer can only perform well if it is made of a waterproof insulation, which basically means that it should be made of cellular glass [12].

This system of adding one layer of FOAMGLAS® cellular glass on a hot line with such a thickness that the permeable insulation has a temperature always exceeding 100°C is called overfit or retrofit, the second word being chosen to indicate that it can be applied on already existing lines. The joints between the FOAMGLAS® cellular glass pipe coverings of the overfit system are left slightly open to allow the water possibly present in the permeable insulation to escape in the form of water vapour. Similarly, the aluminium jacket is generally used to protect the FOAMGLAS® cellular glass layer from exterior aggression and should not have sealed laps. Water can already be present in the insulation before the installation of the overfit on it or water can make its way into the system, for instance during an idling period. Actually the overfit system can dry out several times after different periods of water intrusion [13].

A very interesting point in relation to the overfit system consists in the possibility of applying it when the hot line is in service. This application without shut down is obviously interesting for plants in which any loss of production has to be avoided [14].

Since water vapour ingress and condensation into water is really the typical problem of many insulation systems, the overfit system constitutes the answer to it: water vaporises at high temperature and escapes due to high pressure.

---

[12] Mineral wool and calcium silicate can contain substantial amounts of water. Cellular plastic foams, some of which like extruded polystyrene present a reasonable resistance against water absorption, are not considered for high temperature applications.
[13] Generally a few days of continuous operations are needed to dry out a very wet insulation.
[14] If the overfit system appears very challenging, it should not be specified for cases where the non absorption of water or combustible liquids should be achieved for the total insulation thickness: for instance in the case of organic fluid lines.

V.4.6 INSULATING MATERIALS AND WATER VAPOUR TRANSMISSION

If it is generally accepted that mineral wool ($\mu = 1$) and calcium silicate do not offer much resistance to water vapour transmission, some people think that non protected cellular plastic insulating materials are "just good enough" for the majority of the applications. To clarify this issue, we may present a few statements made by plastic foam manufacturers or by independent experts.

Koppers engineers [15] state:

> "Phenolic cells breathe water vapour... the moisture permeability of polyurethane, isocyanurate and polystyrene is relatively low... Once moisture has entered it is difficult to remove, even where humidity conditions are low".

Regarding polyurethane foams, the *Donald S. Gilmore Research laboratories* state in a report on cryogenic insulation [16]:

> "Protection to exclude moisture from the insulant is also required. Although isocyanurate-based cellular plastics are closed cell structures, they do have the ability to "breathe" or transmit moisture vapour in the same manner as wood, a naturally occurring cellular product. This is a slow process when compared to fibrous materials but must still be accounted for in the design of the system".

and J.H. Saunders and F.C. Frisch [17] write in *Polyurethanes, Chemistry and Technology*:

> "Permeability studies of various gases through a urethane film, indicate that the diffusion of atmospheric gases into the foam, with resulting dilution of the fluorocarbon, cause the drift in the k factor".

---

[15] "Design Advantages of Phenolic Foam Roof Insulation", Edward Kifer and Clyde Henry, *Roof Design,* May/June (1985), pp. 48-54.
[16] "Cryogenic Insulation", Donald S. Gilmore Research Laboratories, *Insulation Outlook,* November (1981), pp. 23-28.
[17] "Polyurethanes, Chemistry and Technology, Part II Technology", J.H. Saunders, F.C. Frisch, *Interscience Publishers, II,* p. 258.

They also added a table of permeability of gases. In this table, water vapour has the highest value.

	Gas permeability $\dfrac{\text{CC gas (stp)} \times \text{mm}}{\text{s} \times \text{cm}^2 \times \text{cm Hg}}$
Trichlorofluoromethane	0.029×10^{-10}
Nitrogen	0.270×10^{-10}
Air (calc)	0.430×10^{-10}
Oxygen	1.070×10^{-10}
Carbon dioxide	4.000×10^{-10}
Water vapour	1100.000×10^{-10}

stp = standard temperature and pressure.

Conclusion: This shows that polyurethane is about 2500 times more permeable to water vapour than to air. That means the permeability of water vapour is an intrinsic property of polyurethane, whereby open cells are not necessary.

The *Oak Ridge National Laboratory* published a study on thermal insulation [18]. In this study they wrote:

> "For intermediate low-temperature service applications, there is considerable discontent with expanded polyurethane because of its tendency to be permeated by water vapor, which subsequently deposits as ice at the cold surface. This progressively destroys the insulation properties of the urethane, although the warmer surface appears normal and dry. The vapor barriers available for low-temperature foamed urethane and polystyrene insulation systems are generally inadequate, and need to be greatly improved, and their performance needs to be verified in service".

Among the cellular plastics, extruded polystyrene foam shows a better resistance to water and water vapour than many others with the possible exception of expanded rubber.

Nevertheless it shows in certain applications a "water absorption" and a "water vapour diffusion" which cannot be neglected. Since it is less used for

---

[18] "Industrial Thermal Insulation – An Assessment", R.G. Donnelly, V.J. Tennery, D.L. McElroy, *Oak Ridge National Laboratory, TID-27120,* August (1976), p. 32.

industrial applications than for instance polyurethane, we will take an analysis of a building application. Calvin C. Hay [19] indicates that extruded polystyrene insulation applied in an inverted roof system may absorb as much as 20-30% moisture (by volume).

As already indicated, FOAMGLAS® cellular glass has a µ value which is practically infinite, 70,000 being a value often considered for calculation purposes when the infinite value appears difficult to introduce in the calculation. It corresponds to a thick layer of bitumen, reflecting the fact that this material if often used to fill the joints.

As a matter of general interest, some tests run by *Pittsburgh Corning Corporation* using a mass spectrometer resulted in a µ value close to 5,000,000 instead of 70,000. At this level, the difference of values presents a scientific interest but has no influence on practical applications.

V.4.7 CONCLUSION

The advantages of using a water vapour tight material like FOAMGLAS® cellular glass have been pointed out by several authors. It may be sufficient to quote again the *Oak Ridge National Laboratory*:

> "Foamed glass was cited by several contacts as a superior insulation for use at intermediate low temperatures. Most of these low-temperature systems require a periodic heating and drying because of the inevitable moisture permeation and ice formation. The high-temperature stability of foamed glass relative to urethane or polystyrene foam is another significant advantage for the glass material".

---

[19] "Moisture Absorption and its Effect on the Thermal Performance of Extruded Polystyrene", C.C. Hay, *Graduate Paper Mechanical Engineering,* 4 May (1984), pp. 1-56.

ANNEX

TABLE V.2
Absolute humidity ψ of water vapour saturated air in g/m^3 as a function of air temperature

τ (°C)	ψ (g/m^3)	τ (°C)	ψ (g/m^3)	τ (°C)	ψ (g/m^3)	τ (°C)	ψ (g/m^3)	τ (°C)	ψ (g/m^3)
-20	0.88	0	4.84	20	17.29	40	51.14	60	130.2
-19	0.97	1	5.19	21	18.33	41	53.77	61	135.9
-18	1.06	2	5.56	22	19.42	42	56.50	62	141.9
-17	1.16	3	5.95	23	20.56	43	59.35	63	148.1
-16	1.27	4	6.36	24	21.77	44	62.33	64	154.5
-15	1.39	5	6.79	25	23.04	45	65.40	65	161.1
-14	1.51	6	7.25	26	24.37	46	68.67	66	168.0
-13	1.65	7	7.74	27	25.76	47	72.03	67	175.2
-12	1.80	8	8.26	28	27.23	48	75.33	68	182.6
-11	1.96	9	8.81	29	28.76	49	79.18	69	190.2
-10	2.14	10	9.39	30	30.36	50	82.98	70	198.1
-9	2.33	11	10.00	31	32.04	51	86.93	71	206.2
-8	2.53	12	10.65	32	33.80	52	91.03	72	214.6
-7	2.75	13	11.34	33	35.65	53	95.30	73	223.4
-6	2.99	14	12.07	34	37.58	54	99.74	74	232.4
-5	3.25	15	12.84	35	39.60	55	104.30	75	241.8
-4	3.53	16	13.65	36	41.71	56	109.1	76	251.4
-3	3.83	17	14.50	37	43.92	57	114.1	77	261.4
-2	4.15	18	15.39	38	46.22	58	119.3	78	271.7
-1	4.49	19	16.32	39	48.63	59	124.7	79	282.3

TABLE V.3
Dewpoint temperature

Ambient air temperature (°C)	Relative humidity											
	40	45	50	55	60	65	70	75	80	85	90	95	100
	Dewpoint temperatures (°C)											
-50	-57.2	-56.3	-55.5	-54.7	-54.1	-53.4	-52.9	-52.3	-51.8	-51.3	-50.9	-50.4	-50
-49	-56.2	-55.3	-54.5	-53.8	-53.1	-52.5	-51.9	-51.3	-50.8	-50.3	-49.9	-49.4	-49
-48	-55.3	-54.4	-53.6	-52.8	-52.1	-51.5	-50.9	-50.3	-49.8	-49.3	-48.9	-48.4	-48
-47	-54.4	-53.5	-52.6	-51.9	-51.2	-50.5	-49.9	-49.4	-48.8	-48.3	-47.9	-47.4	-47
-46	-53.4	-52.5	-51.7	-50.9	-50.2	-49.6	-49.0	-48.4	-47.9	-47.4	-46.9	-46.4	-46
-45	-52.5	-51.6	-50.7	-50.0	-49.2	-48.6	-48.0	-47.4	-46.9	-46.4	-45.9	-45.4	-45
-44	-51.6	-50.6	-49.8	-49.0	-48.3	-47.6	-47.0	-46.4	-45.9	-45.4	-44.9	-44.4	-44
-43	-50.6	-49.7	-48.8	-48.0	-47.3	-46.7	-46.0	-45.5	-44.9	-44.4	-43.9	-43.4	-43
-42	-49.7	-48.7	-47.9	-47.1	-46.4	-45.7	-45.1	-44.5	-43.9	-43.4	-42.9	-42.4	-42
-41	-48.8	-47.8	-46.9	-46.1	-45.4	-44.7	-44.1	-43.5	-42.9	-42.4	-41.9	-41.5	-41
-40	-47.8	-46.9	-46.0	-45.2	-44.4	-43.7	-43.1	-42.5	-42.0	-41.4	-40.9	-40.5	-40
-39	-46.9	-45.9	-45.0	-44.2	-43.5	-42.8	-42.1	-41.5	-41.0	-40.4	-39.9	-39.5	-39
-38	-46.0	-45.0	-44.1	-43.3	-42.5	-41.8	-41.2	-40.6	-40.0	-39.5	-38.9	-38.5	-38
-37	-45.0	-44.0	-43.1	-42.3	-41.5	-40.8	-40.2	-39.6	-39.0	-38.5	-38.0	-37.5	-37
-36	-44.1	-43.1	-42.2	-41.3	-40.6	-39.9	-39.2	-38.6	-38.0	-37.5	-37.0	-36.5	-36
-35	-43.2	-42.1	-41.2	-40.4	-39.6	-38.9	-38.2	-37.6	-37.0	-36.5	-36.0	-35.5	-35
-34	-42.2	-41.2	-40.3	-39.4	-38.7	-37.9	-37.3	-36.6	-36.1	-35.5	-35.0	-34.5	-34
-33	-41.3	-40.3	-39.3	-38.5	-37.7	-37.0	-36.3	-35.7	-35.1	-34.5	-34.0	-33.5	-33
-32	-40.4	-39.3	-38.4	-37.5	-36.7	-36.0	-35.3	-34.7	-34.1	-33.5	-33.0	-32.5	-32
-31	-39.4	-38.4	-37.4	-36.6	-35.8	-35.0	-34.4	-33.7	-33.1	-32.5	-32.0	-31.5	-31
-30	-38.5	-37.4	-36.5	-35.6	-34.8	-34.1	-33.4	-32.7	-32.1	-31.6	-31.0	-30.5	-30
-29	-37.6	-36.5	-35.5	-34.7	-33.9	-33.1	-32.4	-31.8	-31.1	-30.6	-30.0	-29.5	-29
-28	-36.6	-35.6	-34.6	-33.7	-32.9	-32.1	-31.4	-30.8	-30.2	-29.6	-29.0	-28.5	-28
-27	-35.7	-34.6	-33.6	-32.8	-31.9	-31.2	-30.5	-29.8	-29.2	-28.6	-28.0	-27.5	-27
-26	-34.8	-33.7	-32.7	-31.8	-31.0	-30.2	-29.5	-28.8	-28.2	-27.6	-27.0	-26.5	-26
-25	-33.8	-32.7	-31.8	-30.8	-30.0	-29.2	-28.5	-27.8	-27.2	-26.6	-26.1	-25.5	-25
-24	-32.9	-31.8	-30.8	-29.9	-29.1	-28.3	-27.6	-26.9	-26.2	-25.6	-25.1	-24.5	-24
-23	-32.0	-30.9	-29.9	-28.9	-28.1	-27.3	-26.6	-25.9	-25.3	-24.6	-24.1	-23.5	-23
-22	-31.1	-29.9	-28.9	-28.0	-27.1	-26.3	-25.6	-24.9	-24.3	-23.7	-23.1	-22.5	-22
-21	-30.1	-29.0	-28.0	-27.0	-26.2	-25.4	-24.6	-23.9	-23.3	-22.7	-22.1	-21.5	-21
-20	-29.2	-28.1	-27.0	-26.1	-25.2	-24.4	-23.7	-23.0	-22.3	-21.7	-21.1	-20.5	-20
-19	-28.3	-27.1	-26.1	-25.1	-24.3	-23.4	-22.7	-22.0	-21.3	-20.7	-20.1	-19.5	-19
-18	-27.3	-26.2	-25.1	-24.2	-23.3	-22.5	-21.7	-21.0	-20.3	-19.7	-19.1	-18.5	-18
-17	-26.4	-25.2	-24.2	-23.2	-22.3	-21.5	-20.8	-20.0	-19.4	-18.7	-18.1	-17.5	-17
-16	-25.5	-24.3	-23.2	-22.3	-21.4	-20.6	-19.8	-19.1	-18.4	-17.7	-17.1	-16.6	-16
-15	-24.6	-23.4	-22.3	-21.3	-20.4	-19.6	-18.8	-18.1	-17.4	-16.8	-16.1	-15.6	-15
-14	-23.6	-22.4	-21.4	-20.4	-19.5	-18.6	-17.8	-17.1	-16.4	-15.8	-15.1	-14.6	-14
-13	-22.7	-21.5	-20.4	-19.4	-18.5	-17.7	-16.9	-16.1	-15.4	-14.8	-14.2	-13.6	-13
-12	-21.8	-20.6	-19.5	-18.5	-17.5	-16.7	-15.9	-15.2	-14.5	-13.8	-13.2	-12.6	-12
-11	-20.9	-19.6	-18.5	-17.5	-16.6	-15.7	-14.9	-14.2	-13.5	-12.8	-12.2	-11.6	-11
-10	-19.9	-18.7	-17.6	-16.6	-15.6	-14.8	-14.0	-13.2	-12.5	-11.8	-11.2	-10.6	-10
-9	-19.0	-17.8	-16.6	-15.6	-14.7	-13.8	-13.0	-12.2	-11.5	-10.8	-10.2	-9.6	-9
-8	-18.1	-16.8	-15.7	-14.7	-13.7	-12.8	-12.0	-11.3	-10.5	-9.8	-9.2	-8.6	-8
-7	-17.2	-15.9	-14.8	-13.7	-12.8	-11.9	-11.0	-10.3	-9.5	-8.9	-8.2	-7.6	-7
-6	-16.2	-15.0	-13.8	-12.8	-11.8	-10.9	-10.1	-9.3	-8.6	-7.9	-7.2	-6.6	-6
-5	-15.3	-14.0	-12.9	-11.8	-10.8	-9.9	-9.1	-8.3	-7.6	-6.9	-6.2	-5.6	-5
-4	-14.4	-13.1	-11.9	-10.9	-9.9	-9.0	-8.1	-7.3	-6.6	-5.9	-5.2	-4.6	-4
-3	-13.5	-12.2	-11.0	-9.9	-8.9	-8.0	-7.2	-6.4	-5.6	-4.9	-4.2	-3.6	-3
-2	-12.5	-11.2	-10.0	-9.0	-8.0	-7.1	-6.2	-5.4	-4.6	-3.9	-3.3	-2.6	-2
-1	-11.6	-10.3	-9.1	-8.0	-7.0	-6.1	-5.2	-4.4	-3.7	-2.9	-2.3	-1.6	-1

TABLE V.3 (*continued and the end*)

Ambient air temperature (°C)	\multicolumn{12}{c}{Relative humidity}
	40	45	50	55	60	65	70	75	80	85	90	95	100
	\multicolumn{12}{c}{Dewpoint temperatures (°C)}
0	-10.7	-9.4	-8.2	-7.1	-6.1	-5.1	-4.3	-3.4	-2.7	-2.0	-1.3	-0.6	0
1	-9.9	-8.5	-7.3	-6.2	-5.2	-4.3	-3.4	-2.6	-1.8	-1.1	-0.4	0.3	1
2	-9.1	-7.7	-6.5	-5.4	-4.4	-3.4	-2.6	-1.7	-1.0	-0.2	0.5	1.3	2
3	-8.3	-6.9	-5.7	-4.6	-3.5	-2.6	-1.7	-0.9	-0.1	0.7	1.5	2.3	3
4	-7.4	-6.1	-4.9	-3.7	-2.7	-1.7	-0.9	0.0	0.9	1.7	2.5	3.3	4
5	-6.6	-5.3	-4.0	-2.9	-1.9	-0.9	0.0	0.9	1.8	2.7	3.5	4.3	5
6	-5.8	-4.5	-3.2	-2.1	-1.0	-0.1	0.9	1.9	2.8	3.7	4.5	5.3	6
7	-5.0	-3.6	-2.4	-1.3	-0.2	0.9	1.9	2.9	3.8	4.7	5.5	6.3	7
8	-4.2	-2.8	-1.6	-0.4	0.7	1.8	2.9	3.8	4.8	5.6	6.5	7.2	8
9	-3.4	-2.0	-0.8	0.4	1.7	2.8	3.8	4.8	5.7	6.6	7.4	8.2	9
10	-2.6	-1.2	0.1	1.4	2.6	3.7	4.8	5.8	6.7	7.6	8.4	9.2	10
11	-1.8	-0.4	1.0	2.3	3.5	4.7	5.7	6.7	7.7	8.6	9.4	10.2	11
12	-1.0	0.4	1.9	3.2	4.5	5.6	6.7	7.7	8.7	9.6	10.4	11.2	12
13	-0.2	1.4	2.8	4.2	5.4	6.6	7.7	8.7	9.6	10.5	11.4	12.2	13
14	0.6	2.3	3.7	5.1	6.4	7.5	8.6	9.6	10.6	11.5	12.4	13.2	14
15	1.5	3.2	4.7	6.0	7.3	8.5	9.6	10.6	11.6	12.5	13.4	14.2	15
16	2.4	4.1	5.6	7.0	8.2	9.4	10.5	11.6	12.6	13.5	14.4	15.2	16
17	3.3	5.0	6.5	7.9	9.2	10.4	11.5	12.5	13.5	14.5	15.3	16.2	17
18	4.2	5.9	7.4	8.8	10.1	11.3	12.5	13.5	14.5	15.4	16.3	17.2	18
19	5.1	6.8	8.4	9.8	11.1	12.3	13.4	14.5	15.5	16.4	17.3	18.2	19
20	6.0	7.7	9.3	10.7	12.0	13.2	14.4	15.4	16.4	17.4	18.3	19.2	20
21	6.9	8.6	10.2	11.6	12.9	14.2	15.3	16.4	17.4	18.4	19.3	20.2	21
22	7.8	9.5	11.1	12.6	13.9	15.1	16.3	17.4	18.4	19.4	20.3	21.2	22
23	8.7	10.4	12.0	13.5	14.8	16.1	17.2	18.3	19.4	20.3	21.3	22.2	23
24	9.6	11.3	12.9	14.4	15.8	17.0	18.2	19.3	20.3	21.3	22.3	23.1	24
25	10.5	12.3	13.9	15.3	16.7	18.0	19.1	20.3	21.3	22.3	23.2	24.1	25
26	11.4	13.2	14.8	16.3	17.6	18.9	20.1	21.2	22.3	23.3	24.2	25.1	26
27	12.3	14.1	15.7	17.2	18.6	19.9	21.1	22.2	23.3	24.3	25.2	26.1	27
28	13.2	15.0	16.6	18.1	19.5	20.8	22.0	23.2	24.2	25.2	26.2	27.1	28
29	14.0	15.9	17.5	19.0	20.4	21.8	23.0	24.1	25.2	26.2	27.2	28.1	29
30	14.9	16.8	18.4	20.0	21.4	22.7	23.9	25.1	26.2	27.2	28.2	29.1	30
31	15.8	17.7	19.4	20.9	22.3	23.6	24.9	26.0	27.1	28.2	29.2	30.1	31
32	16.7	18.6	20.3	21.8	23.3	24.6	25.8	27.0	28.1	29.2	30.1	31.1	32
33	17.6	19.5	21.2	22.8	24.2	25.5	26.8	28.0	29.1	30.1	31.1	32.1	33
34	18.5	20.4	22.1	23.7	25.1	26.5	27.7	28.9	30.1	31.1	32.1	33.1	34
35	19.4	21.3	23.0	24.6	26.1	27.4	28.7	29.9	31.0	32.1	33.1	34.1	35
36	20.3	22.2	23.9	25.5	27.0	28.4	29.7	30.9	32.0	33.1	34.1	35.1	36
37	21.2	23.1	24.8	26.5	27.9	29.3	30.6	31.8	33.0	34.0	35.1	36.1	37
38	22.0	24.0	25.8	27.4	28.9	30.3	31.6	32.8	33.9	35.0	36.1	37.1	38
39	22.9	24.9	26.7	28.3	29.8	31.2	32.5	33.7	34.9	36.0	37.1	38.0	39
40	23.8	25.8	27.6	29.2	30.7	32.1	33.5	34.7	35.9	37.0	38.0	39.0	40
41	24.7	26.7	28.5	30.1	31.7	33.1	34.4	35.7	36.8	38.0	39.0	40.0	41
42	25.6	27.6	29.4	31.1	32.6	34.0	35.4	36.6	37.8	38.9	40.0	41.0	42
43	26.5	28.5	30.3	32.0	33.5	35.0	36.3	37.6	38.8	39.9	41.0	42.0	43
44	27.4	29.4	31.2	32.9	34.5	35.9	37.3	38.6	39.8	40.9	42.0	43.0	44
45	28.2	30.3	32.1	33.8	35.4	36.9	38.2	39.5	40.7	41.9	43.0	44.0	45
46	29.1	31.2	33.0	34.8	36.3	37.8	39.2	40.5	41.7	42.9	43.9	45.0	46
47	30.0	32.1	34.0	35.7	37.3	38.7	40.1	41.4	42.7	43.8	44.9	46.0	47
48	30.9	33.0	34.9	36.6	38.2	39.7	41.1	42.4	43.6	44.8	45.9	47.0	48
49	31.8	33.9	35.8	37.5	39.1	40.6	42.0	43.4	44.6	45.8	46.9	48.0	49
50	32.7	34.8	36.7	38.4	40.1	41.6	43.0	44.3	45.6	46.8	47.9	49.0	50

TABLE V.4
Saturation water vapour pressure in Pa (N/m^2), according to Goff formulas

Air temperature (°C)	0.0	0.1	0.2	0.3	0.4	0.5	0.6	0.7	0.8	0.9
-50	3.9	3.9	3.8	3.8	3.7	3.7	3.7	3.6	3.6	3.5
-49	4.4	4.4	4.3	4.3	4.2	4.2	4.1	4.1	4.0	4.0
-48	5.0	5.0	4.9	4.8	4.8	4.7	4.7	4.6	4.6	4.5
-47	5.7	5.6	5.5	5.5	5.4	5.3	5.3	5.2	5.2	5.1
-46	6.4	6.3	6.2	6.2	6.1	6.0	5.9	5.9	5.8	5.7
-45	7.2	7.1	7.0	6.9	6.9	6.8	6.7	6.6	6.6	6.5
-44	8.1	8.0	7.9	7.8	7.7	7.6	7.6	7.5	7.4	7.3
-43	9.1	9.0	8.9	8.8	8.7	8.6	8.5	8.4	8.3	8.2
-42	10.2	10.1	10.0	9.9	9.8	9.6	9.5	9.4	9.3	9.2
-41	11.4	11.3	11.2	11.1	10.9	10.8	10.7	10.6	10.4	10.3
-40	12.8	12.7	12.5	12.4	12.3	12.1	12.0	11.9	11.7	11.6
-39	14.4	14.2	14.0	13.9	13.7	13.6	13.4	13.3	13.1	13.0
-38	16.1	15.9	15.7	15.5	15.4	15.2	15.0	14.9	14.7	14.5
-37	17.9	17.7	17.5	17.4	17.2	17.0	16.8	16.6	16.4	16.2
-36	20.0	19.8	19.6	19.4	19.2	19.0	18.8	18.5	18.3	18.1
-35	22.3	22.1	21.9	21.6	21.4	21.1	20.9	20.7	20.5	20.2
-34	24.9	24.6	24.4	24.1	23.8	23.6	23.3	23.1	22.8	22.6
-33	27.7	27.4	27.1	26.8	26.5	26.3	26.0	25.7	25.4	25.1
-32	30.8	30.5	30.1	29.8	29.5	29.2	28.9	28.6	28.3	28.0
-31	34.2	33.9	33.5	33.2	32.8	32.5	32.1	31.8	31.5	31.1
-30	38.0	37.6	37.2	36.8	36.4	36.0	35.7	35.3	34.9	34.6
-29	42.1	41.7	41.3	40.8	40.4	40.0	39.6	39.2	38.8	38.4
-28	46.7	46.2	45.7	45.3	44.8	44.3	43.9	43.5	43.0	42.6
-27	51.7	51.2	50.7	50.2	49.6	49.1	48.6	48.1	47.7	47.2
-26	57.2	56.6	56.1	55.5	54.9	54.4	53.8	53.3	52.8	52.2
-25	63.2	62.6	62.0	61.4	60.8	60.2	59.5	59.0	58.4	57.8
-24	69.8	69.2	68.5	67.8	67.1	66.5	65.8	65.2	64.5	63.9
-23	77.1	76.3	75.6	74.8	74.1	73.4	72.7	72.0	71.3	70.6
-22	85.0	84.2	83.4	82.6	81.8	81.0	80.2	79.4	78.6	77.9
-21	93.7	92.8	91.9	91.0	90.1	89.3	88.4	87.6	86.7	85.9
-20	103.2	102.2	101.2	100.3	99.3	98.3	97.4	96.5	95.5	94.6
-19	113.5	112.5	111.4	110.3	109.3	108.2	107.2	106.2	105.2	104.2
-18	124.8	123.7	122.5	121.3	120.2	119.1	117.9	116.8	115.7	114.6
-17	137.1	135.9	134.6	133.3	132.1	130.9	129.6	128.4	127.2	126.0
-16	150.6	149.2	147.8	146.4	145.1	143.7	142.4	141.1	139.7	138.4
-15	165.2	163.7	162.2	160.7	159.2	157.7	156.3	154.8	153.4	152.0
-14	181.1	179.4	177.8	176.2	174.6	173.0	171.4	169.8	168.3	166.7
-13	198.4	196.6	194.8	193.0	191.3	189.6	187.8	186.1	184.4	182.8
-12	217.2	215.2	213.3	211.4	209.5	207.6	205.7	203.9	202.0	200.2
-11	237.6	235.5	233.4	231.3	229.2	227.2	225.1	223.1	221.1	219.1
-10	259.7	257.4	255.2	252.9	250.7	248.4	246.2	244.0	241.9	239.7
-9	283.8	281.3	278.8	276.3	273.9	271.5	269.1	266.7	264.4	262.0
-8	309.8	307.1	304.4	301.8	299.1	296.5	293.9	291.4	288.8	286.3
-7	338.0	335.1	332.2	329.3	326.4	323.6	320.8	318.0	315.3	312.5
-6	368.5	365.3	362.2	359.1	356.0	352.9	349.9	346.9	343.9	340.9
-5	401.5	398.1	394.7	391.3	388.0	384.7	381.4	378.1	374.9	371.7
-4	437.2	433.5	429.9	426.2	422.6	419.0	415.5	411.9	408.4	405.0
-3	475.8	471.8	467.8	463.9	460.0	456.1	452.3	448.5	444.7	440.9
-2	517.4	513.1	508.8	504.6	500.4	496.2	492.0	487.9	483.9	479.8
-1	562.3	557.7	553.1	548.5	544.0	539.5	535.0	530.5	526.1	521.8
-0	610.8	605.8	600.8	595.9	591.0	586.1	581.3	576.5	571.8	567.0

TABLE V.4 (*continued and the end*)

Air temperature (°C)	0.0	0.1	0.2	0.3	0.4	0.5	0.6	0.7	0.8	0.9
0	611	615	620	624	629	633	638	643	647	652
1	657	661	666	671	676	681	686	691	696	701
2	706	711	716	721	726	731	736	742	747	752
3	758	763	768	774	779	785	790	796	802	807
4	813	819	825	830	836	842	848	854	860	866
5	872	878	884	890	897	903	909	916	922	928
6	935	941	948	954	961	968	974	981	988	995
7	1001	1008	1015	1022	1029	1036	1043	1051	1058	1065
8	1072	1080	1087	1094	1102	1109	1117	1125	1132	1140
9	1148	1155	1163	1171	1179	1187	1195	1203	1211	1219
10	1227	1236	1244	1252	1261	1269	1278	1286	1295	1303
11	1312	1321	1330	1339	1347	1356	1365	1374	1384	1393
12	1402	1411	1421	1430	1439	1449	1458	1468	1478	1487
13	1497	1507	1517	1527	1537	1547	1557	1567	1577	1588
14	1598	1608	1619	1629	1640	1651	1661	1672	1683	1694
15	1705	1716	1727	1738	1749	1760	1772	1783	1795	1806
16	1818	1829	1841	1853	1865	1877	1889	1901	1913	1925
17	1937	1949	1962	1974	1987	1999	2012	2025	2038	2050
18	2063	2076	2089	2103	2116	2129	2143	2156	2170	2183
19	2197	2211	2224	2238	2252	2266	2280	2295	2309	2323
20	2338	2352	2367	2382	2396	2411	2426	2441	2456	2471
21	2487	2502	2517	2533	2548	2564	2580	2596	2612	2627
22	2644	2660	2676	2692	2709	2725	2742	2759	2775	2792
23	2809	2826	2843	2861	2878	2895	2913	2930	2948	2966
24	2984	3002	3020	3038	3056	3075	3093	3112	3130	3149
25	3168	3187	3206	3225	3244	3263	3283	3302	3322	3342
26	3362	3382	3402	3422	3442	3462	3483	3503	3524	3545
27	3566	3587	3608	3629	3650	3672	3693	3715	3737	3758
28	3780	3803	3825	3847	3869	3892	3915	3937	3960	3983
29	4006	4030	4053	4076	4100	4124	4148	4171	4196	4220
30	4244	4268	4293	4318	4342	4367	4392	4418	4443	4468
31	4494	4519	4545	4571	4597	4623	4650	4676	4703	4729
32	4756	4783	4810	4838	4865	4892	4920	4948	4976	5004
33	5032	5060	5089	5117	5146	5175	5204	5233	5262	5292
34	5321	5351	5381	5411	5441	5471	5502	5532	5563	5594
35	5625	5656	5688	5719	5751	5782	5814	5846	5879	5911
36	5944	5976	6009	6042	6075	6109	6142	6176	6210	6244
37	6278	6312	6347	6381	6416	6451	6486	6521	6557	6592
38	6628	6664	6700	6736	6773	6809	6846	6883	6920	6958
39	6995	7033	7071	7109	7147	7185	7224	7262	7301	7340
40	7380	7419	7459	7498	7538	7579	7619	7659	7700	7741
41	7782	7823	7865	7907	7948	7991	8033	8075	8118	8161
42	8204	8247	8290	8334	8378	8422	8466	8510	8555	8600
43	8645	8690	8735	8781	8827	8873	8919	8965	9012	9059
44	9106	9153	9200	9248	9296	9344	9393	9441	9490	9539
45	9588	9637	9687	9737	9787	9837	9888	9938	9989	10041
46	10092	10144	10196	10248	10300	10353	10405	10458	10512	10565
47	10619	10673	10727	10781	10836	10891	10946	11001	11057	11113
48	11169	11225	11282	11339	11396	11453	11511	11568	11626	11685
49	11743	11802	11861	11920	11980	12040	12100	12160	12221	12282
50	12343	12404	12466	12528	12590	12652	12715	12778	12841	12905

The saturation pressure over liquid water for the temperature range from –50°C to 100°C is

$$\log_{10}(p_{ws}) = 10.79586(1-\theta) + 5.02808\log_{10}(\theta) + 1.50474 \times 10^{-4}(1 - 10^{-8.29692[(1/\theta)-1]})$$
$$+ 0.42873 \times 10^{-3}(10^{4.76955(1-\theta)} - 1) - 2.2195983$$

p_{ws} = saturation vapour pressure in atmospheres. (Note: 1 atm = 760 mm Hg)
θ = 273.16/T
T = absolute temperature in Kelvin.

TABLE V.5
N values

°C	N.h/(kg.m) 10^6	°C	N.h/(kg.m) 10^6	°C	N.h/(kg.m) 10^6	°C	N.h/(kg.m) 10^6	°C	N.h/(kg.m) 10^6	°C	N.h/(kg.m) 10^6	°C	N.h/(kg.m) 10^6	°C	N.h/(kg.m) 10^6	°C	N.h/(kg.m) 10^6	°C	N.h/(kg.m) 10^6
-50.0	1.792	-49.5	1.789	-49.0	1.785	-48.5	1.782	-48.0	1.779	-47.5	1.776	-47.0	1.773	-46.5	1.769	-46.0	1.766		
-45.0	1.760	-44.5	1.757	-44.0	1.754	-43.5	1.751	-43.0	1.748	-42.5	1.745	-42.0	1.741	-41.5	1.738	-41.0	1.735		
-40.0	1.729	-39.5	1.726	-39.0	1.723	-38.5	1.720	-38.0	1.717	-37.5	1.714	-37.0	1.712	-36.5	1.709	-36.0	1.706		
-35.0	1.700	-34.5	1.697	-34.0	1.694	-33.5	1.691	-33.0	1.688	-32.5	1.686	-32.0	1.683	-31.5	1.680	-31.0	1.677		
-30.0	1.671	-29.5	1.669	-29.0	1.666	-28.5	1.663	-28.0	1.660	-27.5	1.658	-27.0	1.655	-26.5	1.652	-26.0	1.650		
-25.0	1.644	-24.5	1.641	-24.0	1.639	-23.5	1.636	-23.0	1.633	-22.5	1.631	-22.0	1.628	-21.5	1.626	-21.0	1.623		
-20.0	1.618	-19.5	1.615	-19.0	1.613	-18.5	1.610	-18.0	1.607	-17.5	1.605	-17.0	1.602	-16.5	1.600	-16.0	1.597		
-15.0	1.592	-14.5	1.590	-14.0	1.587	-13.5	1.585	-13.0	1.582	-12.5	1.580	-12.0	1.577	-11.5	1.575	-11.0	1.573		
-10.0	1.568	-9.5	1.565	-9.0	1.563	-8.5	1.561	-8.0	1.558	-7.5	1.556	-7.0	1.553	-6.5	1.551	-6.0	1.549		
-5.0	1.544	-4.5	1.542	-4.0	1.539	-3.5	1.537	-3.0	1.535	-2.5	1.532	-2.0	1.530	-1.5	1.528	-1.0	1.526		
0.0	1.521	0.5	1.519	1.0	1.517	1.5	1.514	2.0	1.512	2.5	1.510	3.0	1.508	3.5	1.505	4.0	1.503		
5.0	1.499	5.5	1.497	6.0	1.495	6.5	1.492	7.0	1.490	7.5	1.488	8.0	1.486	8.5	1.484	9.0	1.482		
10.0	1.477	10.5	1.475	11.0	1.473	11.5	1.471	12.0	1.469	12.5	1.467	13.0	1.465	13.5	1.463	14.0	1.461		
15.0	1.457	15.5	1.455	16.0	1.452	16.5	1.450	17.0	1.448	17.5	1.446	18.0	1.444	18.5	1.442	19.0	1.440		
20.0	1.436	20.5	1.434	21.0	1.432	21.5	1.430	22.0	1.429	22.5	1.427	23.0	1.425	23.5	1.423	24.0	1.421		
25.0	1.417	25.5	1.415	26.0	1.413	26.5	1.411	27.0	1.409	27.5	1.407	28.0	1.405	28.5	1.404	29.0	1.402		
30.0	1.398	30.5	1.396	31.0	1.394	31.5	1.392	32.0	1.390	32.5	1.389	33.0	1.387	33.5	1.385	34.0	1.383		
35.0	1.379	35.5	1.378	36.0	1.376	36.5	1.374	37.0	1.372	37.5	1.370	38.0	1.369	38.5	1.367	39.0	1.365		
40.0	1.362	40.5	1.360	41.0	1.358	41.5	1.356	42.0	1.355	42.5	1.353	43.0	1.351	43.5	1.349	44.0	1.348		
45.0	1.344	45.5	1.343	46.0	1.341	46.5	1.339	47.0	1.337	47.5	1.336	48.0	1.334	48.5	1.332	49.0	1.331		
50.0	1.327	50.5	1.326	51.0	1.324	51.5	1.322	52.0	1.321	52.5	1.319	53.0	1.317	53.5	1.316	54.0	1.314		
55.0	1.311	55.5	1.309	56.0	1.308	56.5	1.306	57.0	1.304	57.5	1.303	58.0	1.301	58.5	1.300	59.0	1.298		
60.0	1.295	60.5	1.293	61.0	1.292	61.5	1.290	62.0	1.289	62.5	1.287	63.0	1.286	63.5	1.284	64.0	1.283		
65.0	1.279	65.5	1.278	66.0	1.276	66.5	1.275	67.0	1.273	67.5	1.272	68.0	1.270	68.5	1.269	69.0	1.267		
70.0	1.264	70.5	1.263	71.0	1.261	71.5	1.260	72.0	1.258	72.5	1.257	73.0	1.255	73.5	1.254	74.0	1.252		
75.0	1.250	75.5	1.248	76.0	1.247	76.5	1.245	77.0	1.244	77.5	1.242	78.0	1.241	78.5	1.239	79.0	1.238		
80.0	1.235	80.5	1.234	81.0	1.232	81.5	1.231	82.0	1.230	82.5	1.228	83.0	1.227	83.5	1.225	84.0	1.224		
85.0	1.221	85.5	1.220	86.0	1.218	86.5	1.217	87.0	1.216	87.5	1.214	88.0	1.213	88.5	1.212	89.0	1.210		
90.0	1.208	90.5	1.206	91.0	1.205	91.5	1.204	92.0	1.202	92.5	1.201	93.0	1.200	93.5	1.198	94.0	1.197		
95.0	1.194	95.5	1.193	96.0	1.192	96.5	1.190	97.0	1.189	97.5	1.188	98.0	1.186	98.5	1.185	99.0	1.184		

Formula

$$N = \frac{0.022414 \times T_m \times p}{0.018 \times 273 \times 0.083 \times \left(\frac{T_m}{273}\right)^{1.81}}$$

where:
p = Atmospheric pressure in Pa or N/m^2
T_m = mean temperature in K
N = in N·h/(kg·m)
δ = 0.083 in m^2/h = diffusivity of water vapour in air.

V.5

THERMAL EFFICIENCY IN REAL LIFE

V.5.1 THE PROBLEM

Reliable and long term insulation efficiency is a prerequisite to safe and economical plant operation. It must be built into the facility through the use of proper materials and systems. If the goal of long term efficiency is compromised, severe technical and economic problems will result. These problems include:

- Increased operating costs
- Possible loss of process control
- Compromised personnel safety
- Mechanical damage
- Chemical corrosion.

V.5.1.1 Operating Costs

Thermal insulation that has been degraded leads to increased operating costs. In contrast, with constant insulation efficiency a short payback period can be achieved and operating costs over the lifetime of the project can be reduced.

Mr. Sam Moy of *Hercules* determined that a FOAMGLAS® insulation system on the sides of boiling vats/tubs reduced energy costs by 18% and provided an acceptable payback time of 3.75 years. An additional 29% would be saved by insulating the tops of these vats [1].

---

[1] "Insulation Cuts Steam Use, Payback for Insulating Nitrocellular Boiling Tubs is Less than Four Years", *Manufacturing Technology Note,* U.S. Army Material Development and Readiness Command, Project Number MMT57942, March (1984).

PART V. WHY TO CHOOSE FOAMGLAS® INSULATION

The overfit system of FOAMGLAS® insulation may have simple payback periods of the order of 2 to 5 years and retain the revitalised efficiency.

Stanley Kastanas of *Colonial Gas* points to the economic consequences of thermal inefficiency for the life of an operation, particularly in low temperature and cryogenic processes. Of particular concern is the importance of keeping water vapour out of the system in order to retain the thermal efficiency, thus FOAMGLAS® insulation can be used both as a vapour barrier and as an impermeable insulation [2].

Moak of *DuPont* states [3]:

"Water is the worst enemy of thermal insulation systems because it increases the heat loss".

This loss of thermal efficiency directly leads to higher operating costs.

V.5.1.2 Process Control

Degradation of insulation efficiency can cause severe problems with process control and interfere with both the quality and quantity of the manufactured product. Examples of such problems include excessive boil-off or product loss, shut-down of chemical processes, interruption of fluid flow by freezing or solidification or undesired changes in viscosity. In particular, materials such as sulphur, residual oils, asphalt, waxes, etc., can solidify or increase in viscosity.

V.5.1.3 Personnel Protection

Thermal insulation that has been degraded after installation will not provide the proper personnel protection, allowing burns from hot insulation and may even cause skin or tissue damage in the case of cold insulation as well.

As explained later in this paper, even a small degradation due to mechanisms such as moisture, ageing, shrinkage, cracks, etc. can lead to a safety problem in regard to personnel protection, due to unexpected surface temperatures.

---

[2] "Updating Piping and Vessel Insulation with Composite Systems", S. Kastanas, *AGA 1987 Operating Section Proceedings*, (1987), pp. 446-470.
[3] "An Owner's View of Industrial Thermal Insulation", H. Moak, *Insulation*, June (1983), pp. 9-11.

V.5.1.4 Damage

At low temperatures, degraded thermal insulation can lead to ice formation resulting in damage both to the vessel or equipment and/or the insulation system that will in turn accelerate the degradation and damage.

Dr. Hal Wesson points out that such situations in refrigerated liquefied gas storage tanks can lead to freezing of the subsoil and an ice formation that may cause serious damage to the tank foundation system [4].

V.5.1.5 Corrosion

Degradation of thermal insulation by absorption/adsorption of water vapour or water can lead to corrosion problems. In a similar way, loss of efficiency in the insulation systems of chimneys can lead to excessive temperatures, resulting in mechanical damage and degradation of the corrosion resistant barriers, which would accelerate additional structural damage.

V.5.2 MECHANISMS FOR LOSS OF EFFICIENCY

V.5.2.1 Loss of Efficiency in the Insulation Material

A. The Effect of Water and Water Vapour

The reasons for which water and water vapour tend to penetrate insulation systems have been explained in details in V.3 and V.4. Their influences on thermal conductivity are dramatic. This can be understood by comparing the following figures: even though the dry thermal conductivity of the main insulating products ranges from 0.020 W/(m.K) to 0.050 W/(m.K) and is often around 0.035 W/(m.K), the thermal conductivity of water is 0.55 W/(m.K) or about 15 times the average conductivity of insulating products and the conductivity of ice at freezing point is about 2.2 W/(m.K), about 60 times the average thermal conductivity of dry insulating products.

---

[4] "Refrigerated Liquefied Gas Storage Tank: Summary of Verifiable Incidents, Accidents and Failures", H. Wesson, (1987), pp. 1-17.

At low temperatures, things get even worse as the conductivity of ice rapidly increases with decreasing temperature. For instance, the thermal conductivity of ice at –130°C is about seven times that of water at 24°C instead of about four times at 0°C. It can be easily seen that it is a question of different order of magnitude and not of a moderate increase.

As a summary:
- Typical dry thermal conductivity of insulating products   0.020 to 0.050 W/(m.K)
- Average............................   0.035 W/(m.K)

- Water...............................   0.55 W/(m.K)
- Ice at 0°C ....   2.20 W/(m.K)

It can be easily appreciated that the influence of any sizeable amount of water vapour, liquid water or especially ice on the thermal conductivity of insulating materials is substantial. But the distribution of the water in the insulating material and eventually other factors may modify its influence on the thermal conductivity of insulating materials (Fig. V.16). As an example, we could consider a slab of insulation applied on a cold tank wall. If a moderate amount of water has condensed near the cold face of the slab, without affecting the main part of the slab, the deterioration of the thermal conductivity will be less pronounced than if some columns of water link the warm and the cold faces creating thermal bridges. This is probably the main reason why various scientists, having run different tests and investigations, found somewhat different values to evaluate the thermal conductivity increases.

For instance, C.P. Hedlin of *National Research Council of Canada* investigated the heat flow in moist insulations and found 100 to 300% increase in heat flow for only 20 per cent by volume of moisture dependant upon on the type of porous insulation [5].

Batty et al report wet fibrous insulation with a heat transmission 24 times that of the dry state and, when frozen, rising to between 50 and 100 times [6].

---

[5] "Effect of Moisture on Thermal Resistance of Some Insulation in a Flat Roof Under Field-Type Conditions", C. Hedlin, *ASTM STP 789* (1983), pp. 602-625.
[6] "Apparent Thermal Conductivity of Glass-Fibre Insulation: Effects of Compression and Moisture Content", W. Batty, P. O'Callaghan, S. Probert, *Applied Energy*, 9, (1981), pp. 55-57.

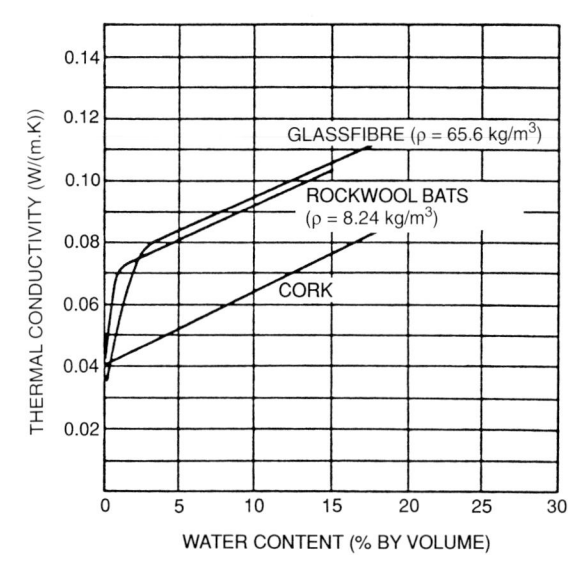

Fig. V.16. Effects of moisture content on conductivity for various materials (Ludwig Adams).

L. Adams (fellow ASHRAE) of *Pittsburgh Des Moines, Inc.*, gives moisture intensification factors for both water and ice [7]. As little as one volume per cent moisture can make a very significant increase in the conductivity (30% increase).

Horst Zehendner of *FIW München* has experimentally shown the severe degradation of moist cellular plastic insulations and the sharp step increase at the ice point [8]. The amount of degradation depends to some degree on the type of insulation and is related to the amount of moisture as well as the temperature. However, in general Zehendner's results show that the presence of even small levels of moisture causes substantial degradation of insulation efficiency.

As shown by these various references, even small amounts will degrade the thermal insulation, destroying its efficiency. At temperatures below freezing, a situation becomes many times worse.

---

[7] "Thermal Conductivity of Wet Insulations", Ludwig-Adams, *ASHRAE, Journal*, October (1974), pp. 61-62.
[8] "Einfluss von Feuchtigkeit auf die Wärmeleitfähigkeit von Schaumkunststoffen im Bereich von –30°C bis +30°C", H. Zehendner, *Kunststoffe im Bau, 11*, January (1979), pp. 3-7.

B. Ageing of Materials

Plastic thermal insulations are prone to degradation by "ageing". The ageing can be affected by time, by exposure to temperature variations, by exposure to chemicals, by exposure to nuclear radiation and by exposure to UV (ultraviolet radiation).

The most commonly known effect is the time ageing of polyurethanes and polyisocyanurates, extruded polystyrenes and phenolic foams caused by gas diffusion. This effect can be accelerated by temperature variations that occur daily and by exposure to elevated temperatures.

Although this type of ageing has received widespread discussion in trade literature, it is important to point out that the effect on thermal performance from water or ice intrusion can easily be many times more important. Nevertheless, ageing effects should not be neglected.

As an example, work by S.P. Muhlenkamp and S.E. Johnson of *Owens-Corning Fiberglas Corporation* [9] indicate that the thermal conductivity of polyurethane foams could double. In addition, these data indicate that the RIC/TIMA value may understate the ageing effect by over 25% dependant upon the specific foam.

Kugler et al indicate even greater ageing effects [10]. They found changes up to almost 30% during only three years storage in a warehouse after the initial "aged" tests.

As pointed out by *Owens-Corning Fiberglas Corporation* [11], this ageing mechanism is diffusion of air into the foam and diffusion of the fluorocarbon out of the foam. The rate of diffusion depends upon the operating temperature to which insulation is exposed. Thus exposure to a more severe industrial environment would accelerate this ageing.

The speed of diffusion of water vapour in polyurethane is much higher than for oxygen or nitrogen, thus ageing in a moist relative humidity environment

---

[9] "In-Place Aging of Polyurethane Foam Roof Insulations", S. Muhlenkamp, S. Johnson, *Seventh Conference on Roofing, 11*, April (1983), pp. 49-55.
[10] "Are 'Aged' Values Adequate for Rating Roof Insulations", W. Kugler, L. Morris, J. McCorkle, *RSI*, July (1986), pp. 38-62.
[11] "The 'Aged K' Phenomenon – A New Urethane Factor", *RSI*, October (1976).

of the real industrial world would be much greater than the mild conditions generally recommended for testing purpose.

C. Absorption of Liquid Chemicals

Another mechanism for degradation of thermal insulation and loss of thermal efficiency is the absorption of spilled or leaking liquids or chemicals [12]. The conductivity of the absorbed liquid chemical directly increases the conductivity of the wet insulation. In addition, there is a possible destruction of the insulation material and/or binder by chemical attack. Both factors degrade thermal efficiency.

D. Compression and/or Vibrations

Thermal insulation efficiency can be degraded by compression and/or vibrations. Fibrous insulations are particularly susceptible to this type of problem.

Field testing of high temperature piping thermal insulation systems by K. Horr of *Pabco Division* revealed significant problems with some mineral wools. Over a hundred tests were conducted at five major petrochemical facilities in the USA. Horr states [13] that:

> "All mineral wool installations tested had less than the original thickness on the top side. Even new mineral wools appear to compress from five to ten percent due to their own weight. Older mineral wool had ten to fifty percent compression on the top. Additional factors influencing the poor dimensional stability of mineral wool include settling due to vibration and the volatilisation of organic binders used in some products".

V.5.2.2 Loss of Efficiency in the Insulating System

A. Convection

Air movement both within the thermal insulation and particularly within the thermal insulation system can cause significant loss of thermal resistance.

---

[12] "Common In-Plant Problems Encountered with Insulation", G. Lang, *Chemical Processing,* January (1984), pp. 38-39.
[13] "Field Testing of High Temperature Insulation Using the Pabco Heat Flow Meter", K. Horr, *Pabco Technical Bulletin, PTB-48-1A,* August (1978), pp. 1-13.

J. Sullivan investigated the thermal performance of insulated piping systems [14]. Insulation parameters included insulation material, thickness, air gaps, seams and joints (Fig. V.17). Pipe system parameters included pipe size, hangers, supports, and operating temperature. He found that the existence of cracks or joints that measure 2.5 mm or more in width significantly affects the thermal performance, with a linear increase in heat loss with increasing joint width for a particular set of operating parameters. An opening of 6.4 mm caused a degradation of 15%.

Air flow around the insulation

Air flow through the insulation

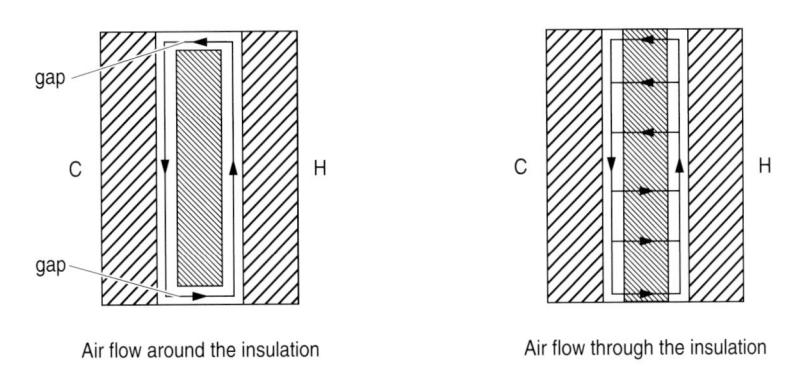

Fig. V.17. Heat losses due to air flows around and through an insulation layer.

J.G.J. Lecompte of *Katholieke Universiteit Leuven* [15] has shown the importance of airtightness within the insulation system. Openings within the insulation system that allow convection around the insulation or through it can significantly degrade the effectiveness. These openings may be caused by dimensional stability problems or by poor workmanship in the application. Degradation of up to approximately 200% was experimentally determined for situations duplicating actual application parameters.

---

[14] "Thermal Performance of Insulated Pipe Systems", J. Sullivan, *ASTM, STP 789*, (1983), pp. 778-795.
[15] "The Influence of Natural Convection in an Insulated Cavity on the Thermal Performance of a Wall", J. Lecompte, *ASTM Symposium*, (1987), pp. 1-14.

This can be especially significant in the case of cryogenic vessels and piping. Work by Musgrave of *Owens-Corning Fiberglas Corporation* [16] indicated that induced natural convection at the joint inside a piping system increases the heat loss since the air inside increases in density, possibly condensing, along with thermal contraction of the insulation widening the gaps and the heat transfer mechanism significantly increases. A double layer urethane system had an effective heat gain 33% higher than expected for a pipe operating at –186°C. The heat gain increase reached 174% when operating this insulation system at –196°C since the air started to condense at this temperature.

This gap problem due to shrinkage has also been observed with polystyrene foams. R. Fricklas, concerning such gaps of only 1% by area in a roofing system, states [17]:

> "That becomes a 10 per cent loss in thermal efficiency which cannot and should not be ignored".

B. Thermal Bridges

Another source of increased heat loss or gain are thermal bridges that may be either inclusions of a high conductivity material (for example, a metal that can provide a thermal short circuit) or paths for direct radiation losses plus convective heat transfer (Fig. V.18).

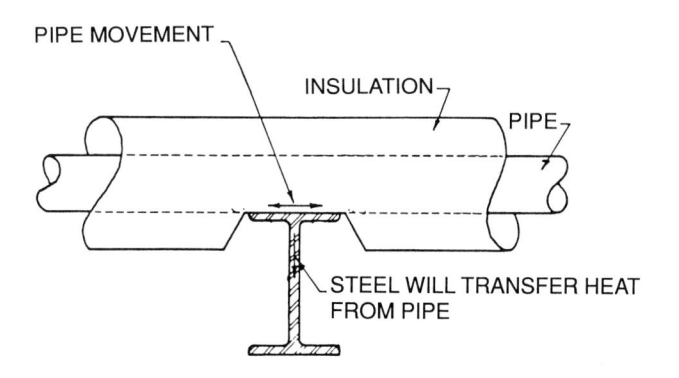

Fig. V.18. Example of thermal bridges.

---

[16] "Thermal Performance of Urethane Foam Pipe Insulation at Cryogenic Temperatures", D. Musgrave, *Journal of Thermal Insulation 3*, July (1979), pp. 3-21.
[17] "Still Room to Improve EPS Usage, 3 Years After Study", R. Fricklas, *RSI*, October (1987), pp. 58-62.

The test program of J.M. Sullivan, Jr. [18], also included investigation of the influence of pipe hangers and supports. He found an additional 14 to 40% heat loss, depending on the type of hanger, compared to that of an uninterrupted insulated pipe.

V.5.2.3 Misleading Concepts

The previously explained physical phenomena, although well established by many researchers and plant operators, have often been neglected or not considered seriously enough. This situation can be explained at least partially by a number of misleading concepts, among which we can quote:

- The dry out concept myth
- The vapour barrier myth
- The waterproofing agent myth.

These concepts have been analysed previously in V.4 and their fallacy has been outlined.

V.5.3 CONCLUSION

As a conclusion it can be said that the real efficiency of many insulation systems is often far from being the expectations, during the design stage. Achieving efficiency comparable to the design can only be obtained when insulation material is properly chosen and the system well designed and built.

---

[18] "Thermal Performance of Insulated Pipe Systems", J. Sullivan, *ASTM, STP 789*, (1983), pp. 778-795.

V.6

CORROSION UNDER INSULATION

V.6.1 THE PROBLEM

George Lang of *DuPont* has observed that:

> "wet thermal insulation can greatly accelerate the rate of corrosion of unprotected metals that are in contact with it" [1].

Moreover Thomas Sikes of *Dow* has indicated that:

> "...the corrosion rate under insulation can be 10 to 20 times greater than the corrosion rate of the surrounding atmosphere" [2].

Rates of corrosion of carbon steel under insulation have been reported to be as high as 2 to 4 mm/year [3, 4]. In the case of austenitic stainless steel there is a risk of stress corrosion cracking.

Corrosion under wet insulation is a particularly dangerous problem because it cannot be visually detected until metal failure occurs. It has been described as the biggest corrosion problem faced by the chemical industry [5]. The two major consequences of corrosion under insulation are related to economics and safety.

---

[1] "Common In-Plant Problems Encountered with Insulation", George Lang, *Chemical Processing,* January (1984), pp. 38-39.
[2] "Problem Solving Forum", Thomas Sikes, *Journal of Protective Coatings and Linings,* January (1985), p. 10.
[3] "Experience with Corrosion under Insulation", R.W. Gregory (Esso Engineering (Europe) Ltd.), pp. 1-5.
[4] "L'Entreprise d'Isolation Face au Problème de la Corrosion sous Calorifuge", Peduzzi, *C.U.I. Conference,* Paris, France, juin (1985).
[5] "Pijpleidingen Aangetast", Pieter den Hollander, *Ad Binnenland,* 17 juni (1986).

V.6.1.1 Economics

Corrosion under insulation has been described as a one billion dollar problem [6]. For *DuPont* the cost just to replace corroded pipes and equipment is in the millions of dollars annually [7]. To the replacement costs must also be added the cost of lost production.

(See colour photo 33 and 34 between pages 456 and 457).

V.6.1.2 Safety

Corrosion under insulation can proceed undetected until a leak finally occurs. Such leaks can be particularly hazardous, if they involve facilities that contain combustible liquids or gases, if they involve materials at very high temperatures, or if they involve pressure vessels.

In one such case:

> "an 8" dia. carbon steel heavy fuel oil pipeline reportedly operating at 250°F (121°C) was externally insulated with calcium silicate blocks and protected with metal weather jacket. Corrosion occurred on this pipeline causing a hydrocarbon leak. Ignition of oil escaping from the corroded pipeline lead to an extremely large fire that, before it was extinguished, resulted in many hundreds of thousands of dollars in damage..." [8].

V.6.2 CONDITIONS

Conditions for corrosion under insulation include the following:

- Water must be present
- The rate of corrosion will depend on the chemical content of the water
- The rate will depend upon temperature.

---

[6] "Understanding Substrate Metal Corrosion", Ted Stanley, *Insulation Outlook*, January (1987), pp. 12-14.
[7] "Reducing the Need for Maintenance on Insulation Systems", John W. Kalis, *Insulation Specifiers' and Buyers' Guide* (1986), pp. 6-14.
[8] "Corrosion Control under Thermal Insulation and Fireproofing", J.F. Delahunt, *Insulation*, February (1982), pp. 10-18.

V.6.2.1 Water

Water must somehow or other intrude into the insulation system in order to cause metal corrosion under insulation. This water will generally contain dissolved oxygen. Water may intrude during storage or installation of insulation materials, through internal leaks, or most commonly through ineffective weatherproofing and improper maintenance.

V.6.2.2 Chemistry

The rate of corrosion under thermal insulation is critically dependent on the chemical content of the absorbed water.

For **carbon steel** the corrosion rate is very dependent on pH as shown in the figure V.19 [9]. It shows that the rate of corrosion climbs dramatically in acid

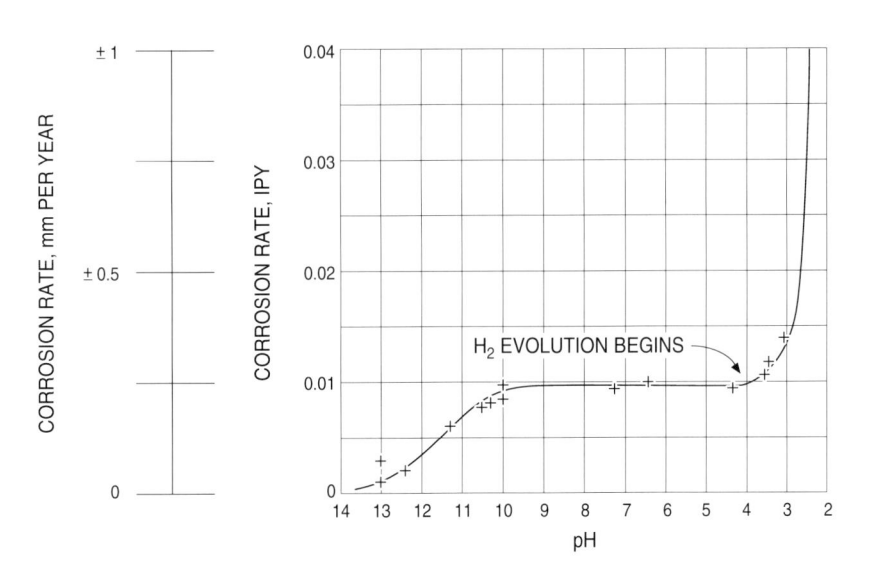

Fig. V.19. Corrosion rate in inches per year (IPY) vs pH.

---

[9] "Corrosion and Corrosion Control", Herbert H. Uhlig, (New York: John Wiley & Sons, Inc., 1971), p. 99 (see Fig. 1).

solution when the pH drops below 4. Acid contamination from external sources of water is quite possible particularly in an industrial environment. However, as we will presently see, acid contamination can also derive from certain thermal insulations.

For **austenitic stainless steel** the primary concern is a combination of mechanical stress and free chloride content. Quality assurance procedures for insulations to be used in contact with stainless steel provide that soluble chloride and fluoride contents should be held to the lowest possible levels and, at least in the US and several other countries, that the acceptable levels be determined in relation to the leachable sodium and silicate ions of the insulation [10]. Leachable chloride ions may be present in the insulation, as manufactured, or may intrude from external sources of water. High contents of chloride ion may be found in rain water, in chemical plant atmospheres, near cooling towers, or even in drinking water: this may occasionally be used for fire fighting, deluge testing or wash down procedures. In addition to the restriction on the chloride content, the pH of the insulation should also be contained in a given range [11].

V.6.2.3 Temperature

Although metal corrosion can occur from cryogenic temperatures up to temperatures in the thousands of degrees, the main problem of corrosion under insulation is found between 0 and 100°C where water can be present. For facilities that are continuously in operation at low temperatures, the water will be present as ice. Similarly, if facilities are in continuous operation at temperatures substantially in excess of 100°C, water will evaporate from the piping and equipment surfaces. Corrosion can occur, however, at operating temperatures that are somewhat in excess of 100°C due to localised cooling by water trapped in or under insulation [12]. Corrosion can also occur at protrusions where the temperature may fall below 100°C. Within the temperature range 0 to 100°C the rate of corrosion will increase with temperature, doubling every 15 to 20°C [13].

---

[10] "Nonmetallic Thermal Insulation for Austenitic Stainless Steel", U.S. Atomic Energy Commission, *Regulatory Guide, 1.36,* 23 February (1973), pp. 1-3.

[11] "Standard Specification for Wicking-Type Thermal Insulation for Use over Austenitic Stainless Steel", *ASTM C 795-77* (reapproved 1983), pp. 1-4.

[12] "Factors Affecting Corrosion on Carbon Steel under Thermal Insulation", Peter Lazar, Paper presented at the symposium on *Corrosion of Metals Under Thermal Insulation,* 11-13 Oct. 1983, edited by Warren I. Pollock, Jack M. Barnhart (Philadelphia, Pennsylvania: ASTM, August, 1985), pp. 11-26.

[13] "L'Entreprise d'Isolation Face au Problème de la Corrosion sous Calorifuge", Peduzzi, *C.I.U. Conference,* Paris, France, juin (1985).

Thus in practical sense, the maximum corrosion potential is more or less between 60°C and 100°C. At temperatures below 60°C insulation may not be necessary while above 100°C the corrosion rate will decrease due to evaporation of the water.

Two temperature conditions require special comments. They are:

- Cyclic temperature systems that will condense moisture at low temperatures and then provide high temperatures that will accelerate metal corrosion [14]. This obviously includes occasional high temperature purges of low temperature equipment
- Plant shut-downs for extended periods of time where liquid water may continue to accumulate without freezing or evaporating. For such plants, it has been recommended, but not often done, that all thermal insulation be removed in order to minimise the risk of corrosion [15].

V.6.3 SOLUTIONS

There are three general solutions for preventing or minimising corrosion under insulation.

The most obvious place to start is to prevent water entry into the insulation system through the use of suitable weather or vapour barriers.

A second approach is to assume that water will enter the system, and deal with this by providing a physical barrier to prevent the water from contacting the piping and equipment. Such barriers are applied to piping and equipment that will operate, at least occasionally, in the temperature range where corrosion can occur – slightly below 0°C to perhaps +200°C. These barriers are generally paints or mastics, however, it is critical that close attention is paid to surface preparation and workmanship. A continuous, defect free coating is required. Such coatings are frequently silicones, epoxy phenolics, coal tar epoxies, or bitumen.

---

[14] "Investigation of Nondestructive Testing Techniques for Detecting Corrosion of Steel under Insulation", George J. Falkenbach, John R. Fox, Robert P. Meister, *MTI Technical Report, 4*, (St. Louis, Missouri: Materials Technology Institute of the Chemical Process Industries, Inc., July, 1981), pp. 1-87.
[15] "Preventing Corrosion During Mothballing", Ronald J. Twigg, *Chemical Engineering*, 16 September (1985), pp. 91-94.

An alternative to a paint or mastic corrosion resistance barrier that can be used for stainless steel, is an aluminium foil barrier. The foil provides both a physical barrier to chloride containing water as well as a cathodic protection layer [16, 17, 18].

The third approach to minimising corrosion under insulation, which is not incompatible with the first two, is to select an insulation that will:

- Minimise water intrusion
- Not retain water
- Not accelerate metal corrosion.

This is the subject of the next section.

V.6.4 INSULATIONS

V.6.4.1 Cellular Plastics

Cellular plastics are frequently chosen for low temperature industrial applications. The most common of these are the polyurethanes and polyisocyanurates but more recently phenolics have been introduced. Unfortunately, both the polyurethane type of foams and the phenolics "are found to form aggressive solutions. pH's associated with solutions are frequently measured at levels 2 to 3... Of the two, phenolic foams are by far the most corrosive" [19].

A. Polyurethane

The problem of corrosion under polyurethane was explained in a report by *Exxon* that described a problem at their refinery in Hamburg, Germany [20]. The report states:

---

[16] "Understanding Substrate Metal Corrosion", Ted Stanley, *Insulation Outlook,* January (1987), pp. 12-14.
[17] "A Review of the European Meeting on Corrosion under Lagging Held in England, November 1980", James A. Richardson, Paper presented at the symposium on *Corrosion of Metals Under Thermal Insulation,* 11-13 Oct. 1983, edited by Warren I. Pollock, Jack M. Barnhart (Philadelphia, Pennsylvania: ASTM, August, 1985), pp. 42-59.
[18] "Thermal Insulation – Specifications and Materials: Applications to Stainless Steel Surfaces", J.D. Nicholson, *Insulation,* May (1982); pp. 22-27.
[19] "Corrosion Control under Thermal Insulation and Fireproofing", J.F. Delahunt, *Insulation,* February (1982), pp. 10-18.

"At this facility, a hot-tank polyurethane foam-insulation system was removed after several years' service. There was severe corrosion on the steel surface beneath the polyurethane foam. Water that had got into the polyurethane foam caused the severe steel corrosion. Our explanation of the corrosion was that the water had combined with halogens (i.e., chlorine, fluorine, bromine ions) in the polyurethane foam causing an acidic environment and accelerating metal corrosion".

A follow-up of this *Exxon* work, "indicated pH values of collected water as low as 1 to 2" [21].

Where do these halogens come from? They were introduced on purpose into the polyurethane foam as fire retardant additives. But these additives can hydrolyse and release strong acids [22, 23], such as HCl, which can easily give a pH of 1 to 2.

The *Exxon* position has been summarised by Lazar who states:

"Although extensively used in the past, polyurethane foam (PUF) is not a popular insulation type in our plants at this time for moderate or cold services".
He also cites specifically "corrosivity of water because of hydrolysis of halogenated flame retardants needed to make the insulation safe in the plant"[24].

Fire retardant additives are not the only sources of chloride in polyurethanes and polyisocyanurates. Both of these materials are made from isocyanurates that are produced from phosgene (a chlorine containing compound). This results in small quantities of hydrolysable chlorine compounds remaining in the polyurethane or polyisocyanurate product [25].

---

[20] "Design Key to Good Insulation Systems", Thomas A. Cross, *Oil and Gas Journal*, 17 March (1975), pp. 126-131.
[21] "Requirements for Optimum Foamed-in-Place Polyurethane Insulating Systems", J.F. Delahunt, Preprint *API Refining Department*, 25-76, 12 May (1976), pp. 1-28.
[22] "Preventing Corrosion under Insulation", V.M. Liss, *Chemical Engineering*, March (1987), pp. 97-100.
[23] "A Review of the European Meeting on Corrosion under Lagging Held in England, November 1980", James A. Richardson, Paper presented at the symposium on *Corrosion of Metals Under Thermal Insulation*, 11-13 Oct. 1983, edited by Warren I. Pollock, Jack M. Barnhart (Philadelphia, Pennsylvania: ASTM, August, 1985), pp. 42-59.
[24] "Factors Affecting Corrosion on Carbon Steel under Thermal Insulation", Peter Lazar, Paper presented at the symposium on *Corrosion of Metals Under Thermal Insulation*, 11-13 Oct. 1983, edited by Warren I. Pollock, Jack M. Barnhart (Philadelphia, Pennsylvania: ASTM, August, 1985), pp. 11-26.
[25] "Isocyanate without Phosgene", Shinsuke Fukuoka, Masazumi Chono, Masashi Kohno, *Chemtech*, November (1984), pp. 670-676.

Manufacturers of polyurethane and polyisocyanurate recognise this problem and stress that "metal corrosion is of particular concern; metal surfaces must be protected with a corrosion inhibiting coating" [26] [27].

There are many examples of corrosion under polyurethane insulation [28], [29], [30]. Reference has already been made to some of them.

(See colour photo 34 between pages 456 and 457).

B. Phenolics

Phenolic foams are manufactured using either mineral acid or organic sulfonic acid catalysts. Thus, they are acidic yielding pH values that can be as low as 2. In addition they are very absorptive to water and water vapour. This combination of characteristics results in extremely corrosive behaviour. In fact, a table from one prominent manufacturer of phenolic foams gives data indicating that this material is between 1.7 and 5.7 times more corrosive to bare steel roof deck than polyurethane [31].

V.6.4.2 High Temperature Insulations

High temperature insulations such as calcium silicate, perlite and mineral wool are inherently porous which facilitates the entry and retention of water. One survey of corrosion under insulation indicates that "the types of insulation giving most problems were fibrous types and calcium silicate" [32].

---

[26] "Trymer 9501 Rigid Foam Insulation", Technical Information, The Dow Chemical Company, *109-766-86*.
[27] "Systèmes d'Isolation", Brochure; Ouest Isol France, juin (1986).
[28] "The Manner of Treating Outdoor Storage Tanks to be Insulated with Urethane Foams", Japan Fire Defense Agency, 3 September (1976).
[29] "Detecting Wet Insulation on Pipelines in Alaska", Edward Feit, *Insulation Outlook*, April (1988), pp. 6-9.
[30] "Shell and Jacket Corrosion of a Foamed in Place Thermally Insulated Liquefied Petroleum Gas Tank", Donald O. Taylor, Rodney D. Bennett, Paper presented at the symposium on *Corrosion of Metals Under Thermal Insulation*, 11-13 Oct. (1983), edited by Warren I. Pollock, Jack M. Barnhart (Philadelphia, Pennsylvania: ASTM, August, 1985), pp. 114-120.
[31] "Corrosion and Koppers Closed-Cell Phenolic Foam", Rx. Bulletin; Koppers Company, Inc., *91*, 4 September (1986).

A. Calcium Silicate

Lazar of *Exxon* tells us that "calcium silicate is highly absorbent, and as such has contributed to much of our corrosion problems...". Elsewhere, Krystow of *Exxon* tells us that:

> "Insulation types such as calcium silicate, which can absorb and retain substantial amounts of water and leach chlorides, can lead to more problems than other insulations such as foam glass, that is not normally subject to moisture absorption and does not contain significant leachable material" [33].

Monsanto's experience is similar. They reported that "the closed cell type is best to use, while calcium silicate gives the most trouble" [32].

A report on a European meeting on corrosion under lagging indicates:

> "there was a consensus that calcium silicate has unfavourable 'wicking' properties, and that closed cell foam glass is relatively impermeable" [34].

Reference was previously made to the oil fire that resulted from corrosion through a calcium silicate insulated pipeline [35].

B. Fibrous Insulation

It has been reported by *Exxon* that "water absorption is a concern with fibrous insulation" [36]. Elsewhere, we read that "at *DuPont* the type of insulation said

---

[32] "Investigation of Nondestructive Testing Techniques for Detecting Corrosion of Steel under Insulation", George J. Falkenbach, John R. Fox, Robert P. Meister, *MTI Technical Report, 4,* (St. Louis, Missouri: Materials Technology Institute of the Chemical Process Industries, Inc., July, 1981), pp. 1-87.

[33] "Controlling Carbon Steel Corrosion Under Insulation", Paul E. Krystow, Paper presented at the symposium on *Corrosion of Metals Under Thermal Insulation,* 11-13 Oct. 1983, edited by Warren I. Pollock, Jack M. Barnhart (Philadelphia, Pennsylvania: ASTM, August, 1985), pp. 145-154.

[34] "A Review of the European Meeting on Corrosion under Lagging Held in England, November 1980", James A. Richardson, Paper presented at the symposium on *Corrosion of Metals Under Thermal Insulation,* 11-13 Oct. 1983, edited by Warren I. Pollock, Jack M. Barnhart (Philadelphia, Pennsylvania: ASTM, August, 1985), pp. 42-59.

[35] "Corrosion Control under Thermal Insulation and Fireproofing", J.F. Delahunt, *Insulation,* February (1982), pp. 10-18.

[36] "Factors Affecting Corrosion on Carbon Steel under Thermal Insulation", Peter Lazar, Paper presented at the symposium on *Corrosion of Metals Under Thermal Insulation,* 11-13 Oct. 1983, edited by Warren I. Pollock, Jack M. Barnhart (Philadelphia, Pennsylvania: ASTM, August, 1985), pp. 11-26.

to give the most problems was glass fibre because of the poultice effect from wetting".

Workers at *Gulf* state that "below 121°C blanket insulation is avoided so as to eliminate possible sagging and moisture pick-up. Consideration should be given to the use of cellular glass" [37].

C. Stress Corrosion Cracking under High Temperature Insulation

The *Institution of Chemical Engineers* in England has addressed the problem of stress corrosion cracking of stainless steel [38]. They tell us that:

> "The rigid open cell types of insulation such as calcium silicate can enhance the risk of stress corrosion due to wetting of the hot metal surface caused by their wicking characteristics.
> The flexible glass fibre or mineral wool insulation materials, due to their open texture, will allow water to the metal surface that can lead to stress corrosion. The closed-cell foam glass types will not normally be a problem".

V.6.4.3 FOAMGLAS® Insulation

At the outset of this section we must state the obvious: water will cause corrosion of steel. Therefore, it is essential that every effort be made to prevent water from intruding behind FOAMGLAS® insulation, which as such, should not be considered as an anti-corrosion protection. However, FOAMGLAS® insulation:

- Will help to minimise water intrusion and retention since it is impermeable (Fig. V.20)
- Will not accelerate the corrosion of carbon steel or the stress corrosion of stainless steel
- Can be used as a component of a corrosion resistant barrier.

As such it is an excellent choice where material corrosion is a concern.

---

[37] "Prevention of Chloride Stress Corrosion Cracking Under Insulation", Louis C. Sumbry, E. Jean Vegdahl, Paper presented at the symposium on *Corrosion of Metals Under Thermal Insulation,* 11-13 Oct. 1983, edited by Warren I. Pollock, Jack M. Barnhart (Philadelphia, Pennsylvania: ASTM, August, 1985), pp. 165-177.
[38] "Guide Notes on the Safe Use of Stainless Steel in Chemical Process Plants", prepared by the Working Group on the Use of Stainless Steel in Process Plant of the International Study Group on Hydrocarbon Oxidation (Rugby, England: The Institution of Chemical Engineers, 1978), pp. 11-15.

Fig. V.20. FOAMGLAS® cellular glass is impermeable.

A. Chemical Properties

The chemical properties of FOAMGLAS® insulation do not contribute to accelerated metal corrosion. To provide evidence to support this statement we must consider the chemistry of the cell gas and the glass matrix, both individually and in combination although it is known that no gas will escape from the cells if they are not broken. But even in this extreme case, the danger of corrosion due to the cell gas is fairly low.

The major cell gas in FOAMGLAS® cellular glass is CO_2, or carbon dioxide. Depending upon the type of FOAMGLAS® insulation under consideration, the CO_2 content can vary in the range 70% to nearly 100%. Carbon dioxide is a weak acid. Theoretically a pH as low as 4.2 could be developed if a cell is damaged, water intrudes and no evaporation of CO_2 occurs. As can be seen in Fig. V.19, this pH does not result in accelerated corrosion. The only other cell gas that exhibits any degree of acidity is H_2S. This gas can be detected at very low concentrations because of its characteristic odour. However, in terms of both its acidity and its quantity CO_2, not H_2S, controls the theoretical cell gas acidity. How dangerous CO_2 is as an acid can be put into perspective by noting that we exhale CO_2 and water vapour with every breath.

Actually we find that the reaction between FOAMGLAS® insulation and water does not create weakly acid conditions but instead is weakly alkaline with a pH in the range of 9 to 10 [39]. This comes about from the interaction of water with the soda-lime-silica glass network which results in the leaching of very small quantities of alkali that effectively neutralise the weak acid cell gases.

Thus for all practical purposes, we can conclude that FOAMGLAS® insulation does not contribute to the acceleration of metal corrosion. This is stated explicitly in a German norm that tells us "The behaviour of cellular glass with metal is neutral" [40].

FOAMGLAS® insulation has also been certified as acceptable for use in contact with austenitic stainless steel according to tests conducted in the US [41] and Belgium [39]. Again the German norm confirms this with the statement: "Cellular glass does not contain water soluble chlorides which can cause stress corrosion cracking with austenitic stainless steel" [40].

When FOAMGLAS® insulation is to be used in contact with austenitic stainless steel, it is of course necessary to choose compatible accessory materials. For such applications we recommend Hydrocal® B-11 or PC® HIGH TEMPERATURE ANTI-ABRASIVE as both are fabricating adhesives as well as bore coatings. These materials have been demonstrated to be compatible with stainless steel [42].

B. Carbon Steel Compatibility

In a paper titled "Preventing Corrosion Under Insulation", Liss has stated that:

---

[39] Test Reports from the University de Liège – Institut du Génie Civil, Liège, Belgium; 10 avril (1986), *P/S.5 N° 39.512*, pp. 1-3, and *P/S.3 N° 40.385/1*, pp. 1-3.

[40] "Cellular Glass as Insulation Material for Industrial Equipment Applications", AGI Work Sheet; Working Association Industrial Constructions e.V., *Q137*, August (1985), pp. 1-12.

[41] "Report from Professional Service Industries, Inc., Pittsburgh Testing Laboratory Division, Pittsburgh, Pennsylvania; *831-73080*, December (1987), pp. 1-2.

[42] Test Report from Lehigh Testing Laboratories, Inc., Wilmington, Delaware; *J575*, 19 August (1975).

> "Insulations such as calcium silicate, glass fibre and, to some extent cellular plastic foam, absorb and retain liquids and vapours... The only fully non-absorbent insulation is cellular glass... Cellular glass should be used where corrosive or flammable liquids are present" [43].

In the paper "Factors Effecting Corrosion of Carbon Steel Under Insulation" Lazar of *Exxon* states that:

> "Cellular glass has been widely adapted by our plants for use from 150°C down, including low temperature requirements. The main advantage is zero water absorption and reasonable installation cost" [44].

In a survey of corrosion of steel under insulation [45], it is reported that:

> "The closed cell type is the best to use, while calcium silicate gives the most trouble" – *Monsanto*
> "Corrosion problems were worst under the polyurethane and least under the foam glass" – *Exxon*

C. Stainless Steel Compatibility

As mentioned earlier, FOAMGLAS® insulation has been found suitable for the use with austenitic stainless steel in both the US and Belgium and its compatibility with stainless steel has been mentioned by the *Institution of Chemical Engineers* in England [38] and in a German norm [40]. Elsewhere it has been acknowledged that:

> "Cellular glass is free of soluble chloride" [34].
> "Consideration should be given to the use of cellular glass" [37].

FOAMGLAS® insulation has been used in a number of industries that employ stainless steel.

---

[43] "Preventing Corrosion under Insulation", V.M. Liss, *Chemical Engineering,* March (1987), pp. 97-100.
[44] "Factors Affecting Corrosion on Carbon Steel under Thermal Insulation", Peter Lazar, Paper presented at the symposium on *Corrosion of Metals Under Thermal Insulation,* 11-13 Oct. 1983, edited by Warren I. Pollock, Jack M. Barnhart (Philadelphia, Pennsylvania: ASTM, August, 1985), pp. 11-26.
[45] "Investigation of Nondestructive Testing Techniques for Detecting Corrosion of Steel under Insulation", George J. Falkenbach, John R. Fox, Robert P. Meister, *MTI Technical Report, 4,* (St. Louis, Missouri: Materials Technology Institute of the Chemical Process Industries, Inc., July, 1981), pp. 1-87.

V.7

CHEMICAL DURABILITY

V.7.1 THE PROBLEM

For some industrial applications, the chemical durability of an insulation material is the single most important criterion for selecting the right product for a particular application. As explained by Liss:

> "Selecting a particular type of insulation is not easy.... Decisions must be made that consider the environment (humidity, corrosive chemicals, temperature, etc.) in which the insulation will be placed" [1].

The chemical environment in which insulation materials may be used covers a wide range, from industrial floor applications on one hand to large storage tanks [2] on the other. It is therefore very important to assess what chemicals are present in each potential application.

Absorption of chemical products on or into the insulation can lead to a loss of thermal resistance, increased risk of fire, increased metal corrosion, loss of strength, etc. These detrimental effects are more fully explored in other chapters.

Typically, there are two sources from which the chemical species can attack the insulation.

One is from outside the insulation system, consisting of atmospheric chemical attack or spillage.

The second source of chemical attack is from the process system that is insulated. Here the attack occurs from within the insulation system. Insulated

---
[1] "Selecting Thermal Insulation", V. Mitchell Liss, *Chemical Engineering*, 26 May (1986), p. 105.
[2] "Wärmeisolierung von Grosstanks", H. Klezath, *OEL, 6,* (1969), p. 182.

tanks or piping systems carrying corrosive acids, alkaline solutions, or organic solvents that may leak at joints, valves, flanges, etc. in the piping system must be insulated with materials that are chemically resistant.

Certain insulation products may even be required to be deliberately exposed to aggressive chemicals, although *Pittsburgh Corning* does not promote such applications except in very particular cases.

V.7.2 THE SOLUTION

The solution to the problem of chemical attack, from whatever source, is to select the insulation materials that are resistant to such attack. FOAMGLAS® cellular glass is an insulation product suited to resist a wide range of chemicals. The primary reason for its chemical durability is the fact that it is composed of a unique material... cellular glass (Fig. V.21).

Fig. V.21. Foamglas® cellular glass is resistant to a wide range of chemical products.

V.7.2.1 Glass Durability

FOAMGLAS® cellular glass insulation gets its unique chemical durability because it is manufactured entirely from glass. According to E.B. Shand of *Corning Glass Works*:

> "... glass ranks high in the scale of chemical durability. Glass containers are commonly used for storing and handling food and chemical products, while in the laboratory many reactions involving very corrosive materials are commonly carried out in glassware. Glass equipment is also used for chemical processes in industry where exceptional durability is required. Window glass withstands the effects of weather for years without protection of any kind" [3].

The glass from which FOAMGLAS® cellular glass insulation is manufactured is a soda-lime-alumino-silicate composition. Silica (SiO_2) is the major constituent of this glass composition (Fig. V.22). Shand states that:

> "The chemical durability of silicate glasses results from the inert nature of silica itself. Silica is practically insoluble in water except at high temperatures. This also applies to other neutral solutions. Acids have little effect on silica, with the exception of hydrofluoric acid and, at high temperatures, phosphoric acid. The solubility of silica increases with the alkalinity of the solution,...".

The chemical durability of bulk glass can be characterised by a variety of tests. One typical test is a glass powder solubility test. There are several versions of the test (ISO 719-1981, DIN 12 111, USP-III, etc.). Basically, the test requires crushing and grinding a sample of glass to a specific particle size. Depending on the test procedure, the ground glass is then heated in distilled water at a temperature ranging from 90 to 121°C for periods of 30 minutes to 4 hours. The amount of alkali extracted by this procedure is then measured analytically and expressed as Na_2O (sodium oxide) per unit weight of the glass sample tested.

The FOAMGLAS® insulation glass composition is designated a Class 3 material based on the ISO test, wherein no more than 264 milligrams of Na_2O per gram of glass may be extracted. Tested according to the *USP (United States Pharmacopœia)* procedure, the FOAMGLAS® cellular glass composition is as chemically durable as glass used for packaging pharmaceutical preparations.

---

[3] "Chemical Durability", E.B. Shand, *Glass Engineering Handbook,* 2nd edition, New York, McGraw-Hill Book Company, Inc., (1958), pp. 91-93.

● Si^{+4} ◯ O^{-2} ◯ Na$^+$

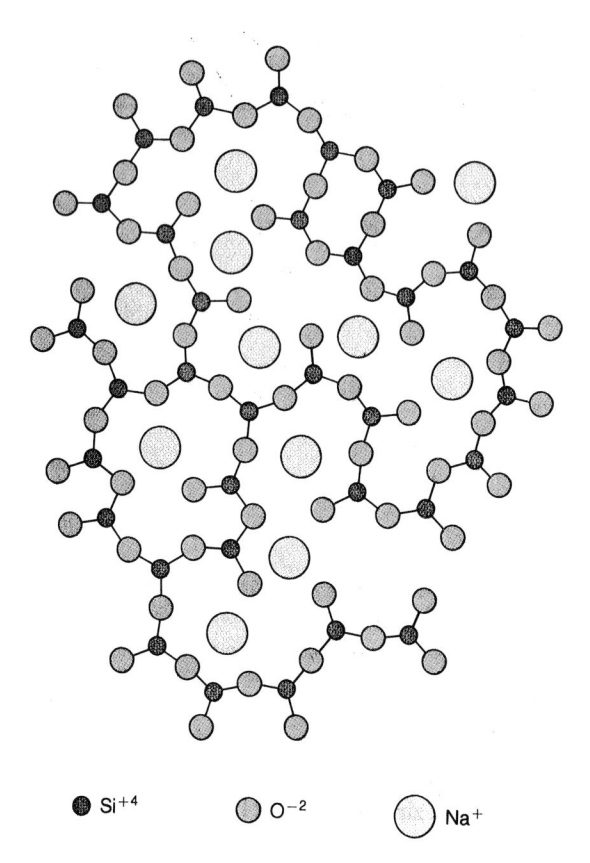

Fig. V.22. Two-dimensional representation of a typical modified glass (Here, sodium silicate glass).

V.7.2.2 FOAMGLAS® Insulation is 100% Glass

Cellular glass insulation is 100% glass. It contains no organic binders or other components of any kind that can be attacked differentially by the chemical environment.

Conversely, many insulation products consist of two or more components (fibrous glass with organic binders, perlite with water repellent agents, etc.). Components such as these are incorporated to improve the physical performance of the end product. However the added component is often chemically or thermally attacked at a faster rate or in a different way than the primary component. This differential attack leads to a premature failure of the insulation product and/or results in a loss of insulation performance.

V.7.2.3 FOAMGLAS® Insulation is Impermeable

Not only is the chemical durability of the cellular glass composition superior, but FOAMGLAS® insulation is also impermeable and non-absorptive. It can neither absorb nor wick chemicals.

V.7.3 MEASURES OF CHEMICAL DURABILITY

The chemical durability of all materials can be characterised by certain tests, but one test will not suffice for every material or even a single material.

Some factors that affect the chemical durability of insulation materials, other than the composition of the material itself, are:

- The temperature of the reagent in contact with the insulation and the possible temperature cycling
- The duration of contact
- The concentration of the reagent
- The atmospheric pressure involved and the possible pressure cycling
- Whether the reagent is static or flowing
- The surface area exposed to the reagent
- Whether the insulation is submerged totally or only partially in the reagent
- Whether the reagent is in a liquid or gaseous state.

V.7.3.1 FOAMGLAS® Cellular Glass Insulation

Examinations of the chemical durability of cellular glass have occurred periodically to characterise the degree of durability, to develop products with improved durability, and to compare durability with other insulation products. These examinations have been made in water, various corrosive chemicals and organic solvents.

A. Water Durability

Water immersion tests to measure durability have been conducted over the years on FOAMGLAS® cellular glass by a number of investigators. For instance, specimens exposed to water were measured for any significant change in volume, weight or strength. These properties were carefully measured for FOAMGLAS® insulation specimens before exposure. The specimens were then submerged for 30 days in 24°C distilled water. At the end of the 30 day period, the specimens were subjected to the same property tests. No significant changes in volume, weight, or compressive strength were found. Other tests were carried out during much longer periods of time. Thus, FOAMGLAS® insulation is not harmed by exposure to water in liquid or vapour form at ambient temperatures.

At temperatures of 66°C and above, changes in sample weight noted by other investigators indicate an interaction with the FOAMGLAS® insulation structure.

Situations where cellular glass insulation is exposed to hot water in liquid or vapour form should be avoided. Appropriate protective measures can then be taken, if necessary. These might include covering the insulation with an efficient moisture barrier and/or the provision of weep holes to rapidly vent accidental water intrusion.

B. Durability in Other Reagents

FOAMGLAS® cellular glass was also exposed to a variety of chemical and organic reagents at room temperature. Volume, weight, and strength measurements were made before and after 30 days submerged in the reagent. No significant deterioration was noted with the exception of the exposure to 10% sodium hydroxide (see Table in II.1.8).

Thus, FOAMGLAS® insulation has been shown to be highly resistant to water, concentrated and diluted acids (except hydrofluoric), salts, hydrocarbons, ketones, alcohols, ethers and esters.

C. The Role of Accessory Materials

FOAMGLAS® cellular glass is rarely used on its own. It is most often used in conjunction with a variety of ancillary products (adhesives, sealers, coatings, anti-abrasives, jackets, etc.). The chemical durability of these materials has to be known to properly design a chemically durable insulation system (see Table in II.2.9.6).

The workmanship used to apply accessory materials to the insulation is also critical to assure a long life, chemically durable insulation system.

D. Insulation Specialist's Recommendations

The excellent chemical durability of cellular glass insulation is widely recognised and insulation specialists recommend its use for a broad range of harsh applications. For instance, Liss writes:

> "If there is potential for absorbing corrosive liquids, then the highly impermeable cellular glass may prove to be the only choice in hot service where vapour or liquid barriers may not withstand the temperature" [1].

In a report of industrial chimneys one can read:

> "The following measures can be taken to avoid damage (attack by acids in chimneys): ... use of a vapour tight insulating material between inner and outer shaft (e.g. 2 layers of foam glass)" [4].

and in the German AGI Q137 documents:

> "Experience has shown that cellular glass is resistant against... all solvents and acids except hydrofluoric acid" [5]

V.7.3.2 Competitive Insulation Materials

A. Polyurethane, Polyisocyanurate, Polystyrene

These products experience various levels of deterioration, when submerged for 30 days in a variety of reagents, including water, held at 24°C.

These results are substantiated by a major manufacturer. *Dow*, the manufacturer of TRYMER polyisocyanurate, states in their product literature that the product:

---

[4] "Report on the Work of the Dutch Working Group on Chimneys – Part 4 – Preservation of Shut-Down Chimneys", F. van Zijl, 57th International Chimney Conference, October 1984, Essen W. Germany, p. 234.
[5] "Cellular Glass as Insulation Material for Industrial Equipment Applications", Working Association Industrial Constructions e.V., AGI Work Sheet Q137, August (1985), p. 9.

"... should not be exposed to liquids, vapours, or residue of any chemicals or solvents that will soften or degrade the foam or affect it in any other manner" [6].

Dow also published product literature [7] listing a wide range of reagents to which polystyrene has low resistance (including aromatic and chlorinated hydrocarbons, olefins, naphtha, ketones, aldehydes, ethers, gasoline, etc.).

Other experts state that PU foam... swells from the action of esters, ketones, alcohols, chlorinated hydrocarbons [8].

B. Phenolics

Two manufacturers of phenolformaldehyde foam, *Kooltherm* and *Saitec*, have published literature listing chemical resistances [9, 10].

Kooltherm, manufacturer of Koolphen, notes that their product is severely attacked by concentrated nitric acid and has only fair to poor resistance to phosphoric acid, concentrated hydrochloric acid, 10% sodium hydroxide, acetone, methylated spirits, and methyl acetate.

C. Mineral Fibre

Mineral fibre products are similar to cellular glass in that their basic composition is essentially a silicate glass. They might therefore be expected to be similar in chemical durability to all silicate-based glasses. However, glass wool fibres lose strength and elasticity when submerged in water at temperatures as low as 40°C. Embrittlement was observed in basalt wool at the same temperature [11].

---

[6] "TRYMER Rigid Foam Insulation", Brochure; The Dow Chemical Co., Form No. 109-764-86, p. 5.
[7] "STYROFOAM Brand Insulation", Brochure; The Dow Chemical Co., Form No. 179-4174-83, p. 5.
[8] "Wärmeisolierung von Speicherbehältern für tiefsiedende Flüssigkeiten", H. Klezath, *Erdöl/Erdgas/Zeitschrift, 85,* [3] (1969), p. 105.
[9] "Koolphen: A Revelation in Insulation", Brochure; Kooltherm Insulation Products Ltd.
[10] "Phenexpan", Brochure; Saitec, p. 4.
[11] "Physical Changes in Basalt and Glass Wool Fibres during Aqueous Corrosion at Different Temperatures", I. Wojnarovits, *Glass Technology, 28,* [5] October (1987), p. 208.

The degree of physical change in mineral fibre products depends on the surface area of the fibres. In this respect it should be noted that mineral fibre products have a much higher surface area than an equivalent volume of FOAMGLAS® cellular glass insulation. Based on specific surface areas for some types of basalt and glass wool given in reference 11, these surface areas are recalculated in the Table V.6 to reflect the total surface area per cubic meter of insulation product and compared to the values for FOAMGLAS® cellular glass. Obviously, a higher rate of chemical attack on the mineral fibre products can be expected because of the very large surface areas exposed.

TABLE V.6

	Surface area	
	(m^2/g)	(m^2/m^3)
Basalt wool [12] (Mineral wool)	0.4	50,100
Glass wool	0.23	18,400
FOAMGLAS® cellular glass [13]	0.00023	30.5

Also, many glass and mineral fibre products are often coated with organic binders, which can be destroyed by heating or chemical attack.

D. Perlite

This product is made of expanded perlite ore, inorganic binders, and reinforced fibres [14]. It also often contains organic-based water repellents.

The organic water repellents can be rapidly deteriorated by heating at temperatures greater than 200°C.

---

[12] Based on basalt wool density of 125 k/m^3 and glass wool density of 80 kg/m^3.
[13] Values based on cell diameter of 1 mm, slab thickness of 130 mm, density of 130 kg/m^3.
[14] "Molded Fitting Covers", Technical Data Sheet; Pamrod Division, Distribution International.

V.7.4 CONCLUSIONS

Especially for industrial applications, the chemical durability of an insulation system is often the deciding factor in the selection of the most suitable insulation product. The inherent chemical durability of FOAMGLAS® insulation in acids, aqueous solutions, and solvents, combined with its impermeability to liquids and vapours of these reagents, makes it the insulation product of choice in many applications.

V.8

DIMENSIONAL STABILITY

V.8.1 THE PROBLEM

Damage caused by insufficient dimensional stability occurs under the influence of temperature and/or moisture variations. It consists in swelling, warping, expansion, shrinkage or other types of distortion.

V.8.1.1 Reversible Changes in the Dimensions of Insulations

Dimensional changes in materials are caused by their thermal expansion coefficients. This is the relative amount by which materials shrink or expand when they are cooled down or heated. Practically all materials have an expansion coefficient that depends on their chemical composition. In general, organic materials have a greater expansion coefficient than inorganic materials. This can be observed with insulation materials. While organic foams [1,2] have linear thermal coefficients ranging from 30 to 90×10^{-6} K^{-1}, the inorganic FOAMGLAS® cellular glass [3] has only a value ranging from 8.5 to 9.0×10^{-6}K^{-1}. In addition, these values can differ for plastic foams, dependant upon the method of production [4,5].

---

[1] "Foamed Plastics Chart", *1971-1972 Modern Plastics Encyclopaedia* (McGraw-Hill, 1971), pp. 627-629.
[2] "Foams", *Guide to Plastics* (New York, N.Y.: McGraw-Hill, 1981), pp. 35-39.
[3] "Lineaire Uitzettingscoëfficiënt van FOAMGLAS® T4", W. van Herk, Rapport; Instituut TNO voor Bouwmaterialen, *B-86-1-318*, 10 september (1986), pp. 1-4.
[4] "Installation Requirements", William C. Turner, John F. Malloy, *Thermal Insulation Handbook*, (New York, N.Y.: McGraw-Hill 1981), pp. 270-271.
[5] "Untersuchung der mechanischen Eigenschaften von Schaumkunststoffen im Bereich von –180°C bis +70°C", H. Zehendner, Sonderdruck *Isolierung*, 5, pp. 1-12.

V.8.1.2 Non Reversible Changes of Dimensions

These can result from various causes, among others ageing. The molecular structure of plastic foams, especially of polyurethane foams directly after their production, are not yet in a stable state; the material itself, depending on the formula used, the density etc., can shrink to a higher or lower extent. Consequently the foam has to be stored for a certain period of time before cutting the large blocks into normal sizes. After production expanded polystyrene boards contain various amounts of the residues of the foaming agent. With time, they diffuse out, and the material shrinks by 0.5 to 2%, depending on density, manufacturing method etc. [6, 7].

Changes in dimension can also be caused by high or low temperatures, through which plastic foams can experience non reversible expansion or contraction. Tests with some plastic foams at high temperatures showed that, depending on temperature, considerable dimensional changes can occur when they are heated up to 120 to 130°C or more [8, 9]. On the other hand, polyurethane foams in cold installations can shrink when their density is not sufficient.

In this case, the low pressure – resulting from the condensation of the gas in the cells – can lead to mechanical damages at the walls of the cells and thereby to shrinkage causing the foam to deteriorate.

For high temperatures over 150°C, inorganic insulation materials must be used. With increasing temperature even calcium silicate begins to shrink, e.g. 1 to 1.5% at 650°C [10, 11, 12].

---

[6] "Anwendung werkgefertigter harter Schaumstoffe", Irmhild Sauerbrunn, Handbuch; IBK-Seminar, 3-4 November (1986), pp. 1-4.
[7] "Still Room to Improve EPS Usage, 3 Years after Study", Richard L. Fricklas, *RSI*, October (1987), pp. 58-62.
[8] "Verhalten von Schaumstoffen unter Druckbeanspruchung bei 100 bis 130°C", H. Zehendner, Sonderdruck *Isolierung, 5,* (1980), pp. 1-9.
[9] "Updating Piping and Vessel Insulation with Composite Systems", Stanley T. Kastanas, *AGA 1987 Operating Section Proceedings* (Arlington, Virginia: American Gas Association, 1987), pp. 466-470.
[10] "Pabco Super Caltemp Block and Pipe Insulation", Brochure; Pabco Division of Fiberboard Corporation, 4-2 B/AD-5, 1 February (1987), pp. 1-2.
[11] "Kaylo Asbestos-Free Pipe Insulation", Brochure; Owens-Corning Fiberglas Corporation, *1-IN-14689,* July (1987), pp. 1-6.
[12] "Calsilite Insulation: Calcium Silicate Pipe Covering and Block Insulation for Installations to 1200°F", Brochure, Calsilite Corporation, 1 September (1985) pp. 1-8.

With FOAMGLAS® cellular glass, problems of shrinkage or expansion are unknown within the temperature range between −200°C to +430°C, except for the normal predictable effect of the coefficient of expansion/contraction.

V.8.1.3 The Combined Effect of Temperature and Humidity

During storage, transport or installation of insulation materials, it is not always possible to protect their surface and they may become wet. When the waterproofing layers are subsequently applied to the insulation, this moisture will be trapped and eventually absorbed by the insulation material.

A similar problem can occur when some plastic foams, which have a relatively high permeability, are subjected to the ingress of water vapour due to an inadequate vapour barrier, which can also be damaged on site.

In all these cases, the insulation material is not only in contact with high humidity but also subject to changes of the outside ambient temperature. Solar radiation can also increase the temperature to more than 75°C, depending on the climatic and geographical situation. Under the combined influence of temperature and humidity, plastic foams can experience substantial changes of dimensions. Shrinkage, warping or bowing are the visible results. Some polyurethane foams tested at 70°C and 100% relative humidity during 14 days according to ASTM D 2126 showed a linear change reaching 2%. When tested at a higher temperature, 150°C, under dry conditions a linear change of 3% was noted [13, 14]. When one realises that a change of 1% means 1 cm per linear meter, the importance of the problem is obvious.

A simple heat lamp and moisture/temperature test demonstrates the dimensional stability of FOAMGLAS® insulation and the irreversible expansion of some other insulation of the same thickness (Fig. V.23).

The situation is similar for phenolic foams.
FOAMGLAS® cellular glass, however, retains its original dimensions under all these conditions.

---

[13] "Trymer Rigid Foam Insulation", Brochure; Dow Chemical Company, *109-764-86* (1986), pp. 1-8.
[14] "Owens Corning Urethane", Brochure; Owens-Corning Fiberglas Corporation, 1-UF-6139, March (1973), pp. 1-8.

PART V. WHY TO CHOOSE FOAMGLAS® INSULATION

Fig. V.23. The exceptional dimensional stability of FOAMGLAS® insulation.

V.8.1.4 The Combined Effect of Temperature and Loading

In addition to the problem of compressive strength as detailed in V.9, the reduction of the thickness of an insulation layer under the influence of load at high temperatures can also be considered as a dimensional change which can have severe consequences. Careful research on plastic foams has been carried out [15] which resulted in recommendations for admissible stresses in relation to the temperature. Safety factors of 4 or more have to be determined in function of the insulation material used and the application temperature [16]. In such conditions, FOAMGLAS® cellular glass is the safest solution because it will not be compressed within temperatures ranging from −200°C to +430°C, if the normal recommended safety factor of 3 is chosen during the design stage and if the material is properly applied.

---

[15] "Verhalten von Schaumstoffen unter Druckbeanspruchung bei 100 bis 130°C", H. Zehendner, Sonderdruck *Isolierung*, 5, (1980), pp. 1-9.
[16] "Untersuchung der mechanischen Eigenschaften von Schaumkunststoffen im Bereich von −180°C bis +70°C", H. Zehendner, Sonderdruck *Isolierung*, 5, pp. 1-12.

For very high temperatures, mineral fibre or calcium silicate are frequently used. The determination of compression under load at high temperatures is somewhat difficult; the results depend on loads and test procedures [17, 18].

V.8.2 CONSEQUENCES OF A LACK OF DIMENSIONAL STABILITY: DIFFERENT TYPES OF PROBLEMS

These can arise when the dimensional stability of the insulation material is not sufficient for the job.

V.8.2.1 Warping and other Types of Distortion

Plastic Foams

Severe damage in the form of warping or buckling can happen in relation to the large coefficient of thermal expansion for plastic foams and the influence of temperature and humidity. If the plastic foam is applied at temperatures ranging from 10 to 20°C with tightly butted joints and if the surface is protected in the usual way from the influence of solar radiation, the exterior surface of the insulation will heat up and the dimensions of the elements will increase, while its lower part will remain cooler and the dimensions will remain practically unchanged. The result can be deformation, warping and buckling, which can occur also with prefabricated elements, e.g. solar collectors [19].

On the other hand, possible blistering of in situ sprayed polyurethane foam on tanks shows that dimensional stability can also be influenced by the applicator, the weather and the mechanical equipment used during application.

---

[17] "Verhalten von Wärmedämmstoffen bei höheren Temperaturen und Ermittlung der oberen Anwendungsgrenztemperatur", H. Zehendner, *Mitteilungen, 12* (München: Forschungsinstitut für Wärmeschutz, e.V., Nachdruck wksb Sonderausgabe, 1985) pp. 1-7.
[18] "Laine de Roche et Laine de Verre Isover", Brochure; Isover Saint-Gobain, November (1986), pp. 1-6.
[19] "Design Guide for Shallow Solar Ponds", A.B. Casamajor, R.E. Parsons, *Lawrence Livermore Laboratory* (Livermore, California: University of California, 6 January, 1978), pp. 1-4.

V.8.2.2 Open Joints or Cracks

These can result from thermal expansion and/or shrinkage of certain types of insulation materials. As an example, the formation of large cracks occurred in the polyurethane insulation of 3 ships owned by *El Paso Co.* in Houston, Texas, USA. Due to the cracks in the insulation of the LNG containment vessels, the ships could no longer be used for the transport of LNG [20].

The joints in polyurethane insulation will open substantially at LNG temperatures. The size of these open joints in a two layer polyurethane insulation system has been established at *Owens-Corning Fiberglas Corporation* [21]. It was found that they were wide enough to permit convection in the open longitudinal, radial and circumferential joints. The result was a substantial increase in heat gain, in the conditions of the test.

Coolant	% increase in heat gain due to opening of the joints
Liquid argon	33
Liquid nitrogen	174

Another example of open joints creating cold bridges is provided when polystyrene foam is used for roof insulation. The reduction of the insulation efficiency through the open joints caused by movements may reach about 10%.

V.8.2.3 Stresses between Insulation and Coatings or Waterproofing Materials

These stresses result from the differential thermal expansion or contraction of the components in an insulation system and from the irreversible dimensional changes of the insulation material. By using computer programs, stresses caused by reversible changes can be predicted [22].

---

[20] "Ship in Tow", *Times-Picayune* (New Orleans), Section 1, 15 October (1980), p. 13.
[21] "Thermal Performance of Urethane Foam Pipe Insulation at Cryogenic Temperatures", Dwight S. Musgrave, *Journal of Thermal Insulation, 3,* July (1979), pp. 3-21.
[22] "Preliminary Stress Evaluation of the Effects of Gaps between Roof Insulation Panels", James E. Lewis, *Journal of Thermal Insulation, 4,* July (1980), pp. 3-25.

V.8.3 FOAMGLAS® CELLULAR GLASS HAS AN EXCEPTIONAL DIMENSIONAL STABILITY (Fig. V.24)

Fig. V.24. FOAMGLAS® cellular glass has a very low thermal expansion coefficient.

V.8.3.1 Reversible Dimensional Changes Due to Temperature Variations Are Minimal

This results from the small linear thermal expansion [23, 24] of only 8.5×10^{-6} K^{-1} to 9.0×10^{-6} K^{-1}. This means that the free expansion/contraction will only range from 0.85 mm/m to 0.9 mm/m for a temperature difference of 100°C.

Compared with other insulation materials, especially plastic foams with 7 to 10 times higher coefficients, one can understand why FOAMGLAS® cellular

---

[23] "Lineaire Uitzettingscoëfficiënt van FOAMGLAS® T4", W. van Herk, Rapport; Instituut TNO voor Bouwmaterialen, *B-86-1-318,* 10 september (1986), pp. 1-4.
[24] "Dämmstoffe für den baulichen Wärmeschutz", Irmhild Sauerbrunn, *Zeitschrift für das Sachverständigenwesen, 6.* [3] Juni (1985), pp. 49-52 [4] August (1985), pp. 73-78.

glass provides the solution for insulation systems working in variable temperature conditions (Fig. V.24). Moreover, the small difference between the expansion coefficient of FOAMGLAS® cellular glass and the one of steel (12×10^{-6} K^{-1} for carbon steel) or concrete (10×10^{-6} K^{-1}) allows "compact" insulation systems to be used, in which all the layers are completely bonded together, all joints being filled. For special constructions in the lower temperature range (cold application or variable temperature) for which tight joints to avoid water vapour penetration are needed, PC® 88 ADHESIVE as a flexible, but strong adhesive and sealer is an excellent solution. Of course, calculations of differential dilatation are recommended in cases when the systems experience considerable temperature variations. In general terms, accessory products having a high modulus of elasticity should not be used with FOAMGLAS® insulation.

V.8.3.2 Non Reversible Changes Due to the Influence of Temperatures in the Range of +430°C to –200°C Are not Known

Several tests as well as practical experience over many years show that non reversible dimensional changes with FOAMGLAS® cellular glass have never been observed. Even for temperatures of about 400°C FOAMGLAS® cellular glass has no shrinkage but only a coefficient of expansion, while other inorganic insulation materials shrink to some extent. This excellent behaviour of FOAMGLAS® cellular glass is one of the reasons for its application in industrial plants.

V.8.3.3 Humidity Does not Cause Dimensional Changes to FOAMGLAS® Cellular Glass

Testing and practical experience have shown that FOAMGLAS® cellular glass remains dimensionally stable when it is in contact with humidity or water. During tests at 20°C and 95% relative humidity, no changes could be observed. Consequently FOAMGLAS® cellular glass can be used as the insulation material when the ambient conditions involve high humidity.

V.8.4 CONCLUSION

On the basis of the information presented in this chapter, it is fair to say that true dimensional stability constitutes a unique advantage.

V.9

COMPRESSIVE STRENGTH

V.9.1 THE PROBLEM

High compressive strength is necessary to support loads (Fig. V.25).

When considering insulating materials, compressive strength is often overlooked. This is probably caused by the fact that many insulating materials, for instance fibrous materials, generally have a limited compressive strength.

Fig. V.25. High compressive strength is necessary to support loads.

PART V. WHY TO CHOOSE FOAMGLAS® INSULATION

In the case of FOAMGLAS® insulation, compressive strength is one of the most important properties and plays a major role in many industrial applications.

V.9.1.1 Low Temperature Tank Base Insulation

In the case of low temperatures and cryogenic tank base insulation, lack of compressive strength will give unacceptable settlement which would result in loss of thermal insulation, ground heaving and foundations being brought to temperatures which would result in major failures and, in case of uneven settlement, rupture of tank steel bottom and spill of the content.

FOAMGLAS® insulation compressive strength depends on the type of capping applied on the product. The question has been studied and two articles published by R. Gerrish [1,2] provide details on this question. Solutions have been found for the base insulation of various types of cryogenic tanks (e.g. LPG, LNG, LOX, and LIN). For instance, a typical LNG tank base FOAMGLAS® insulation system has been tested, taking care of simulating very precisely the actual water test and service situation.

Figure V.26 shows how the application of LNG tank base installation is carried out.

Although more than 90% of the above ground cryogenic storage tanks have bases insulated with FOAMGLAS® cellular glass, no operational problems due to lack of compressive strength have been reported. Many large tanks in major LNG terminals are in operation and perform satisfactorily.

As examples of these cryogenic tank bases insulated with FOAMGLAS® cellular glass, one can quote Lake Charles LNG, Das Island LNG, Zeebrugge LNG, North West Shelf LNG [3], Kaarstoe/Norway LPG storage tank [4], Mossmoran LPG and LNG tanks [5], *British Gas* tanks and many others.

---

[1] "Tank Base Insulation Systems for the Insulation of LNG and LOX", Gerrish, *Conference Papers Gastech 76*, 5-8 October (1976), Session 6, p. 13.
[2] "Cellular Glass Insulation for Load Bearing Application on the Storage of Cryogenic Fluids", R.W. Gerrish, *Advances in Cryogenic Engineering, 22*, p. 250.
[3] "There's More to LNG Tanks Than Meets the Eye", *North West Shelf Report, 3,* July (1987), p. 5.
[4] "The LPG storage terminal at Kaarstoe approx. 4 months before completion of the tanks", Noell GmbH, p. 2.
[5] "Wrapping up at Mossmoran", *Processing,* November (1984).

Fig. V.26. Low temperature tank base insulation.

V.9.1.2 Semi-underground Conical Digesters. Base Insulation

In the case of semi-underground conical digesters working at medium temperature, lack of compressive strength will give settlement, loss of thermal efficiency, and interference with the stable thermal environment.

V.9.1.3 High Temperature Tank Base Insulation

For high temperature tanks, insufficient compressive strength will lead to settlement, loss of thermal performance, possibly high viscosity and solidification of the contents. Fortunately, these potential problems have been overcome with adequate systems, like those shown in Fig. V.27 [6].

Basically three FOAMGLAS® insulation systems have been developed for hot tank base insulation, one for operating temperatures up to 80°C, one for operating temperatures up to 150°C and one for operating temperatures up to

---

[6] "Ervaringen met FOAMGLAS® Cellulaire Glasisolatie: Bodemisolatie voor verwarmde Opslagtanks", Brochure; *PCN-12/86,* p. 3.

Fig. V.27. High temperature tank base insulation.

250°C. It has also been shown that hot tank base insulation leads to substantial energy saving when the wall and the roof of the tank have already been insulated.

V.9.1.4 Industrial Floors

In the same way, FOAMGLAS® cellular glass has been applied extensively for the insulation of large industrial floors [7].

V.9.1.5 Underground Pipe and Vessel Insulation

For underground pipe and vessel insulation, compressive strength is needed to allow direct burial without additional structural protection. Such application permits above ground traffic with proper design. Prestigious jobs like the underground heating system of the Personnel Rapid Transit System at Morgan-

---

[7] "Industriefussböden", I. Sauerbrunn, Paper presented at *International Colloquium Esslingen,* 13-15 January (1987), p. 333.

town, West Virginia, USA or the steam-heat distributing line of *Norenco* at St. Paul, Minnesota, USA [8] illustrate these challenging possibilities.

V.9.1.6 Pipe Supports and Hangers

For insulated pipe supports and hangers, the choice of an insulating material such as FOAMGLAS® cellular glass, which has a high compressive strength, enables the designer to avoid direct short circuits or at least to reduce them very substantially.

Should one or several insulated pipe supports settle considerably or fail, it will create geometrical changes in the steel pipeline which may result in significant stresses in the pipeline, particularly near nozzles, flanges and fittings. Calculation methods to determine the cradle dimensions have been developed and enable the designer to choose the correct dimensions in relation to other factors, such as the distance between the cradles, the load, and the type of capping materials [9].

Rather recently, pipe covering has been cut out of FOAMGLAS® cellular glass HLB 150 slabs to increase the load bearing possibilities. This material was applied in 1985 on the *Esso Flexicoker* plant [10].

V.9.1.7 Living Loads

In the case of pipe insulation, excessive deformation under load can result in either open gaps in metal jackets or damage to vapour barrier materials, which permits water and water vapour to enter the insulation system. Insulation saturation can result, except in the case of cellular glass. This will lead to reduced thermal efficiency and possible corrosion problems (see V.5 and V.6). The same applies to the top of spheres, domed tanks and vessels.

---

[8] "FOAMGLAS® Insulation: Industrial Applications", Case study of Norenco Corporation, St. Paul, Minnesota; Pittsburgh Corning Corporation, *FI-192*, March (1987), pp. 1-2.
[9] "Cradle Supports for Insulated Pipes", Union Carbide, Engineering Standard; *P-82*, September (1965), p. 2.
[10] "FOAMGLAS® Pijpschaalisolatie in Esso-Flexicoker", *Uit Europoortkringen, 25* [10] (1986), pp. 14-15.

Insulated piping should not be walked on or subjected to concentrated loads such as ladders or used as scaffolding, but unfortunately, it is frequently subjected to this severe mechanical abuse. This goes against good practice but unfortunately corresponds to real life and has been widely acknowledged by outstanding authorities like Wiaco [11].

Figure V.28 shows what is a more usual situation than expected. As mentioned by Stanley Kastanas [12]:

> "Essentially, an insulation should have enough compressive strength to "fend-off" that element of personnel ("the climbing apes") that insist on using insulated pipes as stepladders. Depending on density, fibreglass and mineral fibres tend to have the lowest compressive strength (i.e., 5-15 psi/0.035-0.105 N/mm ^{2}), while inorganic cellular foam glass has the highest (i.e. 100 psi/0.7 N/mm^2) compressive strength".

Fig. V.28. Workmen use often horizontal insulated pipes as walkways or scaffoldings.

---

[11] "Nuclear Plant Thermal Insulation", Charles C. Lindsay, Paper presented at the *Congrès Mondial de l'Isolation Thermique et Acoustique*, p. 65.
[12] "Updating Piping and Vessel Insulation with Composite Systems", S.T. Kastanas, *AGA 1987 Operating Section Proceedings*, 87-DT-101, (1987), p. 468.

A similar opinion is expressed by Mrs. Simonetti in an article on *Polysar*'s new plant where the need of high compressive strength is pointed out [13].

V.9.1.8 Industrial Metal Decks

For large industrial metal decks, a similar problem has been identified in a case in which an insulation material with inadequate strength was applied.

This danger is clearly warned against in the documentation of a phenolic foam manufacturer who writes [14]:

> "Metal deck roofs shall be designed so that they will not have a deflection exceeding 1/240 of a clear span under all types of loads (live & dead) when subjected to 300 lbs. (135 kg) concentrated on a 1 sq. ft. (0,093 m^2) area at midspan. Designer should take into account roof traffic conditions, all roofs mounted mechanical equipment, as well as equipment used in the application of the roof".

V.9.1.9 Self Supporting FOAMGLAS® Insulation Walls

In the case of self supporting walls, such as in some types of chimney construction, mechanical strength is obviously needed and its absence will result in the collapse of the system. Similar remarks apply to other self supporting walls such as those of cold stores that may have to carry the pressure of the wind between the supporting girds. Flexural strength is necessary for this type of construction.

V.9.2 DEFINITION OF COMPRESSIVE STRENGTH CHARACTERISTICS

When comparing the compressive strength of various insulating materials, one has to really appreciate the meaning of the indicated values. Here again, a substantial difference does exist between FOAMGLAS® cellular glass and the vast majority of other insulating materials.

---

[13] "Cellular-glass insulation protects equipment efficiently", M.C. Simonetti, *Process Industries Canada,* December (1986), p. 9.
[14] "Koppers RX All-Purpose Roofboard Insulation", *Koppers Buyline, 2281,* September (1986), p. 7.

PART V. WHY TO CHOOSE FOAMGLAS® INSULATION

V.9.2.1 Definition of Compressive Strength Characteristics

Usually, when testing materials, one records the stress-strain curve.

In the case of FOAMGLAS® cellular glass like for many other building materials, for instance insulating concrete, the compressive strength is defined by the stress at which the material main failure occurs. In the case of cellular glass, the deformation at failure is very small and basically only concerns the capping material (bitumen and thin roofing felt when following ASTM C 240-91 for instance). The deformation may represent 1 or 2 mm and is independent from the sample thickness.

After the main failure, FOAMGLAS® cellular glass still keeps a certain load bearing capacity, although often somewhat lower than the compressive strength.

V.9.2.2 The Deformation Issue

For many other insulating materials, such as plastic foams, mineral wool and calcium silicate, the compressive strength is typically measured when the deformation reaches 5 or 10% of the thickness. ASTM C 165-83 even considers a deformation of either 10 or 25% of the thickness (cf. paragraph 7.2.5 of the reference [15]).

For the compressive strength test of foamed plastics, ISO 844-1978 recommends the selection of the higher stress obtained either with a relative deformation lower than 10% or at 10% [16].

The references [17, 18, 19, 20, 21, 22, 23] show typical compressive strengths at 10% for various insulating materials such as phenolic foam, polyisocyanurate foam, polyurethane foam, fibre glass, and extruded polystyrene foam.

---

[15] "Standard Recommended Practice for Measuring Compressive Properties of Thermal Insulations", ASTM C 165-83, (1983).
[16] "Plastiques alvéolaires – Essai de compression des matériaux rigides", Norme Internationale, *ISO 844-1978 (F)*, September (1978), p. 2.
[17] "Koppers RX All-Purpose Roofboard Insulation", *Koppers Buyline, 2281,* September (1986), p. 7.
[18] "Cellobond K: Low conductivity phenolic foam systems", British Petroleum, *BP Chemicals,* p. 4.

Reference [24] gives the compressive strength under 5% deformation of a typical calcium silicate insulation. The Table V.7 is taken from the article published several years ago by Klezath [25] that gives compressive strength of polystyrene of various densities under 5%, 10% and 30% deformation.

TABLE V.7

Compressive strength of polystyrene

	Density (volume weight) (kg/m^3)			
	15	20	30	60
Compressive strength in N/mm^2				
5% deformation....................	0.055-0.065	0.07-0.09	0.14-0.22	0.43-0.53
10% deformation....................	0.06-0.08	0.09-0.12	0.15-0.24	0.47-0.57
30% deformation....................	0.08-0.12	0.12-0.16	0.18-0.28	0.53-0.63

In a relatively recent article [26], Zehendner reports the result of compressive strength tests run in pushing the deformation to higher values than 10%. He went to 25%. The stress-strain graphs generally show that in the 10 to 25% deformation range, the compressive strength moderately increases for expanded polystyrene insulation, slightly increases for one type of polyurethane foam and basically remains constant for extruded polystyrene foam, PVC foam and two other polyurethane foams (Figs. V.29 and V.30).

---

[19] "Carpenter FP 2500", Carpenter Insulation Company, Technical Data Sheet, *301*, December (1986), p. 1.
[20] "Delta Industrial Insulations", Brochure; Rockwool Manufacturing Company, November (1986), p. 9.
[21] "Styrofoam Plan", Brochure; Dow Construction Products, *Dow Eu, 4600-E-583*, April (1985), p. 23.
[22] "Koolphen: A Revelation in Insulation", Data sheet; Kooltherm Insulation Products, *Kooltherm*, p. 6.
[23] "Systèmes d'Isolation", Brochure; Ouest Isol, p. 3.
[24] "Pabco Super Caltemp Block and Pipe Insulation", Data sheet; Pabco Division of Fiberboard Corporation, *4-2-b and 4-2-D,* January (1987), p. 2.
[25] "Wärmeisolierung von Speicherbehältern für tiefsiedende Flüssigkeiten", H. Klezath, *Erdoel-Erdgas-Zeitschrift, 3,* March (1969), p. 101.
[26] "Untersuchung der mechanischen Eigenschaften von Schaumkunststoffen im Bereich von –180°C bis +70°C", H. Zehendner, Reprint from *Isolierung, 5,* p. 9.

PART V. WHY TO CHOOSE FOAMGLAS ® INSULATION

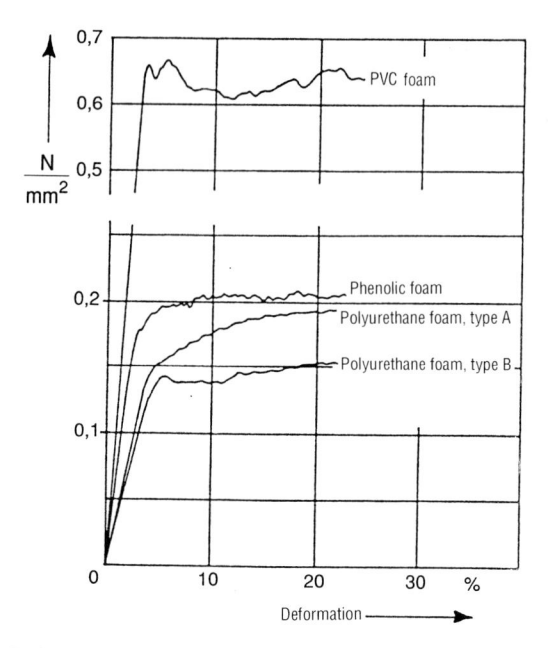

Fig. V.29. Stress-strain diagram of PVC, phenolic and polyurethane foams.

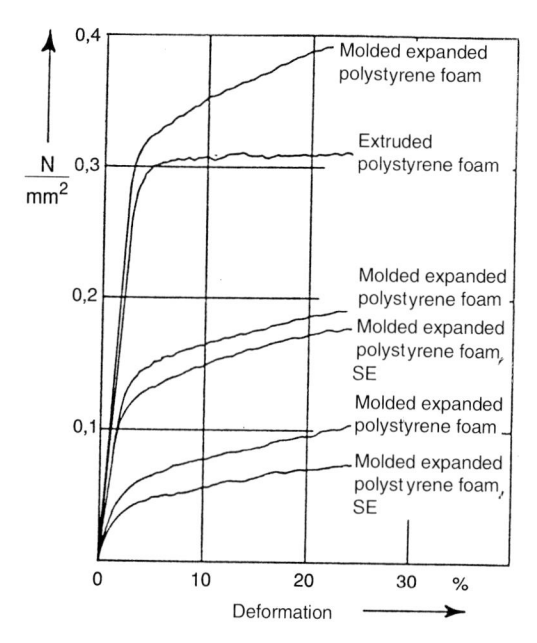

Fig. V.30. Stress-strain diagram of polystyrene foam.

446

V.9.2.3 Testing Procedure for FOAMGLAS® Cellular Glass

In the case of FOAMGLAS® cellular glass, the procedure for testing compressive strength has been precisely defined in at least three different norms: ASTM C 240-91, DIN 18 174 and ÖNorm B 6041. Fortunately the three procedures are very similar and lead to comparable results. They also have the merit to simulate the most common method of applying FOAMGLAS® insulation, when using bitumen as an adhesive.

When using FOAMGLAS® cellular glass cappings different from the one foreseen in ASTM C 240-91, the compressive strength does vary.

V.9.2.4 The Capping Influence

The proper capping should be selected consistent with the specific engineering needs of the application, such as for instance a liquid oxygen or a molten sulphur tank base.

These alternate cappings give different but well-defined compressive strengths. In almost all cases, the deformations are very limited and homogeneous.

In the case of insulation materials, like cellular plastics, for which compressive strength is measured under 5 or 10% deformation, special engineering considerations must be taken into account to cope with the substantial deformation.

V.9.2.5 Consequences of Important Deformation

It is obvious that deformation can create very serious problems when designing insulation systems, especially with the increasing insulation thickness specified today that can result in higher deformation. Connections, inlets, outlets, flanges, etc. can be subjected to concentrated stresses creating major dangers of rupture and causing potentially hazardous liquid spillage.

The degree of recovery after the deformation of plastic foam also varies from product to product.

V.9.2.6 Influence of Temperature and Time

In addition to the normally reported compressive strength data, the insulation specifier should be cognisant of the potential adverse effects of long term creep and strength variations with temperatures.

The influence of high temperatures on the compressive strength of foamed plastics has been studied by H. Zehendner who published several interesting articles [27]. His graphs indicate for instance that the tested phenolic foam and the polyurethane foam had the compressive strength under 10% deformation reduced by more than 50% when the temperature increased from 20 to 130°C (Fig. V.31).

The reduction is even higher for the tested PVC foam.

Regarding the combined influence of temperature and time, samples of polyurethane foam bearing only a moderate load of 0.02 N/mm^2 at 130°C provide a thickness deformation ranging from about 5 to 10% after 75 days.

Zehendner also investigated the AGT (Anwendungsgrenztemperaturen or application limit temperatures) of fibrous insulating materials and produced many graphs indicating the thickness reduction in the test conditions at various elevated temperatures [28].

In contrast to this behaviour, FOAMGLAS® cellular glass compressive strength is practically unaffected by temperature increase, at least up to its given limit of 430°C. The creep of FOAMGLAS® cellular glass under load at 430°C remains fairly low.

The AGI brochure Q 137 on cellular glass [29] reads as follows (paragraph 5.7):

> "The compressive strength within the application limiting temperature area is independent of time and temperature".

---

[27] "Verhalten von Schaumstoffen unter Druckbeanspruchung bei 100 bis 130°C", H. Zehendner, *Isolierung, 5/80,* May (1980), p. 3.
[28] "Verhalten von Wärmedämmstoffen bei höheren Temperaturen und Ermittlung der oberen Anwendungsgrenztemperatur", H. Zehendner, *Forschungsinstitut Mitteilungen,* (1985), p. 3.
[29] "Cellular Glass as Insulation Material for Industrial Equipment Applications", AGI Work Sheet, *Q 137,* August (1985), p. 4.

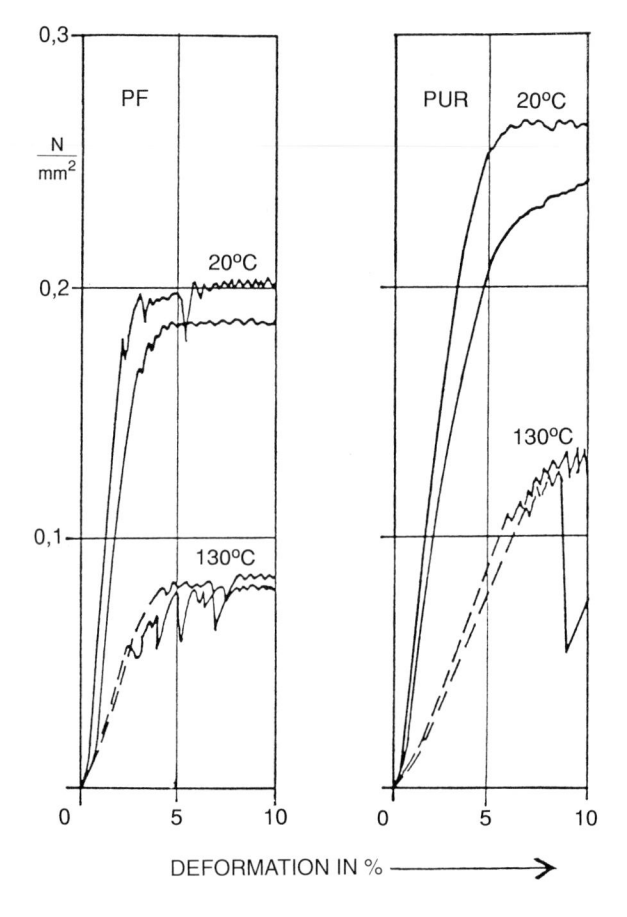

Fig. V.31. Stress-strain diagram of PF foam block and PUR foam block – type B1 at 20 and 130°C test temperatures.

And report [30] indicates average compressive strength of 0.64 N/mm² when testing FOAMGLAS® cellular glass at 250°C.

---

30 "Staatliches Materialprüfungsamt Nordrhein-Westfalen", Test Report, PCE Technical Sales Letter, *E 188,* 14 October (1975), p. 2.

V.9.2.7 Influence of the Load Direction

Regarding other influences that affect an insulating material compressive strength, one should also quote the direction of the load in relation to the foaming direction.

For instance a report of the *Gilmore Research Laboratories* [31] indicates for a polyurethane foam having a density ranging from 28.8 to 36.8 kg/m^3, a compressive strength measured according to ASTM D 1621 of 0.235 N/mm^2 in the direction parallel to the foam rise and of 0.159 N/mm^2 in the perpendicular direction. Similarly, the Kooltherm phenolic foam documentation [32] shows a compressive strength of 0.172 N/mm^2 parallel to rise and 0.1 N/mm^2 perpendicular to rise for a 35 kg/m^3 foam. Similar values have been published for Cellobond phenolic foam.

V.9.3 ENGINEERING DESIGN INFORMATION

Pittsburgh Corning has responded to the various market demands by producing several grades of FOAMGLAS® cellular glass with different strengths and thermal properties for different applications.

V.9.3.1 Types of FOAMGLAS® Insulation

The characteristics of the most usual grades of FOAMGLAS® insulation – type T4, T2 and S3 – are given in II.1 and in FOAMGLAS® cellular glass data sheets.

High compressive strength grades are labelled HLB types (High Load Bearing) and four usual types exist named FOAMGLAS® cellular glass HLB 115, HLB 135, HLB 155 and HLB 175. They are especially aimed at applications requiring high compressive strength like large tank base insulation for which they have constituted the classical solution for many years.

---

[31] "Cryogenic Insulation", Donald S. Gilmore Research Laboratories, *Insulation Outlook,* November (1981), p. 24.
[32] "Koolphen: A Revelation in Insulation", Data sheet; Kooltherm Insulation Products, *Kooltherm,* p. 6.

More specialised types of FOAMGLAS® insulation are also available on request.

Because of the reliable and predictable properties of FOAMGLAS® insulation, it is specified for applications requiring high strength by major engineering companies throughout the world.

V.9.3.2 The Safety Factor

Moreover, as indicated these worldwide engineering companies have generally adopted for FOAMGLAS® insulation [33] a safety factor of three, based on the maximum working stress and the average published compressive strength, which is more or less becoming an international design criterion for industrial FOAMGLAS® cellular glass applications. The AGI Q 137 note reads [29]:

> "In practical application, the compressive strength should only be stressed for a third of its given value".

For other insulating materials, especially cellular plastics, for which deformation and strength reduction due to various factors are prime considerations, significantly higher safety factors are generally chosen.

In a paper presented during the International Colloquium for Industrial Floors in 1987, I. Sauerbrunn lists the usual safety factors [34].

V.9.3.3 Quality Control Procedure: Internal and by Third Party

Pittsburgh Corning has developed stringent internal procedures bearing on several properties with emphasis on compressive strength and has voluntarily agreed to subject routinely its quality control to third party inspection.

It does include audit of internal procedure as well as independent sampling and testing. Due to the international character of the corporation it is actually carried out simultaneously by several third parties.

---

[33] "FOAMGLAS® HLB Cellular Glass for the Insulation of Liquefied Gas Tank Bases", PCE Data Sheet, *PCE-ENG-i22*, June, (1986), p. 2.
[34] "Industriefussböden", I. Sauerbrunn, Paper presented at *International Colloquium Esslingen*, 13-15 January (1987), p. 333.

We can quote for *Pittsburgh Corning Europe* various third parties evaluating the quality control for one or more properties and running their own tests among which are

- *ACERMI*, Paris
- *Forschungsinstitut für Wärmeschutz eV*, München
- *Norwegian Building Institute*, Trondheim
- *Staatliches Materialprüfungsamt Nordrhein-Westfalen*, Dortmund
- *UBAtc* (Union belge pour l'Agrément technique dans la construction) [35].

Moreover, in the case of HLB ware aimed at demanding applications, *Pittsburgh Corning* has developed special quality assurance specifications, which involve customer and third party inspection.

V.9.4 PERFORMANCE OF FOAMGLAS® INSULATION COMPARED TO ALTERNATIVE INSULATION

As previously explained, the performance of many insulation systems depends on compressive strength and deformation in service conditions. These data widely differ between FOAMGLAS® cellular glass and competitive insulation. It is impracticable to provide complete engineering data on every competitive insulation product and brand. As general information, the Table V.8 presented by I. Sauerbrunn [36] shows ranges of compressive strength values for families of competitive products.

They generally correspond to procedures outlined in norms simulating idealised conditions.

In some cases, differences due to test conditions have been noticed. Prudent designers should always verify that available information truly corresponds to actual service conditions.

---

[35] "Verre Cellulaire FOAMGLAS® et FOAMGLAS® BOARD", *UBAtc*, ATG/H 539, 17 September (1991).
[36] "Dämmstoffe für den baulichen Wärmeschutz (part 1 + 2)", I. Sauerbrunn, *Zeitschrift für das Sachverständigenwesen, 6,* June (1985), p. 75.

TABLE V.8
Compressive strength or compressive stress
with 10% deformation of insulation materials
for industrial floors in N/mm² (published values)

Product	Density (kg/m³)	(N/mm²)
Wood-wool building slab	360-570	0.15-0.20
Multiple lightweight building slab		As expanded PS foam
Cork	80	0.05-0.11
Cork	120	0.05-011
Phenolic foam	35	0.10-0.25
Phenolic foam	45	0.15-0.35
Expanded polystyrene foam	15	0.07-0.12
Expanded polystyrene foam	20	0.12-0.16
Expanded polystyrene foam	30	0.18-0.26
Extruded polystyrene foam	30	0.25-0.42
Extruded polystyrene foam	50	0.8
Polyurethane foam	30	0.10-0.20
Polyurethane foam	40	0.20-0.35
Urea formaldehyde resin foam	10	0.01
Cellular glass	125	0.50-0.70
Cellular glass	135	0.70-1.00
Perlite insulation slabs	150	0.3
Perlite insulation slabs	210	0.3
Expanded perlite (bulk)	90	0.01

V.10

VERMIN RESISTANCE OF INSULATION MATERIALS

V.10.1 THE PROBLEM

According to a US study [1], rodents cause important destructions to commercial, industrial and institutional buildings. They come next to fire and Acts of God. The problems are even more severe in agricultural buildings. Heavy damage can also be caused by insects. Finally, materials with an open structure can represent a source of air pollution caused by the absorption of dust and moisture and provide a culture medium for the development of microorganisms.

The attack from different animal species can result in the loss of insulation efficiency, bad hygienic conditions and severe economic loss.

FOAMGLAS® insulation can help to solve these problems. It resists practically to all kinds of rodents and vermin. In addition, it can serve as a sort of "vermin barrier" when it is correctly used.

V.10.2 WHAT TYPES OF CREATURES ARE DANGEROUS TO INSULATING MATERIALS?

V.10.2.1 Rodents

Rats and mice are rodents which like to live near human settlements because of the availability of food. This applies to dwellings and also to agricultural, food processing and pharmaceutical installations.

---

[1] " Rodent proofing in new construction: It pays", G.N. Paterson, *The Construction Specifier,* November (1976), pp. 35-41.

The most severe problem is caused by the breeding habits of mice [2], which can produce 50 or more offspring from one female mouse. Especially when the insulation material is used as a preferred environment which serves as a nest and a home for young mice, the conditions for breeding are optimal. As a result, these insulation materials can be destroyed within in a few months [3]. Two papers [4, 5] show how fast the deterioration can take place.

V.10.2.2 Insects

Insulation materials can also be destroyed by different kinds of insects when they are not tough enough to resist attack from their gnawing and boring action.

Beetles of different kinds, e.g. hide beetles [6, 7] and others, bore tunnels until finally the whole insulation becomes damaged and ineffective.

Termites seem to be an increasing problem [8].

V.10.2.3 Microorganisms

Spores of fungi and other microorganisms are present in the air. In most of the cases, the insulation materials themselves are not a culture medium for these organisms, but the presence of moisture, dust and greasy substances in materials with an open structure can produce a "soil".

---

[2] "Zur Problematik der Zerstörung von Dämmstoffen an landwirtschaftlichen Gebäuden durch Hausmäuse", M. Süss, B. Mittrach, G. Koller, *Bayrisches Landwirt. Jahrbuch, 54,* March, (1977), pp. 333-337.

[3] "Uber die Notwendigkeit und Möglichkeit der völligen Beseitigung von Mäusebefall in Anlagen der industriemässigen Geflügelproduktion", W. Tannert, *Monatsheft der Veterinärmedizin, 31,* (1976), pp. 108-114.

[4] "Die Hausmaus ist ein Risikofaktor für Dämmstoffe, M. Süss, Der Praktische Schädlingsbekämpfer, 35 (1983), pp. 131-134.

[5] "Mäuse zerstören fast jeden Dämmstoff", M. Süss, B. Mittrach, *Landtechnik, 2,* February (1982), pp. 1-3.

[6] "Invading insects cause maintenance headaches in South", J. Chapman, *Professional roofing,* March (1988), pp. 19-21.

[7] "Auch Käfer fressen Hartschaumplatten", K.H. Hoppenbrock, *Top Agrar,* November (1980), pp. 3-4.

[8] "Le développement des termites est lié à l'utilisation croissante des isolants dont ils sont très friands", J.-L. Lafleur, *Le Moniteur,* p. 1.

Photo 29. Fire damage.

Photo 30. Prefabricated insulation.

Photo 31. FOAMGLAS® cellular glass application in industry.

Photo 32. Damage due to water vapour deterioration.

Photo 33. Corrosion problems imply often loss of production.

Photo 34. Damages due to corrosion under insulation.

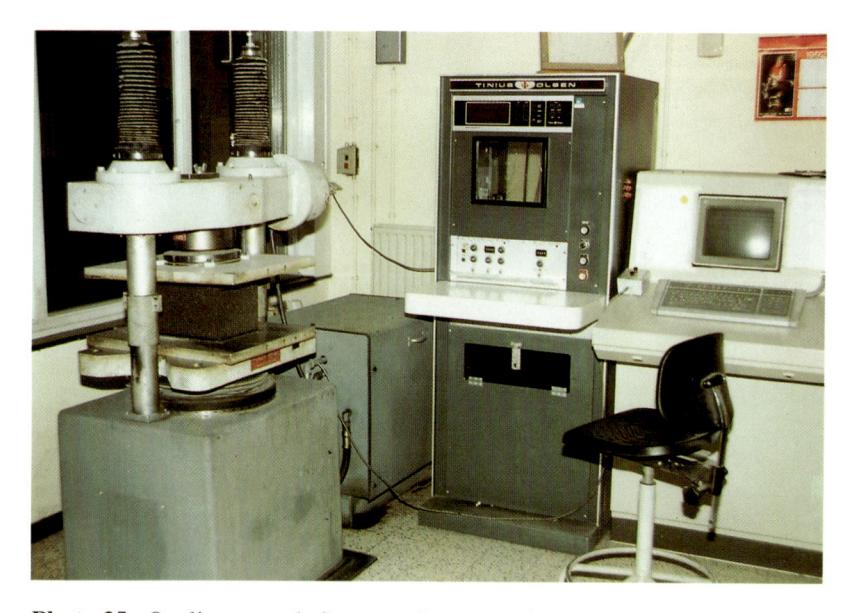

Photo 35. Quality control. Compressive strength test.

Photo 36. FOAMGLAS® insulation is ratproof.

V.10.3 THE DIFFERENT TYPES OF DAMAGE CAUSED BY RODENTS AND VERMIN

V.10.3.1 Insulating Efficiency

When they are partially or totally destroyed by vermin, insulation materials lose partially or totally their efficiency. Several authors inform us that this happens with pipe insulation and also with other types of insulation systems [2, 3, 9].

V.10.3.2 Bad Hygienic Conditions

All kinds of harmful living creatures as described in V.10.2 can transfer most of the infectious diseases to human beings and domestic animals. The excrements and bodies of rats and mice will pollute food and other products and cause various diseases [2, 3, 10].

V.10.3.3 Economic Losses

The losses caused by rodents and vermin are extremely high. They occur in different kinds of buildings including agricultural installations, industrial plants, etc.

Energy consumption rises when the insulation is destroyed. Corrosion occurs in cold installations when the vapour barrier and insulation are gnawed through, and subsequent renovation is very expensive. Rodents multiply rapidly when they nest in warm insulation. They cause deterioration of the thermal insulation and can also damage electrical installations, resulting in shortcircuits.

---

[9] "Wärmedämmung für Industriefussböden", I. Sauerbrunn, Paper presented at the *International Colloquium Industrial Floors,* 13-15 Jan. 1987, edited by P. Seidler (Esslingen, Germany: Technische Akademie), pp. 327-334.

[10] "Fungal Contamination of Fireproofing Material", C.R. Schimpff, B. Arch, *The Construction specifier,* November (1976), pp. 50-52.

V.10.4 DIFFERENT INSULATION MATERIALS SHOW DIFFERENT BEHAVIOUR AGAINST RODENTS AND VERMIN

V.10.4.1 FOAMGLAS® Insulation

To emphasize: FOAMGLAS® insulation is the most vermin and rodent resistant insulation material. Up to now, there have been no reports that FOAMGLAS® insulation has been damaged or destroyed by such life forms.

This experience is confirmed by the results of different tests. As early as 1944 and 1945, tests were executed by independent laboratories. The result reads:

"FOAMGLAS® insulation is ratproof" [11, 12].

Subsequent tests on mice were carried out with different insulation materials in Germany. In these tests, the quantity of insulation material destroyed through the nest building of mice were measured. While FOAMGLAS® insulation remained virtually unattacked, the other materials had been more or less damaged. Because the results are very important for the insulation of agricultural installations they have been described by several authors [4].

After testing the resistance of FOAMGLAS® cellular glass against insects in the USA, it was reported in 1949 that this material showed no evidence of attack after more than 7 months, while wood was very quickly destroyed [13]. These results were confirmed in 1985 by extensive tests in Germany. FOAMGLAS® insulation was resistant against two specimens of termites. At the same time, tests with different beetles also showed that FOAMGLAS® insulation is resistant to these gnawing or boring insects [14].

---

[11] "Field studies with FOAMGLAS® as a rat proofing material", J.F. Welch, United States Department of the Interior, Fish and Wildlife Service, Wildlife Research Laboratory, 19 April (1945).
[12] "Material Examination Report", Armstrong Cork Company, June 26 (1944).
[13] "Termitarium inspection report", Earl M. de Noon, *South Florida Test Service*, 3 September (1949), p. 1.
[14] "Prüfung der Insektenbeständigkeit des FOAMGLAS®-Platten-Musters mit nagenden und bohrenden Insekten", Bundesanstalt für Materialprüfung (BAM), 31 October (1978), pp. 1-15.

V.10.4.2 Other Insulation Materials

As already explained, nearly all insulation materials such as plastic foams, mineral fibres, etc., can be damaged by mice, except FOAMGLAS® insulation. It occurs in spite of the fact that some producers of these insulation materials claim that their products have no nutritive value. At most, they recommend the protection of the material against rodents, e.g. for underground or agricultural insulation [15, 16]. A large number of reports inform us of the damage caused by rodents or insects on plastic foams or mineral wool [2, 3].

V.10.5 INSTALLATIONS USING FOAMGLAS® CELLULAR GLASS AS THE INSULATION MATERIAL

Due to the excellent test results and the practical experience, FOAMGLAS® insulation is more and more used for all cases where the insulation system can be attacked by rodents or vermin and where safety is the main point to be considered.

Facilities where the resistance against rodents is important, especially agricultural installations, such as poultry farms, pipe installations, storage and freezer rooms, should be insulated with FOAMGLAS® insulation as a material which helps to avoid the development of vermin. This property is of prime importance when selecting insulation material for the whole food producing and processing industry.

---

[15] "Coquille Phenexisol", Prospectus, Ouest Isol, March (1984), pp. 1-6.
[16] "Styrofoam TG – Sous-toiture de Bâtiments Agricoles", Brochure, J. Labeille S.A., November (1977), pp. 1-2.

CONCLUSION

In this Handbook, we have presented the applications for which systems based on FOAMGLAS® insulation offer a safe solution for the long term.

We have explained the concepts of the FOAMGLAS® cellular glass systems and given some technical reasons why they should be chosen.

We hope that it will help the owners, designers and contractors to select, design and install insulation systems which will satisfy them for many years.

INDEX

ABB Lummus Crest, 44
Absolute humidity, 370, 389
Absorption
 of combustible liquids, 345
 of water, 353
Accessory products, 95, 108, 118
Acetaldehyde, 69
Acetic acid, 69
Acetone, 69
Acetylenic, 10
Acids, 80, 118
Acrylates, 69
Acrylic acid, 69
Acrylonitrile, 69
Adhesives, 95, 97
Adipic acid, 69
Ageing of materials, 400
Air conditioning, 62
 in mines, 62
Air ducts, 63, 70
Alkenes, 10
Alkylation
 (H_2SO_4), 68
 (HF), 68
Alkynes, 10
Ambergate, 307
America Burning, 329
Amine unit, 68
Ammonia, 4, 32, 69, 202, 223, 229
Aniline, 69
Anti-abrasive coatings, 107
Anti-abrasives, 116
Application guide, 278
Application requirements, 127

Architecture Building, 66
Argon, 29
Aromatics, 69
ASTM C 240-91, 87
ASTM E 136, 328
Atmospheric distillation, 15, 68
Autoignition, 347

Bellow chamber, 272
Bitumen, 53, 96
Bitumen tanks, 70
Boiling temperatures, 12
Breweries, 59, 70
Brewery Courage's Berkshire, 297
BS 476 Part 4, 328
Buildings, 66, 70
Butadiene, 53, 69
Butane, 53, 202, 223, 229

Calculation method for insulation thickness determination, 292
Capping influence, 447
Carbon monoxide, 35, 69
Catalytic reforming, 68
Cell filler, 104
Cheese dairies, 60, 70
Chemical
 resistance requirements, 136
 resistance towards acids, 80
 treatment process, 68
Chilled water lines
 Lurgi, 44, 301
 specification concept, 194

INDEX

Chimneys, 65, 70, 238
CO₂, 53
Coal tar distillates, 347
Coatings, 99
Coking process, 68
Cold applications, 185
Cold stores, 60
Combustion products, 331
Compact roof, 66
Compatibility with liquid oxygen, 133
Composite insulation system, 363
Compressive strength, 437, 444
Compressive strength of FOAMGLAS®cellular glass, 85
Condensation, 143, 369
Cone Segment CSG, 181
Constant thermal efficiency, 78
Contact adhesives, 97
Convection, 401
Corrosion, 405
  conditions, 406
  resistance requirements, 135
Cosmetics, 70
Cryogenic, 4
Cumene, 69

Dairies, 60, 70
Density of FOAMGLAS® cellular glass, 85
Dew point, 371, 390
Digesters, 64, 70, 234
Dimensional stability, 429
Dimersol, 68
DIN 18 174, 87
DIN 4102, 328
DIN 4108, 373
Distillation, 15, 68
DMT, 69
Dowtherm® A, 347
Dragon project, 311
Dual temperature applications, 192

Ecological standpoint, 82
Elbows E90 and E45, 177

Emissivity, 106
Encacel T, 121
Epiradiateur test, 336
Equipment insulation
  cold applications, 185
  dual temperature applications, 192
  high temperature applications, 189
Essential oils, 347
Ethane, 53, 202
Ethylbenzene, 69
Ethylene, 4, 35, 53, 202
Ethylene glycol, 69
Ethylene oxide, 44, 69
Expansion/contraction joints
  insulated pipelines, 249
  vessels and tank walls, 256

Fabricated FOAMGLAS® geometry, 168
Fabrication method of FOAMGLAS® cellular glass, 170
Fabrication shop, 170
Farms, 60
Federal Trade Commission, 330
Fertilizers, 31
Fire
  consequences, 324
  control (pool fire), 55
  epiradiateur test, 336
  examples, 331
  influence of insulating materials, 329
  literature, 333
  protection requirements, 131
  reaction, 326
  resistance, 77
  test methods, 327
  triangle, 326
Flanges, 257
Flanges F15 and F30, 177
Flexural modulus of elasticity of FOAMGLAS® cellular glass, 85
Flexural strength of FOAMGLAS® cellular glass, 85
FOAMGLAS® cellular glass
  chemical durability, 79

464

CO_2 in the cells, 83
coefficient of thermal expansion, 73, 81
compressive strength, 86
dimensional stability, 81, 435
dimensions of the slabs, 75
fire reaction, 76, 335
flexural strength, 89
limitations, 92, 361
pay back in energy, 82
preliminary conditions, 183
production process, 73
shear strength, 90
temperature limits, 74
tensile strength, 91
thermal conductivity, 86, 92
types, 84
water vapour resistance, 78
FOAMSEAL®, 120
Food industry, 7
Foster® 60-25, 120
Foster® 60-75, 120
Foster® 81-33, 120
Foster® 81-80, 117
Foster® 81-82, 117
Foster® 82-10, 120
Foster® 95-50, 120
Foster® Anti-abrasive 30-16, 121
Foster® FOAMSEAL® 30-45, 120
Foster® Lagtone® 30-70, 121
Foster® Monolar®, 121
Freezing lines, 61, 70

Gases
industrial, 29, 68
synthesis, 29
Glass durability, 42

H₂S, CO₂ removal process, 68
Heat traced pipelines, 265
Heat transfer
limitation, 127
lines, 54, 70

Heating, 62, 70
central, 63
district, 63
Helium, 29
High temperature applications, 189
HLB (FOAMGLAS®), 85
Hot box expansion chamber, 271
Hot tank base, 210
Humidity, 370, 389
Hydraulic setting cement adhesives, 97
Hydrocarbon chemistry, 9
Hydrocarbon fire test, 340
Hydrocarbons
saturated, 10
unsaturated, 10
Hydroconversion, 68
Hydrogen, 29
Hydrolytic class, 78
Hydrotreatment process, 68

Ice rinks, 63, 70
Industrial gases, 29, 68
Insects, 456
Inside area of fabricated products, 274
Insulation drying, 358, 364
Insulation thickness determination
based statistically on weather data, 143
for cold applications, 141
for dual temperatures applications, 161
for hot applications, 154
for insulated undergrund pipings, 163
for overfit/retrofit, 162
for tanks, 166
ISO 1182, 328
Isobutene, 69
Isomerisation, 68
Isopropanol, 69

Jackets, 99
Jobsite, 170
Joints, 256
Joints aera of fabricated products, 274

465

Kellogg, M.W., 44

LAB, 69
Lagtone®, 121
Langmuir's law, 150
Limitations to the use of cellular glass, 92, 361
Linde, 44
Line anchors, 273
Liquefaction processes, 68
LNG, 53
   Ambergate, 307
   applications, 202, 225
   liquefaction process, 23
   Mossmorran, 303
   receiving terminal, 27, 68
   storage, 27
Low temperature
   cryogenic tank wall and roof, 216
   sphere, 229
   tank bottom, 202
   tank wall and roof, 221
LPG gasoline fractionation, 15
LPG-NGL recovery, 23, 68
Lurgi, 44, 301

Maintenance, 287
Maleic anhydride, 69
Malt-houses, 59, 70
Mastics, 102
MDI-Diphenyl Methane 4, 4-Diisocyanate, 46, 69
Meat, 60
Meat poultry, 70
Mechanical requirements, 138
Metal jackets, 103
Methane, 4, 202, 225
Methanol, 46, 69
Microorganisms, 456
Mineral oil, 347
Mining, 70
Mobiltherm® 600, 347

Moisture
   content in vol%, 355
   resistance requirements, 134
Monolar®, 121
Mossmorran, 303
MTBE - Methyl Ter Butyl Ether, 52, 69

N values, 374, 394
N-Butanol, 69
Naphtha cracker, 37
Natural gas processing, 21
NEN 3881, 335
Neoprenes, 31
Nitro-paraffins, 53
Nitrobenzene, 69
Nitrogen, 29, 53, 202
Non absorption of flammable liquids, 133
Nuclear, 247
   applications, 133, 247
   plants, 65, 70
Number of layers, 167

Offshore, 54, 70, 242, 319
Oils, 347
Olefins, 10
ONorm B 6041, 87
Outside area of fabricated products, 274
Overfit system, 20, 197, 364, 384
Oxygen, 4, 29, 53, 202, 351

P-Xylene, 44
Paper mills, 64
Paraxylene, 69
Partial pressure of water vapour in the air, 371
PC® 18, 114
PC® 56 ADHESIVE, 110
PC® 74A, 114
PC® 80M MORTAR, 111
PC® 85 POWDER, 114, 207, 215
PC® 86T, 111
PC® 88 ADHESIVE, 110
PC® ANTI-ABRASIVE COMPOUND 1A, 121

PC® ANTI-ABRASIVE COMPOUND 2A, 116
PC® FABRIC 79 G, 116
PC® FABRIC 79 HD, 116
PC® FABRIC 79 P, 115
PC® HIGH TEMPERATURE ANTI-ABRASIVE, 116
Permeability, 372
Permeance, 372
Personnel protection, 396
Petro Wax PA, 315
Petrochemistry, 31
Pharmaceutical, 70
Phenol, 69
Pipe
  hangers, 259
  segments (PSG), 172
  shells (PSH), 171
  supports, 5
Piping insulation
  cold applications, 185
  dual temperature applications, 192
  high temperature applications, 189
PITTCOTE® 300, 113
PITTCOTE® 404, 113
PITTSEAL® 111, 120
PITTSEAL® 444, 112
PITTWRAP®, 115
Pluvex N° 1 Damp Proof Course Class A, 87, 206
Polybutadiene, 46, 69
Polyester polyols, 69
Polyethylene, 31
  HDPE, 69
  LDPE, 69
  linear LDPE, 69
Polypropylene, 46, 69
Polystyrenes, 31, 69
Polyurethanes, 31
Poultry-farms, 60
Power plants, 65, 70
Preliminary conditions before insulation application, 183

Process
  alkylation, 17
  catalytic cracking, 16
  catalytic reforming, 16
  Claus, 18
  coking, 16
  Dimersol, 17
  Fluor Solvent, 21
  hydrocracking, 17
  hydrotreatment, 17
  isomerisation, 16
  Kraft, 64
  purification, 17
  Rectisol, 21
  refining, 18
  separation, 18
  silicone, 311
  steam cracking, 35
  sulphite, 64
  synthesis, 17
  thermal cracking, 16
  visbreaking, 16
Propane, 4, 53, 202, 223
Propylene, 35, 53, 202, 223
  glycol, 69
  oxide, 69
Protection
  from burns, 129
  of structural material against temperature, 130
PVC, 31, 46, 69
Pyrolysis products, 331

Quality control, 451

Reference, 32-year old, 315
Refinery, 15
Reinforcement fabrics, 102, 115
Relative humidity, 370
Resins, 31
Resistance
  to chemical products, 80, 118
  to fire, 326
Retrofit, 197, 364, 384

467

Road tunnels, 66, 70
Rodents, 7, 455
Roofs, 66, 70

S3 (FOAMGLAS®), 85
Safety factor, 89, 451
Saturation water vapour pressure, 392
SBR, 69
Schwedeneck-See, 319
Sealers, 98
Self Ignition Temperature - SIT, 44, 346
Silicone, 311
Site fabrication, 170
Small radius segments SRS, 180
Smoke, 336
Solvent desasphalting, 68
Sources of leaks, 348
Specification concept
   chilled water pipelines, 194
   cold applications, 185
   cryogenic tank walls, 225
   digesters, 234
   dual temperature applications, 192
   expansion/contraction joints
     of vessels and tank walls, 256
     on insulated pipelines, 249
   heat traced pipelines, 265
   high temperature applications, 189
   hot tank bases, 210
   industrial chimneys, 238
   insulation of pipe hanger and support assembly, 259
   insulation of pipeline flanges, 257
   low temperature and cryogenic tank wall and roof, 216
   low temperature sphere, 229
   low temperature tank bottom, 202
   low temperature tank wall and roof, 221
   nuclear, 247
   offshore, 242
   overfit/retrofit, 197
   particular points, 249, 269
   sphere legs, 264
   sphere low temperature, 229

   stainless steel support, 244
   undergrund lines, 199
   vessel supports, 262
Sphere legs, 264
Spherical head segments SHS, 180
Spherical tanks, 70
Standardisation systems, 171
Steam cracking, 69
Stefan-Boltzmann's law, 150
Storage tanks
   bitumen, 54
   spherical, 53
   vertical, 52
Styrene, 53, 69
Suikerunie, 317
Sulphur, 53
   pits, 54, 70
   unit, 68
Supports, 244, 259, 262
Surface
   condensation limitation, 128
   temperature limitation, 129
   water retention, 78
Synthesis gases, 68

T2 (FOAMGLAS®), 85
T4 (FOAMGLAS®), 85
Table of standardisation of fabricated products, 274
Tank
   bottoms, 5, 70, 202, 210, 317, 438
   fixed roof, 220
   segments (TSG), 174
   suspended roof, 220
   wall and roof, 216, 221
TDI - Tolylene Diisocyanate, 46, 69
Temperature
   and humidity effects, 431
   and loading effects, 432
   and time effects, 448
   of autoignition, 347
   process control, 130
   variations, 130

Test
  deluge, 5
  epiradiateur, 336
  hydrocarbon fire, 6, 340
  methods (fire), 327
Textile fibres, 31
Thermal
  bridges, 403
  conductivity of FOAMGLAS® cellular glass, 85
  diffusivity of FOAMGLAS® cellular glass, 85
  requirements, 127
Therminol®, 347
Thickness determination
  based statistically on weather data, 143
  for cold applications, 141
  for dual temperatures applications, 161
  for hot applications, 154
  for overfit/retrofit, 162
  for tank insulation, 166
  for underground pipings, 163
Toxic gases, 336
TPA, 69
Transformer oil, 347
Turbo expander, 23
Two-component adhesives, 97

Underground
  lines, 199
  particular points, 269
  pipes, 70
Urea, 31, 69

Vacuum distillation, 68

Valve chamber, 272
Valves V15 and V30, 177
Vapour
  brake, 100
  cracker, 37
  retarder, 100, 379
Ventilation, 62
Vermin resistance, 455
Vessel
  heads, 180
  supports, 262
Vinyl
  acetate, 69
  chloride, 46, 69
Visbreaking process, 68

Water
  absorption, 353, 381
  absorption data for various insulation materials, 356
  chilled water pipes, 62
  effect, 397
  entry, 358, 365
  treatment, 64, 70
Water vapour
  barrier, 379
  diffusion resistance, 372
  effect, 397
  flow, 375
  retarder, 379
  transport, 369
Waxes, 53
Weather
  barrier, 100
  data, 143

ACHEVÉ D'IMPRIMER EN OCTOBRE 1992
PAR L'IMPRIMERIE LOUIS JEAN, 05000 GAP
DÉPÔT LÉGAL NOVEMBRE 1992, N° 632. IMPRIMÉ EN FRANCE